THIRD INTERNATIONAL CONFERENCE ON COLLECTIVE PHENOMENA

ANNALS OF THE NEW YORK ACADEMY OF SCIENCES

Volume 337

THIRD INTERNATIONAL CONFERENCE ON COLLECTIVE PHENOMENA

Edited by Joel L. Lebowitz, James S. Langer, and William I. Glaberson

The New York Academy of Sciences
New York, New York
1980

The New York Academy of Sciences believes that it has a responsibility to provide an open forum for discussion of scientific questions. The positions taken by the scientists whose papers are presented here are their own and not those of The Academy. The Academy has no intent to influence legislation by providing such forums.

Library of Congress Cataloging in Publication Data

International Conference on Collective Phenomena, 3d, Moscow, 1978.
Third International Conference on Collective Phenomena.

(Annals of the New York Academy of Sciences; v. 337)
Proceedings of the conference sponsored by the New York Academy of Sciences.
1. Mathematical physics—Congresses. 2. Physics —Congresses. I. Lebowitz, Joel Louis, 1930– II. Langer, James S. III. Glaberson, W. I. IV. Series: New York Academy of Sciences. Annals; v. 337.
Q11.N5 vol. 337 [QC19.2] 500s [530.1′5] 80–17323

PCP
Printed in the United States of America
ISBN 0–89766–074–9 (cloth)
ISBN 0–89766–075–7 (paper)
ISSN 0077–8923

ANNALS OF THE NEW YORK ACADEMY OF SCIENCES

VOLUME 337

June 26, 1980

THIRD INTERNATIONAL CONFERENCE ON COLLECTIVE PHENOMENA*

Editors

JOEL L. LEBOWITZ, JAMES S. LANGER,
AND WILLIAM I. GLABERSON

CONTENTS

* This series of papers is the result of a conference entitled Third International Conference on Collective Phenomena held on December 27, 28, and 29, 1978 in Moscow, USSR, sponsored by the New York Academy of Sciences.

FOREWORD

Joel L. Lebowitz*

President, 1979
The New York Academy of Sciences
New York, New York 10021

In December 1978, I participated, together with two other American, seven French, one British, and about thirty Russian scientists, in a most unusual conference in Moscow. Five other American scientists who wished to participate had their visas to the Soviet Union revoked ten days before the meeting. Some Russian scientists living outside Moscow were similarly prevented from attending; other Russians who participated did so in the face of threats and warnings from the Soviet authorities.

Despite these and other serious harassments of some participants, the meeting itself, the Third International Conference on Collective Phenomena, was not disrupted. It took place in a very warm atmosphere in the rather small living room of the apartment of "refusniks" Viktor and Irina Brailovsky on Vernadskova Prospect. (A refusnik is someone who has been officially refused an exit permit from the Soviet Union for Israel.) The scientific level of the meeting was remarkable, as can be seen in the present *Annal* of the New York Academy of Sciences, which officially sponsored the meeting.

The conference was an extension of the Moscow Sunday Seminar, which is no ordinary seminar. Indeed, it is often referred to by Soviet authorities as the Nonexistent Seminar; the present conference was, likewise, flatly stated by the official newspaper *Trud* to be a nonexistent Western fabrication.

The Moscow Sunday Seminar was organized about seven years ago by refusnik scientists who, as is common in the USSR, lost their status in the Soviet scientific community as soon as they applied to emigrate. Some of them were demoted to minor non-research positions, while others lost all employment (and thus became liable to legal prosecution for "parasitism"). This has meant that any continuation of their scientific work has had to be done in isolation at home. It was to overcome this isolation that the weekly seminar was established.

The seminar in Moscow meets every Sunday at noon. The total membership is about 70, of which some 20 or 30 show up each week to hear a presentation by a member or by a Western scientist visiting Moscow on some official or private business. (There is a similar seminar in Leningrad as well as additional specialized seminars in Moscow.)

There are many reasons why a visit to the seminar from a Western scientist is important to its regular participants. First is their need of scientific contacts to keep alive as scientists. Then, too, they need to feel in touch with colleagues in order to keep from despairing over their bleak situation. And, most urgently, they want the Soviet authorities to know that they are not abandoned by scientists in the West. This, they feel, is their primary, perhaps only, protection against disruption of the seminar by the authorities, despite its legality. They also fear, not without reason, that even worse things (the Gulag) may be in store for them if the authorities feel that they can act without evoking a strong response from the West.

It cannot be emphasized enough how much the members of the seminar value

* Department of Mathematics and Physics, Rutgers University, New Brunswick, New Jersey 08903.

Western visits. Since the fields in which its members formerly worked cover practically the whole of science, any topic on which a visiting scientist wishes to speak will attract an appreciative audience. Necessity has made the Sunday Seminar one of the few nonspecialized scientific meeting places. If you are going to the USSR, I urge you not to pass up the opportunity to attend one of these sessions—you will be performing an extremely important good deed and you will find it both personally and scientifically rewarding. The audience is eager and the discussions are stimulating. (Nobel laureate A. Penzias says that it was his best audience.) If you do not wish to give a talk, a visit to listen and discuss is equally well-appreciated.

Even though the Soviet government frowns on this and sometimes tries hard to persuade visiting Western scientists not to mix with refusniks, it is *not* against Soviet law to meet in a private house for lectures and discussion. Some visiting scientists have even used official Soviet cars to go to the Nonexistent Seminar. No one, as far as we know, has ever suffered any ill consequences from such a visit. Indeed, many scientists have visited the seminar on more than one occasion—and two of the five scientists whose visas were revoked have since been there—so participation does not seem to interfere with later travel to the Soviet Union. And the benefits, both scientific and psychological, to our beleaguered fellow-scientists are many.

If you wish more information on participating in a meeting of the seminar or how you might help in other ways, e.g., sending preprints to and corresponding with members of the seminar, please contact me.

A SUNDAY SEMINAR ON LATTICE-GAUGE THEORY*

John B. Kogut

Department of Physics
University of Illinois at Urbana–Champaign
Urbana, Illinois 61801

Static Quark Potential and β Function of SU(3) Gauge Fields

Two important characteristics of SU(3) lattice-gauge theory are its phase diagram and continuum limit.[1] Everyone is familiar with the speculation that the theory exists only in a disordered phase in which static quarks are confined. The theory's only critical point is expected to be at zero coupling, and the vicinity of $g = 0$ can be described by ordinary weak-coupling perturbation theory. These hopes are motivated by a plausibility argument that first notes that the theory is asymptotically free with a β function, calculated in the one-loop approximation,[2]

$$\beta(g) = -\frac{11}{16\pi^2} g^3 - \cdots \qquad \text{(weak coupling)} \tag{1}$$

The famous minus sign in Equation 1 indicates that the theory is more strongly coupled at large distances than at small distances. The second part of the plausibility argument observes that for large coupling, the lattice Hamiltonian theory[3] gives

$$\beta(g) = -g + \cdots \qquad \text{(strong coupling)} \tag{2}$$

This last result corresponds to quark confinement through the formation of an electric-flux tube between a static quark and its antipartner. Recall how this occurs. The lattice Hamiltonian for pure SU(3) gauge fields is

$$\mathscr{H} = \frac{g^2}{2a}\left[\sum_{\text{links}} \vec{E}_l^2 - x \sum_{\text{boxes}} (\text{tr } UUUU + \text{h.c.})\right], \qquad x = \frac{2}{g^4}, \tag{3}$$

where $\vec{E}_l^2$ is the quadratic Casimir operator for SU(3), U is a representation matrix of an SU(3) group element that resides on a link of the lattice, and a is the lattice spacing. Now place a quark and an antiquark on the lattice, as shown in Figure 1. Suppose also that g^2 is so large that only the first term in Equation 3 need be considered when estimating the energy of this state. Gauss' law assures us that the quark (antiquark) is a source (sink) of a unit of electric flux. The flux will distribute itself among the links of the lattice so as to minimize the energy of the state. A little thought shows that this is accomplished when the flux forms a tube between the quarks. Then, Equation 3 implies that the energy, or static quark potential, is

$$E = \frac{g^2}{2a} \cdot \frac{4}{3} \cdot n, \tag{4}$$

* Supported in part by a grant (NSF PHY 79-00272) from the National Science Foundation.

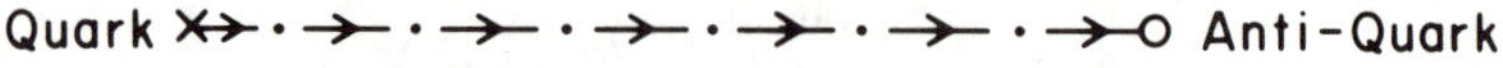

Quark ⟩⟩· → · → · → · → · → · →o Anti-Quark

FIGURE 1. A quark and an antiquark connected by a flux tube ($g^2 \gg 1$).

where n is the number of links separating the static quarks. But n is L/a, where L is the distance between the quarks measured in physical units, so Equation 4 yields the familiar linear confining potential,

$$E = \frac{4}{3} \cdot \frac{g^2}{2a^2} \cdot L. \tag{5}$$

To obtain a strong coupling β-function from these observations, we require that the string tension, $4g^2/6a^2$, be a physical quantity independent of the lattice spacing a. Therefore, the coupling g defined on a lattice of spacing a must grow linearly with a,

$$g = \text{const.} \cdot a, \tag{6}$$

and the β-function becomes

$$\beta(g) \equiv -a\frac{\partial g}{\partial a} = -g, \qquad \text{(strong coupling)} \tag{7}$$

as promised in Equation 2. Roughly speaking, Equations 1 and 2 mean that electric flux spreads out easily at short distances but collimates into quark-confining tubes at large distances.

The serious question is now whether the β-function of weak coupling smoothly attaches onto that of strong coupling. If this is true, the continuum limit of the lattice-gauge theory can be taken; asymptotic freedom and confinement will coexist in the same phase of the theory. Calculations that can deal with the intermediate-coupling region with good accuracy are necessary to handle this problem. I shall discuss the status of Hamiltonian strong-coupling expansions[4] in attacking this problem, although Euclidean strong-coupling expansions[4,5] and Monte Carlo numerical methods[6] are being applied to these questions as well. In addition, instanton calculations[7] in the continuum formulation of the theory also give the β-function in the intermediate-coupling region. Although none of these calculations are very refined as yet, they all indicate that a smooth transition region exists and that non-Abelian gauge fields confine static quarks.

To study the β-function by use of Hamiltonian strong-coupling methods, return to FIGURE 1 and the interquark potential. The energy of this state can be calculated as a power series in $x = 2/g^4$ by means of standard Rayleigh–Schrödinger perturbation theory,

$$E = \frac{g^2}{2a^2} \cdot F(x) \cdot L, \tag{8a}$$

where

$$F(x) = \tfrac{4}{3} + \text{higher-order effects.} \tag{8b}$$

The theory's β-function then follows from the requirement that the string tension be a fixed, physical quantity,

$$\frac{\mathrm{d}}{\mathrm{d}a}\left[\frac{g^2}{2a^2} F(x)\right] = 0. \tag{9}$$

Working out Equation 9 gives a strong-coupling expansion for $\beta(g)/g$ in terms of that for $F(x)$,

$$-\frac{\beta(g)}{g} = \frac{1}{1 - 2xF'(x)/F(x)}. \tag{10}$$

Before discussing the detailed calculation of Equation 10, I shall discuss a few general features of this approach to the confinement-asymptotic freedom problem. We are choosing to renormalize the theory by holding the string tension constant in physical units. This means that we are always dealing with the confining phase of the theory. If the $\beta(g)$ of Equation 10 is a smooth function with no zeros but behaves as $-(11/16\pi^2)g^3$ near the origin, we will have shown that the continuum limit of the lattice theory is asymptotically free and confines quarks. The function $\beta(g)$ is the best way to study this problem because of its relative simplicity. It is a dimensionless function of a dimensionless argument, and its weak- and strong-coupling behaviors are known. We shall see that the strong-coupling series suggests that its structure in the complex coupling-constant plane is also relatively simple. This simplicity should be contrasted with the string tension $F(x)$, which must have an essential singularity $\exp(-c/g^2)$ near the origin.

The strong-coupling expansion for the string tension is

$$F(x) = \tfrac{4}{3} - 0.07190x^2 - 0.03738x^3 - 0.01271x^4 - 0.003067x^5 - \cdots \tag{11}$$

By use of Equation 10, the result for the β-function is

$$-\frac{\beta(g)}{g} = 1 - 0.21569x^2 - 0.16820x^3 - 0.04138x^4 + 0.03444x^5 + \cdots \tag{12}$$

The coefficients of the power series have been computed exactly to $0(g^{-20})$ but are listed here just to several significant figures. It is interesting that the values of the coefficients are very sensitive to the non-Abelian character of the theory. Intermediate states in which the electric flux on links resides in representations as complicated as the **15** and **15′** contribute through this order. Extensive tables of SU(3) $3-j$ (or better, $6-j$) symbols are needed to carry out this calculation.[8]

Now to the physics of Equation 12. Since the series has a finite radius of convergence around $x = 0$, it is sensible to analytically continue it toward weak coupling by means of Padé approximants. By use of the first five terms in the series, the [2, 3] and [3, 2] approximants can be constructed. These extrapolation methods have no knowledge of the weak-coupling result $-\beta(g)/g = (11/16\pi^2)g^2$. We should examine these approximants at small g and see if there is a region, Δg, where the extrapolations and the weak-coupling result overlap. FIGURE 2 shows the [2, 3] and [3, 2] approximants, and one indeed observes matching in the vicinity of $g \approx 0.9$. The fact that the [2, 3] and [3, 2] approximants give essentially identical results for g ranging from infinity to ≈ 0.9 may be an indication that the calculation is stable and reliable. Below $g \approx 0.9$, the extrapolation procedures deviate from one another, indicating that higher-order terms are necessary in Equation 12 to penetrate further into the weak-coupling regime. Of course, for g, this small lattice theory can be analyzed with weak-coupling methods, and the one-loop β-function can be calculated directly by use of lattice propagators, and so on. This awkward exercise gives $\beta(g) = -(11/16\pi^2)g^3$, as it must.†

Let us discuss FIGURE 2 in more detail. Since the lattice calculation of the

† The β-function is universal through two loops.

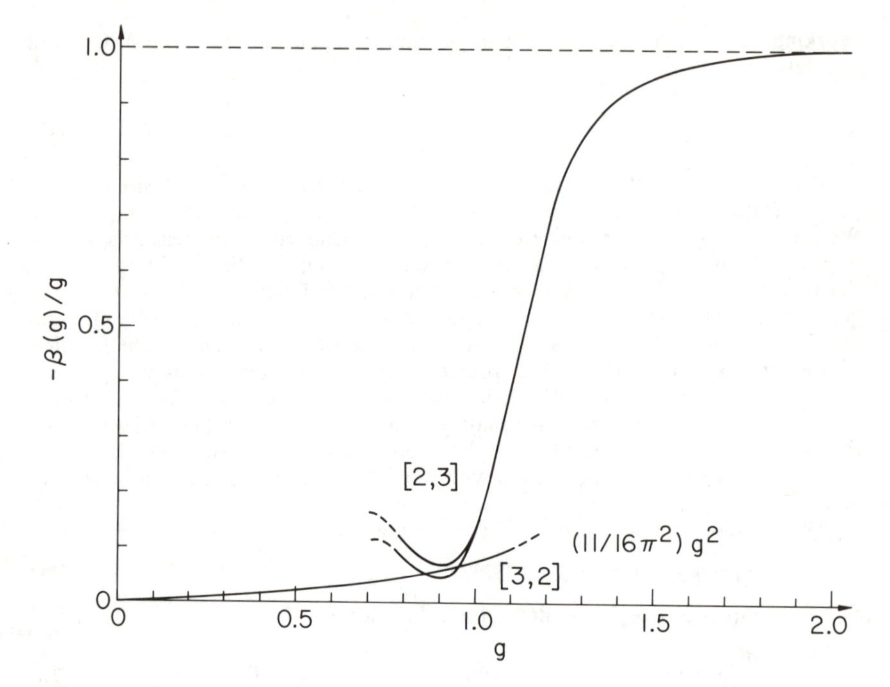

FIGURE 2. Strong- and weak-coupling calculations of the β-function for SU(3) gauge fields on a lattice.

β-function matches onto the weak-coupling continuum result, we have good evidence that the SU(3) Yang–Mills theory in four dimensions exists only in one phase, where confinement and asymptotic freedom coexist. The lattice cutoff is now irrelevant scaffolding and can be forgotten. From FIGURE 2 we learn that the theory has three distinct regions as a function of coupling constant. For $g \lesssim 0.9$, the theory is weakly coupled, with the one-loop calculation of the β-function being an accurate approximation to the theory's coupling constant renormalization. In the region $0.9 \lesssim g \lesssim 1.5$, the theory switches from weak to strong coupling. This intermediate-coupling region has two surprising features: it occurs where $\alpha_s \equiv g^2/4\pi$ is quite small, so that the two-loop contribution to the β-function is a tiny fraction of the one-loop term; and the intermediate region is quite narrow and abrupt. The second feature follows from the character of the coefficients in Equation 12 and the fact that the natural expansion parameter of the strong-coupling series is $x = 2/g^4$, so that small changes in g are amplified into large changes in x. For $g \gtrsim 1.5$, the theory resides in the strong-coupling regime. It is interesting to express these results in terms of a heavy quark potential or, equivalently, the running coupling constant. If the string tension is assigned the value $1/2\pi(\mathrm{GeV})^{-2}$, as suggested by various dual-resonance models and hadronic spectroscopy,‡ one can convert FIGURE 2 into a plot of g versus distance measured in fermi.[4] This exercise produces a curve with three distinct regions: for distances $r < \frac{1}{10}$ fm, the force law between static quarks is

‡ This string tension gives the correct Regge trajectory slope.

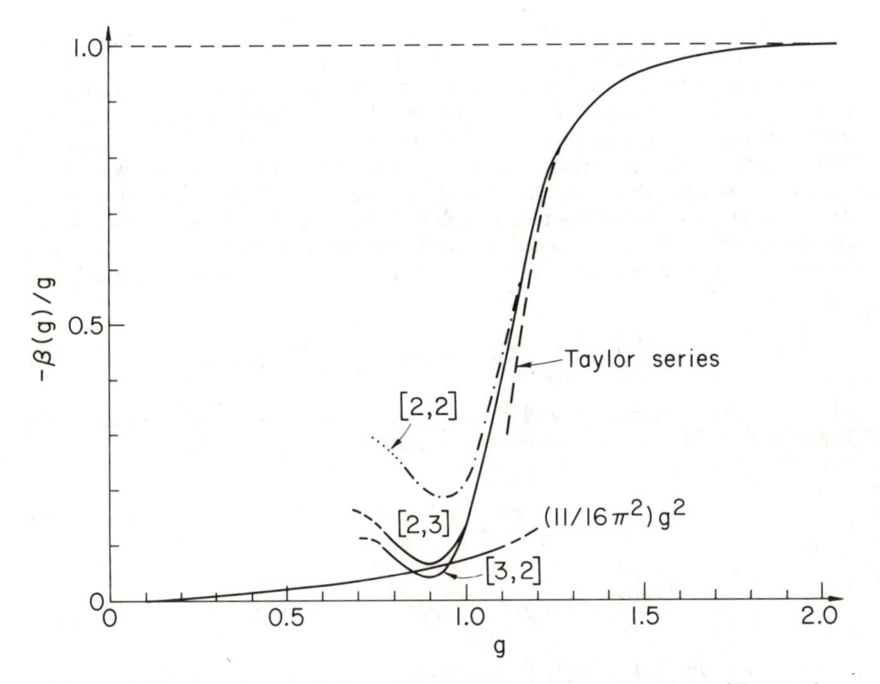

FIGURE 3. Lower-order calculations of the β-function compared to those of FIGURE 2.

governed by asymptotic freedom; if r is between $\frac{1}{10}$ fm and $\frac{1}{2}$ fm, the force law is essentially Coulombic; and if r is greater than $\frac{1}{2}$ fm, the linear potential dominates. It is encouraging that spin-independent potentials of this kind have been used with some success in describing heavy-quark bound states. Of course, this force law does not take into account the light quarks of QCD, so it must be used judiciously.

One of the surprising and pleasing features of this calculation is that the strong-coupling expansion is a well-behaved series, amenable to simple extrapolation methods. To emphasize this point, we plot in FIGURE 3 the [2/2] Padé approximant, which uses only the first four terms in Equation 12 and the Taylor series itself. Note that the prediction that the strong-coupling edge of the intermediate-coupling region lies between g of 1.2 and 1.5 is completely independent of the way the series is dealt with. This makes us quite confident of the fact that the upper edge of the intermediate-coupling region is quite sharp and that the strong-coupling region begins at $g \approx 1.5$. The small coefficients in the expansion Equation 12 suggest that the radius of convergence of the strong-coupling series lies near or perhaps inside the intermediate-coupling region. So it appears that Padé extrapolation procedures are only necessary to continue from $g \approx 1.2$ into the matching region of $g \approx 0.9$. This is a mild extrapolation indeed, so one can hope for an accurate β-function for all g as more terms in the strong-coupling expansion are computed. Further evidence that the extrapolation methods are converging to the true β-function follows from our earlier observation that the [2, 3] and [3, 2] approximants are essentially identical down to $g \approx 0.975$, where they flatten out and match unto the weak-coupling expansion. Hopefully, higher-order approximants will agree with these curves until $g \approx 0.975$ and then will slide along the weak-coupling curve before behaving erratically. It is interesting that just this sort of behavior was observed in similar β-function calculations of two-dimensional $0(n)$ spin systems.[10]

It is interesting to contrast the SU(3) series of Equations 11 and 12 to those of Ising lattice-gauge theory[11] in $2 + 1$ dimensions. This theory is dual to the Ising model, which has a second-order phase transition. The calculation of the string tension maps onto the calculation of the surface tension of a domain wall in the three-dimensional statistical mechanics problem.[12] Alternatively, it can be described as the surface tension of a bubble of vapor inside a liquid medium. As the critical temperature is approached, the surface tension vanishes, and the distinction between the two phases disappears. The Hamiltonian for the Ising lattice-gauge theory is

$$\mathscr{H} = \frac{G^2}{2a}\left(\frac{-1}{2}\sum_{\text{links}}\sigma_1 - x\sum_{\text{boxes}}\sigma_3\sigma_3\sigma_3\sigma_3\right), \qquad x = 4/G^4, \tag{13}$$

where the Pauli matrices σ_1 and σ_3 are placed on the links of a two-dimensional spatial lattice. The string-tension series is relatively easy to calculate and one finds, to $0(G^{-24})$,[4]

$$F(x) = 1 - \tfrac{1}{2}x^2 - \tfrac{69}{288}x^4 - \tfrac{1211}{6912}x^6 - \cdots \tag{14}$$

giving

$$-\frac{\beta(G)}{G} = 1 - 2x^2 + 1.0833x^4 - 0.37326x^6 - \cdots \tag{15}$$

One can form the [2, 1] and [1, 2] approximants to this series and find in both cases zeros on the positive G axis ($G_c = 0.98$ from the [2, 1] and 1.24 from the [1, 2]). Although the calculation is good enough to show that the model has two phases, the series Equation 15 has too few terms to predict the critical properties of the theory accurately. This can be contrasted with the richer β-function series for SU(3) in $3 + 1$ dimensions, where the higher-order approximants showed good evidence for convergence down to $g \approx 0.9$.

Lattice-Gauge Theories with 6-Link Coupling Terms

Lattice-gauge theories are constructed so that their classical continuum limits reduce to familiar Yang–Mills equations. In addition, the lattice theory analyzed with weak-coupling perturbation theory gives the same physics as its continuum-quantized relative. These requirements of the lattice theory still leave considerable freedom in its construction. Other criteria, such as locality, are usually invoked to motivate a simple lattice Hamiltonian. However, when such theories are analyzed via renormalization groups, additional interaction terms are usually generated. In spin systems, one finds, in addition to the nearest-neighbor coupling that might have characterized one's initial Hamiltonian, that next-to-nearest-neighbor coupling terms are generated in the sequence of effective Hamiltonians. All these Hamiltonians give the same continuum physics, but in a practical calculation a more complicated Hamiltonian may expose the interesting continuum physics more easily than its simpler relative.

Consider the coupling term tr $UUUU$ in a quantum Hamiltonian of a lattice gauge theory. In a smooth-continuum limit, this operator reduces to $\mathbf{B}^2$, the magnetic-field term of the continuum-energy density. On a finite lattice, however, $\mathbf{B}^2$ and tr $UUUU$ differ by quantities of the generic form: a power of the lattice

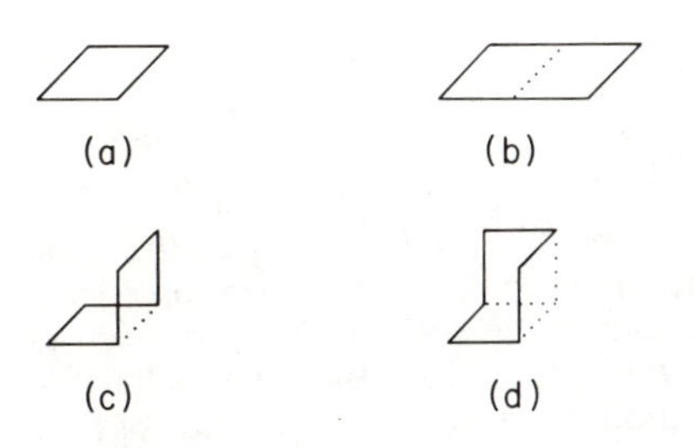

FIGURE 4. a, The links of the 4-link coupling term; b, a rectangular 6-link coupling term; c, a wedge 6-link term; d, an indescribable 6-link term.

spacing times a nonrenormalizable operator. These irrelevant operators do not contribute in the continuum limit, and in a weak coupling renormalization group calculation they iterate to zero. In a practical calculation, only a few iterations of a renormalization group can be accomplished, so it is best if the strength of these irrelevant operators is small. Experience has shown that to control them, one can adjust the lattice Hamiltonian appropriately by including 6-link coupling terms,§

$$\operatorname{tr} UUUU \rightarrow c_4 \operatorname{tr} UUUU + c_6 \operatorname{tr} UUUUUU + \cdots, \tag{16}$$

where the product of the six U-matrices is taken around closed 6-link paths of the spatial lattice (FIGURE 4). One constraint on coefficients c_4, c_6, etc., is that they add up such that the modified Hamiltonian has the usual smooth-continuum limit. In addition, the individual coefficients can be adjusted to reduce the strength of the unwanted irrelevant operators.

One might also use the modified Hamiltonian for strong-coupling expansions. Then, one might choose c_6/c_4 so that the low-lying states have a relativistic energy-momentum relation even when computed only to low orders.

Consider a simple example: Ising lattice-gauge theory in 2 + 1 dimensions. We will write down a version of the theory with a 6-link coupling term and calculate the theory's critical coupling and critical index ν. The Hamiltonian is

$$\mathscr{H} = -\sum \sigma_1 - \lambda \sum \sigma_3\sigma_3\sigma_3\sigma_3\sigma_3\sigma_3, \tag{17}$$

where the second sum extends over 6-link rectangles in the spatial plane. We are not including any 4-link coupling terms just for illustrative purposes. Equation 17 has some simple properties: it is equivalent to a spin system that is self-dual. Recall that Ising lattice-gauge theory formulated with 4-link couplings (Equation 13) is dual to the Ising model in three dimensions.[11] In establishing this, one works within the gauge-invariant subspace of the theory and introduces dual variables, μ_1 and μ_3, on a dual lattice of sites $\mathbf{n}^*$.¶ In particular,

$$\mu_1(\mathbf{n}^*) = \sigma_3\sigma_3\sigma_3\sigma_3 = \prod_{\text{box}} \sigma_3, \tag{18}$$

where the four σ's are those of the box surrounding the dual site $\mathbf{n}^*$. And

$$\mu_3(\mathbf{n}^*) = \prod_{n' \geq 0} \sigma_1(n - n'\hat{y}, \hat{x}). \tag{19}$$

Then, it is easy to see that within the gauge-invariant subspace, μ_1 and μ_3 are Pauli operators and the Hamiltonian Equation 17 is

$$\mathscr{H} = -\sum \mu_3\mu_3 - \lambda \sum \mu_1\mu_1, \tag{20}$$

§ This observation has also been made by K. G. Wilson.[13]

¶ For the details of this construction, see Kogut.[1]

where the sums extend over nearest-neighbor sites. Since the subscripts 1 and 3 can be interchanged, $\mathscr{H}$ is self-dual,

$$\mathscr{H}(\lambda) = \lambda \mathscr{H}(1/\lambda) \tag{21}$$

with the self-dual point $\lambda^* = 1$. Since Equations 17 and 13 belong to the same universality class, $\lambda^* = 1$ should be the critical point of the three-dimensional Ising model. In gauge-theory language, the $1 \rightleftarrows 3$ duality of Equation 20 is akin to electric-magnetic duality of four-dimensional models. Its phases can be labeled as follows. In the small-λ phase of the spin system, the ground-state expectation value of μ_3 is nonzero, so, in gauge-theory language,

$$\left\langle 0 \middle| \prod_{n' \geq 0} \sigma_1(n - n'\hat{y}, \hat{x}) \middle| 0 \right\rangle \neq 0, \qquad (\lambda < 1), \tag{22}$$

which is kink-condensation. In this phase, the gauge-invariant correlation function $\prod_c \sigma_3$ is short-ranged,

$$\left\langle 0 \middle| \prod_c \sigma_3 \middle| 0 \right\rangle \sim \exp(-\text{Area}), \qquad (\lambda < 1), \tag{23}$$

where the contour c is purely spatial, and the Area in Equation 23 is that enclosed by c. In the large-λ phase, μ_1 has a ground-state expectation value, so

$$\left\langle 0 \middle| \prod_{\text{box}} \sigma_3 \middle| 0 \right\rangle \neq 0, \qquad (\lambda > 1), \tag{24}$$

which is boxiton-condensation; the gauge-invariant correlation function is relatively long-ranged,

$$\left\langle 0 \middle| \prod_c \sigma_3 \middle| 0 \right\rangle \sim \exp(-\text{Perimeter}). \qquad (\lambda > 1). \tag{25}$$

Finally, it is interesting to use Equation 17, or equivalently, Equation 20, to calculate some characteristics of the theory's continuum limit. Consider the mass gap in the small-λ phase. The lowest-energy excitation at $\lambda = 0$ is the boxiton at zero momentum,

$$\frac{1}{\sqrt{N}} \sum \sigma_3\sigma_3\sigma_3\sigma_3 \,|0\rangle. \tag{26}$$

This state is manifestly gauge invariant and contains the fewest possible excited links. It is easy to compute its mass as a power series in λ,

$$F(\lambda) = 8 - 4\lambda - \tfrac{5}{3}\lambda^2 - \tfrac{1}{6}\lambda^3 - \cdots \tag{27}$$

Since the theory is self-dual, the mass gap must have the generic form

$$F(\lambda) = (1 + \lambda) f(u), \qquad u = 4\lambda/(1 + \lambda)^2, \tag{28}$$

as implied by Equation 21. Equation 27 gives the expansion for the self-dual function $f(u)$,

$$f(u) = 8 - 3u - \tfrac{41}{48}u^2 - \tfrac{175}{384}u^3 - \cdots \tag{29}$$

In the vicinity of the critical point $u^* = 1$, $f(u)$ should have an algebraic zero,

$$f(u) \sim (u^* - u)^{\nu/2}, \qquad u \approx u^*, \tag{30}$$

where ν is the correlation-length critical index of the three-dimensional Ising model ($\nu \cong 0.64$). Note that $\nu/2$ appears in Equation 30 because $u^* - u \approx \frac{1}{4}(\lambda^* - \lambda)^2$, so the mass gap $F(\lambda)$ behaves as $(\lambda^* - \lambda)^\nu$ in the critical region. We can extract an estimate of ν from the series Equation 29 by noting that the logarithmic derivative of Equation 30 satisfies

$$-(u^* - u)\frac{f'(u)}{f(u)} \sim \frac{\nu}{2}\,. \tag{31}$$

From Equation 30, we have

$$-(1 - u)\frac{f'(u)}{f(u)} = \frac{3}{8}\left(1 - \frac{1}{18}u - \frac{1}{36}u^2\right), \tag{32}$$

which can be extrapolated toward $u^* = 1$ by use of a [1, 1] Padé approximant,

$$-(1 - u)\frac{f'(u)}{f(u)} = \frac{3}{8}\left(\frac{1 - \frac{5}{9}u}{1 - \frac{1}{2}u}\right), \tag{33}$$

which equals $\frac{1}{3}$ at the critical point, implying

$$\nu = \tfrac{2}{3} \qquad (\nu_{\text{exact}} \approx 0.64). \tag{34}$$

This result is quite good for such a simple exercise in three dimensions. Note that the Taylor series in Equation 32 also yields a very good estimate of ν. In all cases, we obtain estimates that are better than similar low-order calculations using the 4-link form of the Hamiltonian. The reader familiar with spin systems can probably site several reasons for the improvement to complement the remarks at the beginning of this section. Consider the graphs of the perturbation theory calculation giving Equation 27. The Hamiltonian with 6-link coupling allows the boxiton state to propagate in first order of λ: applying $\sigma_3\sigma_3\sigma_3\sigma_3\sigma_3\sigma_3$ to $|B\rangle$ can move it by one square. Such effects do not occur in the analogous calculation with the 4-link Hamiltonian until higher order. Although the 6-link Hamiltonian is more difficult to work with, it exposes the continuum physics of the model much more easily than does its 4-link relative. Six-link Hamiltonians of SU(3) gauge fields in 3 + 1 dimensions have similar advantages.

Z_p and Pott's Lattice-Gauge Theories

Ising lattice-gauge theory in four dimensions has been studied by use of a wide variety of methods. Wegner[11] showed that the model is self-dual and that the gauge-invariant correlation function satisfies the area (perimeter) law at high (low) temperatures. By use of expansion methods, Balian *et al.*[14] argued that the theory undergoes a first-order phase transition at the self-dual point. This result means that the theory does not have an interesting continuum limit. Migdal's approximate recursion relation,[15] however, suggested that the transition is second order. In fact, the Migdal recursion relation makes the following correspondences:

$$\begin{array}{ccc} \underline{\textit{2-D Spin System}} & & \underline{\textit{4-D Gauge System}} \\ Z_p & \longleftrightarrow & Z_p \\ \mathrm{U}(1) & \longleftrightarrow & \mathrm{U}(1) \\ \mathrm{SU}(N)\times\mathrm{SU}(N) & \longleftrightarrow & \mathrm{SU}(N) \\ \text{Disorder} & \longleftrightarrow & \text{Confinement} \end{array} \tag{35}$$

of the phase diagrams of two-dimensional spin systems and four-dimensional gauge systems. Unfortunately, one of the primary flaws of the recursion relation[16] is that it misses the first-order character of the phase transitions in simple models, such as the p-state ($p \geq 5$) Pott's models in two dimensions. Recent Monte Carlo studies[17] of Ising gauge theory in four dimensions predict a first-order transition at the self-dual point. Their numerical data are in good agreement with the strong-coupling expansions of the system's free energy.[14] Monte Carlo studies also indicate that Z_3 and Z_4 gauge theories have first-order transitions, whereas Z_p ($p \geq 5$) theories possess three phases.[18]

The three-phase structure of the large-p models was suggested some time ago in studies motivated by the character of Z_p spin systems in two dimensions.[18] Recall the ideas behind this analysis. One can formulate Z_p spin systems by placing an angular variable $\theta(r)$ at the sites of a two-dimensional square lattice and constraining $\theta(r)$ to the p possible values,

$$\theta(r) = \frac{2\pi}{p} n(r), \qquad n(r) = 0, 1, \ldots, p-1. \tag{36}$$

Nearest-neighbor angles can be coupled together in the usual fashion to give an Action,

$$S = -J \sum_{r,\mu} \cos[\Delta_\mu \theta(r)], \tag{37}$$

where μ denotes the set of unit vectors of the lattice. The partition function of the Z_p model is then

$$Z = \sum_{n(r)} \exp\left\{-J \sum_{r,\mu} \left[1 - \cos\left(\frac{2\pi}{p} \Delta_\mu n(r)\right)\right]\right\}. \tag{38}$$

Following developments on the planar model, it proves more convenient to consider the periodic Gaussian form of the model,[18]

$$Z = \sum_{n(r)} \sum_{m_\mu(r)} \exp\left\{-\frac{1}{2T} \sum_{r,\mu} \left[\frac{2\pi}{p} \Delta_\mu n(r) - 2\pi m_\mu(r)\right]^2\right\}, \tag{39}$$

where the link variable $m_\mu(r)$ ranges over all the integers. There are two advantages to this formulation of the model: as $p \to \infty$, it approaches the periodic Gaussian form of the planar model about which much is known;[19] and these Z_p models are self-dual at the temperature[18]

$$T_{\text{dual}}(p) = 2\pi/p. \tag{40}$$

Equation 40 presents an interesting puzzle. The self-duality of the Z_p models implies that if the model has a unique critical point, it must occur at $T_{\text{dual}}(p) = 2\pi/p$. But the planar model has a critical point at[19]

$$T_{\text{KT}} = 1.35, \tag{41}$$

below which the theory's spin–spin correlation function is power behaved. Since $T_{\text{dual}}(5) < T_{\text{KT}}$, the hypothesis that the dual point is an isolated critical point of the Z_5 model would lead us to the absurd conclusion that there is a range of temperatures (between $T_{\text{dual}}(5)$ and T_{KT}) where the 5-state model is more disordered (its spin–spin correlation function falls off exponentially) than the limiting $p \to \infty$ model. In fact, this possibility is rigorously excluded by Ginibré–Griffiths–Kelly–Sherman inequalities,[20] which state in precise terms that increasing p increases the disorder of the system at a given temperature. The resolution of this puzzle is that all the p-state models ($p \geq 5$) have three phases:

1. Low $T(T < T_1)$. The system magnetizes.
2. Intermediate $T(T_1 < T < T_2)$. The system, although based on a discrete group, contains spin waves, and its spin–spin correlation function is power behaved. As $p \to \infty$, $T_1 \to 0$. T_1 and T_2 are dual, and $T_1 < T_{\text{dual}}(p) < T_2$.
3. High $T(T > T_2)$. The theory is disordered, having exponentially decaying correlation functions. As $p \to \infty$, T_2 approaches T_{KT} from above.

The major points of this analysis generalize to Z_p lattice-gauge theories in four dimensions. The periodic Gaussian version of the U(1) gauge model has been studied, a phase transition was found, and the critical temperature was estimated.[21] For $T < T_c$, the theory was predicted to have massless photons, whereas for $T > T_c$, it confines external quarks. The periodic Gaussian form of the Z_p lattice-gauge theories can be shown to be self-dual,[18] with a temperature $T_{\text{dual}}(p)$ that vanishes as $p \to \infty$. The Z_p models for $p \gtrsim$ 4–5 are therefore predicted to have three phases:

1. Low T. The Z_p theories do not confine quarks. They have a mass gap, and the quark potential $V(R) \sim$ const. as R grows. This is a Higgs phase.
2. Intermediate T. Quarks are not confined. The spectrum contains massless particles, "photons."
3. High T. Quarks are confined, the spectrum contains a gap, and the potential $V(R)$ rises linearly with R.

Our insight into Z_p gauge theories for $p \geq 5$ seems quite satisfactory. However, no comparable insight exists into the Z_2, Z_3, and Z_4 models, although the first-order character of the transition for Z_3 has been suspected for some time. The argument for it begins by considering the theory in three dimensions. Here, it is dual to the Z_3 spin system, which has been analyzed by a variety of methods[22] that strongly favor a first-order transition. For example, mean field theory allows a third power of the theory's magnetization to appear in the theory's free energy, and this is sufficient to cause the transition to be first order.‖ Now continue the Z_3 gauge theory from three to four dimensions. One naively expects that the first-order transition would persist and, in fact, become stronger. Unfortunately, this plausibility argument casts no light on the Z_2 gauge case. It is difficult to develop an intuitive understanding of these theories' transition points since a mean field theory that is compatible with local-gauge invariance has not been developed.

Experience with two-dimensional spin systems suggests that lattice-gauge theories based on the p-state Pott's models may be helpful here. In two dimensions, it is natural to consider the p-state Pott's models as a family and treat p as a continuous parameter. The formal limit $p \to 1$ reduces the spin system to the percolation problem,[23] which is a counting problem with a critical point, indices, and so on. Many features of percolation are good guides to the more complicated phase transition in the Ising model. All the Pott's models are self-dual. For $p \leq 4$, they have continuous phase transitions, whereas for $p \geq 5$, the transitions are first order.[24] It is easy to see that Pott's lattice-gauge theories are self-dual in four dimensions. The $p \to 1$ limit is quite simple, and it may shed light on Ising lattice-gauge theory.[25]

‖ This mean field theory prediction may be reliable in three dimensions, but it fails in two dimensions.

SUMMARY

Several topics in lattice theory are discussed briefly. Recent calculations of the Callan–Symanzik β-function of SU(3) gauge fields in 3 + 1 dimensions support the idea that this theory exists in one phase in which asymptotic freedom and quark confinement coexist. These computations suggest that the transition region between weak and strong coupling is smooth but narrow. Effective gauge-field Hamiltonians with 6-link coupling terms in addition to the usual 4-link terms are argued to expose continuum physics particularly efficiently. The present understanding of the phase diagrams of Z_N gauge theories is reviewed, and some directions for additional work are suggested.

REFERENCES

1. KADANOFF, L. P. 1977. Rev. Mod. Phys. **49:** 267; KOGUT, J. B. 1979. Rev. Mod. Phys. **51:** 659.
2. 'T HOOFT, G. 1972. Unpublished observations; POLITZER, H. D. 1973. Phys. Rev. Lett. **30:** 1346; GROSS, D. J. & F. WILCZEK. 1973. Phys. Rev. Lett. **30:** 1343.
3. KOGUT, J. & L. SUSSKIND. 1975. Phys. Rev. D **11:** 395.
4. KOGUT, J., R. PEARSON & J. SHIGEMITSU. 1979. Phys. Rev. Lett. **43:** 484.
5. DUNCAN, A. & H. VAIDYA. 1979. Columbia University preprint, CU-TP-149.
6. WILSON, K. G. 1979. Unpublished observations.
7. CALLAN, C., R. DASHEN & D. J. GROSS. 1979. In preparation.
8. FREDRICKSON, R. 1979. Ph.D. thesis. Cornell University.
9. HAMER, C. J., J. KOGUT & L. SUSSKIND. 1978. Phys. Rev. Lett. **41:** 1337.
10. WEGNER, F. 1971. J. Math. Phys. **12:** 2259.
11. WATSON, P. G. 1972. *In* Phase Transitions and Critical Phenomena. C. DOMB & M. S. GREEN, Eds. Vol. 2: 101–159. Academic Press. London.
12. SHENKER, S. 1979. Personal communication.
13. BALIAN, R., J. M. DROUFFE & C. ITZYKSON. 1975. Phys. Rev. D **11:** 2098.
14. MIGDAL, A. A. 1975. Zh. Eksp. Teor. Fiz. **69:** 810, 1457.
15. KADANOFF, L. P. 1976. Ann. Phys. **100:** 359.
16. CREUTZ, M., L. JACOBS & C. REBBI. 1979. Phys. Rev. Lett. **42:** 1390; 1979. Personal communication.
17. ELITZUR, S., R. B. PEARSON & J. SHIGEMITSU. 1979. Phys. Rev. D **19:** 3698.
18. KOSTERLITZ, J. M. 1974. J. Phys. C **7:** 1046.
19. GINIBRÉ, J. 1970. Commun. Math. Phys. **16:** 310.
20. BANKS, T., R. MYERSON & J. B. KOGUT. 1977. Nucl. Phys. B **129:** 493.
21. OPPERMANN, R. 1975. J. Phys. A **8:** L43; ENTING, I. G. & C. DOMB. 1975. J. Phys. A **8:** 1228.
22. KASTELEYN, P. W. & C. M. FORTUIN. 1969. J. Phys. Soc. Japan Suppl. **26:** 11.
23. BAXTER, R. J. 1973. J. Phys. C **6:** L445.
24. KOGUT, J. B. & D. K. SINCLAIR. 1979. Unpublished observations.

YUKAWA QUANTUM FIELD THEORY IN THREE DIMENSIONS (Y_3)*

J. Magnen and R. Seneor

Centre de Physique Théorique de l'Ecole Polytechnique
91128 Palaiseau, France

Motivations

This paper is a step in the way of constructing QED in three dimensions and deals essentially with a proof of the ultraviolet stability of the Yukawa model in three dimensions.

The usual way to prove ultraviolet stability is to perform a phase space expansion à la Glimm–Jaffe.[5] This method consists of performing a truncated Taylor expansion on parts of the interaction that are simultaneously localized in space and momentum. Such an expansion can be convergent because of the superrenormalizability of the theory. Typically also, this allows for an interaction with upper cutoff M, located in cubes of size M^{-3}, to apply a classical argument (the positivity of the classical Hamiltonian) to bound the cutoff Hamiltonian.

Description of the Model

In Yukawa, the partition function is given by (in volume Λ):

$$Z_\Lambda = \int e^{\int_\Lambda \bar\psi (i\not p + m + \lambda\Gamma\Phi)\psi} \mathscr{D}\bar\psi \mathscr{D}\psi \, d\mu/\text{normalization},$$

where $d\mu$ is the Gaussian measure of the boson field, $\mathscr{D}\bar\psi\mathscr{D}\psi$ is the "normalized fermion measure," λ is a coupling constant, and $\Gamma = 1$ or γ_5, according to whether the theory is scalar or pseudoscalar

$$Z_\Lambda = \int \frac{\det(i\not p + m + \lambda\Gamma\Phi\Lambda)}{\det(i\not p + m)} \, d\mu$$

$$= \int \det[1 + (i\not p + m)^{-1}\Gamma\Phi\Lambda] \, d\mu = \int \det(1 + K) \, d\mu.$$

This is only formal since we need to renormalize the theory. The divergent diagrams are: Tr K = [diagram], [diagram], the infinite parts of Tr K^2 [diagram], [diagram], [diagram], vacuum diagrams up to the sixth order. All these diagrams are finite if we cut off the fermion momenta, and all except the first and third ones are finite if we cut off the boson momenta. By the choice of a symmetric cutoff, the fermion mass renormalization diagram [diagram] is finite. The last point simplifies the development of the expansion as compared to φ_3^4 (see Reference 9).

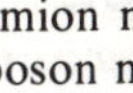

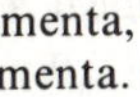

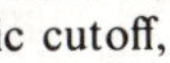

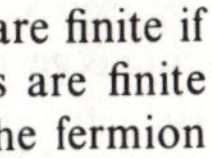

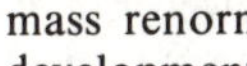

* Supported by a grant (174) from the Conseil Nationale de Recherche Scientifique.

About Expansions

Since Yukawa or QED is expressed in terms of two kinds of fields, bosons or fermions, there are two possible kinds of expansions: α, a perturbative expansion on the boson fields, and β, a perturbative expansion on the fermion lines. Both will produce Yukawa or QED vertexes but with, in case α, a well-localized boson field instead of, in case β, a well-localized fermion line. Because of that, we will need in case α a "classical" bound on the interaction valid for cutoff bosons and no cutoff fermions, and in case β a "classical" bound valid for cutoff fermions and no cutoff bosons.

Let us first look at these "classical" bounds. There are two types of bounds in class α. Seiler's[14] bound is

$$|\det(1+K)e^{-\operatorname{Tr} K}e^{1/2 \operatorname{Tr} K^2}|^2 \leq e^{1/2 \operatorname{Tr}(K+K^*)^2}.$$

This bound is very good for Yukawa theory because the most divergent part of K: $-i\not{p}\Gamma\Phi/(p^2+m^2)$ is anti-Hermitian, and it follows that $\operatorname{Tr}(K+K^*)^2$ is Hilbert Schmidt.[14] This is, however, no more true in QED or in Yukawa for λ complex.

The effective potential bound is the second type in class α. If we take the boson field as a constant external field, in Yukawa

$$|\det(1+K)e^{-\text{divergent part of } 1/2 \operatorname{Tr} K^2}| \sim e^{-\operatorname{Re} \lambda^3 |\Phi^3|\,|\Lambda|}$$

for $\operatorname{Re} \lambda^3 > 0$. The effective potential is $V_{\text{eff}}(\Phi) = \operatorname{Re} \lambda^3 |\Phi^3|$. In QED, $V_{\text{eff}} \equiv 0$. In practice, we will use the effective potential in cubes with periodic boundary conditions.

There is essentially one bound in class β. One has the divergent part of

$$\tfrac{1}{2} \operatorname{Tr} K^2 \simeq -\rho \int_\Lambda \Phi^2(x)\, dx, \qquad (*)$$

with ρ the fermion momentum cutoff.

From

$$|\det(1+K)| \leq e^{\operatorname{Tr}|K|} \leq e^{\rho^2 \int_\Lambda |\Phi(x)|\, dx},$$

it follows that

$$|\det_{\text{ren}}(1+K)| \leq e^{\text{constant} \times \rho^3 |\Lambda|}.$$

This bound is a consequence of the good sign (negative) for the mass counterterm in (*), and, moreover, (*) is responsible for the possibility of an arbitrary mass renormalization for Yukawa theory in two and three dimensions.[7,15] It should be easy to prove, as Frölich and Seiler[4] did for the two-dimensional Higgs model, that with the wrong sign, the theory could not have been ultraviolet stable.

For Yukawa, we will use an α-expansion. In fact, an expansion of type β cannot converge because β-vertexes may have not well-localized boson fields; that is, for the space and momentum localizations that are not compatible, compatibility means $M|\Delta|^{1/3} > 1$, leading to a divergent Gaussian integration. However, a β-expansion can be performed after an α-expansion has produced enough convergent factors to control these divergences. This combination will be necessary in QED because $V_{\text{eff}} \equiv 0$.

A last technical remark is that in each step of the Taylor expansion (or perturbation step), one produces a vertex that has (in an α-expansion) a well-localized boson field, but one needs to perform a renormalization to cancel out the divergent diagrams, which will produce new vertexes, possibly with badly localized boson fields. These fields will be dominated by the effective potential if they are connected with well-localized fermions and by the mass counterterm (bound of type β) otherwise.

Results

We mainly prove that, for $|\text{Arg}\,\lambda| < \pi/6 - \varepsilon$,

$$|Z_\Lambda| \leq e^{K_1|\Lambda|} \text{ for some } K_1(\varepsilon) > 0 \qquad (**)$$

uniformally in the cutoff.

Notice that the analyticity domain is larger (in fact, optimal since it corresponds to the one given by the effective potential[11]) than what one would have obtained by use of a three-dimensional extension of the method of Renouard[12] linked to Seiler's bound.

The existence of Z_Λ, the Schwinger functions, and the corresponding relativistic field theory follows as usual. We also prove the Borel summability of the theory.

Description of the Paper

The main part of the paper is PHASE SPACE EXPANSION, where we prove the bound (**) according to the ideas and notations in References 9 and 10 for the case of a pseudoscalar theory. In BOREL SUMMABILITY, we sketch the proof of this summability, establishing the necessary bound on the nth derivative of the Schwinger functions. The remainder of the argument is left to the reader since it is the same as that for Φ_2^4 and Φ_3^4.[2,9] The extension to the scalar case is discussed in Appendix 3. In Appendix 2, we sketch the arguments that lead to the present formulation of a phase space expansion for Y_3.

PHASE SPACE EXPANSION

Notations

The theory is given by the Matthews–Salam formula with spatial cutoff Λ and ultraviolet cutoff on the fermion momenta M_ζ.

Let $\Phi(x)$ be the Gaussian field of mean zero and covariance

$$C(x, y) = \int d^3p \frac{e^{ip(x-y)}}{p^2 + m^2}$$

and $d\mu$ be the associated measure.

Let K be the operator of L^2 of kernel $K(x, y)$:

$$K_\Lambda(x, y) = \lambda \int S_\Lambda(z)(x, y)\Phi(z)\, d^3z$$

$$S_\Lambda(z)(x, y) = \int \frac{-i\not{k} + M}{(k^2 + M^2)^{3/4}} e^{ik(x-z)}\eta\left(\frac{k}{M_\zeta}\right) d^3k\,\Gamma\Lambda(z) \frac{e^{il(z-y)}}{(l^2 + M^2)^{1/4}}\, d^3l,$$

with the cutoff function η defined below, $\not{p} = \sum_{i=0}^{2} p_i\gamma_i$, γ_i being 3-Hermitian matrices such that $\{\gamma_i, \gamma_j\} = 2\,\delta_{ij}$, $i, j = 0, 1, 2$, $\gamma_5 = \gamma_0\gamma_1\gamma_2$, $\Gamma = 1$ in the scalar case, and $\Gamma = \gamma_5$ in the pseudoscalar case. From now on we restrict ourselves to the pseudoscalar case and omit the Λ-dependence when unnecessary.

Let $f = (f_1, \ldots, f_n)$, $g = (g_1, \ldots, g_{N'})$, $h = (h_1, \ldots, h_{N'})$ with $\{f_i\}$, $\{g_j\}$, $\{h_k\}$ in suitable spaces, and define

$$\hat{g}_k = (k^2 + M^2)^{-1/4} g_k \qquad \text{and} \qquad \hat{h}_k = \frac{-i\not{k} + M}{(k^2 + M^2)^{3/4}} h_k;$$

the unnormalized Schwinger functions are then:

$$S(f, g, h) = \tilde{S}(f, \hat{g}, \hat{h}) = \int \prod_1^n \Phi(z_i) f_i(z_i)\, dz_i \left\{ \det_{1 \le j, l \le N'} \left[\frac{1}{1 + K} \right] (x_j, y_l) \right\}$$
$$\times \prod_{k=1}^{N'} \hat{g}_k(x_k) \hat{h}_k(y_k)\, dx_k\, dy_k\, \det_{\text{ren}}(1 + K)\, d\mu, \tag{1}$$

with

$$\det_{\text{ren}}(1 + K) = \det(1 + K) \exp\{\tfrac{1}{2}{:}(\text{Tr } K^2 - \text{Tr}_{\text{ren}} K^2){:}$$
$$+ \tfrac{1}{4}{:}(\text{Tr}^{(2)} K^4 - \text{Tr}^{(2)}_{\text{ren}} K^4){:} + \delta E_1 + \delta E_2\}$$

$$\text{Tr } K^2 - \text{Tr}_{\text{ren}} K^2 = \lambda^2 \text{ Tr} \int S(x - z) S(z - x)\, dz \Phi^2(x)\, d^3x$$

$$:\text{Tr}^{(2)} K^4: = \text{Tr } K^4 - :\text{Tr } K^4: - \int \text{Tr } K^4\, d\mu$$

$$\text{Tr}^{(2)} K^4 - \text{Tr}^{(2)}_{\text{ren}} K^4 = \lambda^4 \text{ Tr} \int S(x_1 - x_2) S(x_2 - x_3) S(x_3 - x_4)$$
$$\times S(x_4 - x_1) \left[4 \int \Phi(x_2)\Phi(x_3)\, d\mu \right.$$
$$\left. + 2 \int \Phi(x_2)\Phi(x_4)\, d\mu \right] d^3x_2\, d^3x_3\, d^3x_4 \Phi(x_1)^2\, d^3x_1 \tag{2}$$

$$\delta E_1 = \frac{1}{2} \int \text{Tr } K^2\, d\mu + \frac{1}{4} \int \text{Tr } K^4\, d\mu + \frac{1}{6} \int \text{Tr } K^6\, d\mu$$

$$\delta E_2 = -\frac{1}{4 \cdot 2} \int :\text{Tr}_{\text{ren}} K^2::\text{Tr}_{\text{ren}} K^4:\, d\mu - \frac{1}{4 \cdot 2}$$
$$\times \int (:\text{Tr}_{\text{ren}} K^2:)^2\, d\mu + \frac{1}{8 \cdot 3!} \int (:\text{Tr}_{\text{ren}} K^2:)^3\, d\mu.$$

The above equations define $\text{Tr}_{\text{ren}} K^2$ and $\text{Tr}^{(2)}_{\text{ren}} K^4$. We write the exponent in Equation 2 as $F(S, \Phi)$ with S understood as the internal fermion lines of the counterterms.

The partition function is therefore $Z_\Lambda = \int \det_{\text{ren}}(1 + K_\Lambda)\, d\mu$.

Result

Theorem. Let $\text{Re } \lambda^3 > 0$; then, $\exists K_1 > 0$ such that $|Z_\Lambda| \le e^{K_1 |\Lambda|}$.

Cutoff and t_Δ *Dependence*

For given $\varepsilon > 0$ and $M_1 > 0$, to be chosen later, we define the sequence of cutoff

$$M_o = 0, \qquad M_1, \ldots, M_i = M_1^{(1+\varepsilon)^{i-1}}, \ldots, M_\zeta$$

and the sequence of lattices

$$\mathscr{D}_1, \mathscr{D}_2, \ldots, \mathscr{D}_i, \ldots, \mathscr{D}_\zeta,$$

where $\mathscr{D}_1$ is a unit lattice (Λ being taken as the union of cubes of $\mathscr{D}_1$), and $\mathscr{D}_{i+1}$ is obtained from $\mathscr{D}_i$ by successive division in eight of the cubes of $\mathscr{D}_i$. The size $|\Delta|$ of a cube $\Delta \in \mathscr{D}_i$, $i > 1$ is such that:

$$M_i^{-3} < |\Delta| \leq 8M_i^{-3}.$$

For $\Delta \in \cup \mathscr{D}_i$, we define $i(\Delta)$ by $\Delta \in \mathscr{D}_{i(\Delta)}$.

Let us now define the cutoff fermion propagators

$$\begin{aligned} S_{ij}(z)(x, y) &= U_i(x, z)\Gamma\Lambda(z)V_j(z, y) \\ U_i(x, z) &= \int e^{ik(x-z)} \frac{-i\not{k} + M}{(k^2 + M^2)^{3/4}} \eta_i(k)\eta(k/M_\zeta)\, d^3k \\ V_j(z, y) &= \int e^{ik(z-y)} \frac{1}{(k^2 + M^2)^{1/4}} \eta_j(k)\eta(k/M_\zeta)\, d^3k; \end{aligned} \tag{3}$$

thus, $K = \lambda \sum_{i,j} \int S_{ij}(z)\Phi(z)\, d^3z$. We will specify later the exact definition of the cutoff function. For the moment, it is enough to know that $\eta_i(k) \neq 0$ for $M_i \leq |k| \leq M_{i+1}$ and $\sum_{i=0}^{\infty} \eta_i(k) = 1$.

We also decompose the boson field Φ in two different ways. We write

$$\Phi(x) = \sum_0^\infty \Phi_i(x).$$

The first decomposition is

$$\begin{aligned} \Phi^l(x) &= \int \Phi(x - y)\tilde{\rho}_l(y)\, dy \\ \rho_i(k) &= \prod_{j=0}^{2} g(k_j/M_{i+1}) - \prod_{j=0}^{2} g(k_j/M_i) \qquad i = 0, 1, \ldots \end{aligned} \tag{4}$$

and $g(k_j/M_o) = 0$, where

$$\begin{aligned} g(x) &= 1 & 0 &\leq |x| \leq \tfrac{1}{2} \\ 0 \leq g(x) &\leq 1 & \tfrac{1}{2} &\leq |x| \leq \tfrac{3}{2} \\ g(x) &= 0 & |x| &\geq \tfrac{3}{2} \end{aligned}$$

Φ^l with $l \geq k$ is well localized in cubes of $\mathscr{D}_k$.

The second decomposition is a weaker one

$$\begin{aligned} \Phi_o(x) &= \sum_{\Delta \in \mathscr{D}_1} \chi_\Delta(x) \frac{1}{|\Delta|} \int_\Delta \Phi(x)\, d^3x \\ \delta\Phi_j(x) &= \Phi(x) - \sum_{i=0}^{j-1} \Phi_i(x) \qquad j \geq 1 \\ \Phi_j(x) &= \sum_{\Delta \in \mathscr{D}_{j+1}} \chi_\Delta(x) \frac{1}{|\Delta|} \int_\Delta \delta\Phi_j(x)\, d^3x \qquad 1 \leq j < \zeta \\ \Phi_\zeta(x) &= \delta\Phi_\zeta(x); \end{aligned} \tag{4'}$$

therefore,

$$\Phi(x) = \sum_{i=0}^{\zeta} \Phi_i(x)$$

$$\delta\Phi_j(x) = \Phi(x) - \sum_{\Delta \in \mathscr{D}_j} \chi_\Delta(x) \frac{1}{|\Delta|} \int_\Delta \Phi(x)\, \mathrm{d}^3 x.$$

In this way, $\Phi_j(x)$ is weakly well localized in the cubes of $\mathscr{D}_j$ and constant in the cubes of $\mathscr{D}_{j+1}$ (roughly speaking, the integration over $\Delta \in \mathscr{D}_i$ acts as a momentum cutoff of order $|\Delta|^{-1/3}$, but for $j = \zeta$).

Now we introduce the expansion parameters $\{t_\Delta\}$, $0 \le t_\Delta \le 1$. Only the weakly well-localized part of Φ will be t_Δ-dependent; for counterterms, only the ones for which all internal fermion lines are well localized will be t_Δ-dependent.

Ignoring $t_{\Delta'}$ in our previous notation, we define for $\Delta \in \mathscr{D}_{i(\Delta)}$ the t_Δ-dependence by T_Δ such that on fields:

$$T_\Delta \Phi(x) = [1 - \chi_\Delta(x)]\Phi(x) + \chi_\Delta(x)\left[\sum_{j=0}^{i(\Delta)-1} \Phi_j(x) + \sum_{j=i(\Delta)}^{\zeta} t_\Delta \Phi_j(x)\right], \tag{5}$$

and on counterterms

$$T_\Delta F(S, \Phi) = F\left(\sum_{i,\, j \ge i(\Delta)+1} S_{ij},\, T_\Delta \Phi\right) + F(S, \Phi)$$

$$- F\left(\sum_{i,\, j \ge i(\Delta)+1} S_{ij},\, \Phi\right). \tag{6}$$

If $G := \det_{1 \le k,\, j \le N'}(1/(1+K))(x_k, y_j)\det_{\text{ren}}(1+K)$, its t_Δ-dependence is given by

$$T_\Delta G = \det_{1 \le k,\, j \le N'}\left(\frac{1}{1 + T_\Delta K}\right)(x_k, y_j)\det(1 + T_\Delta K)\exp\{T_\Delta F(S, \Phi)\}. \tag{7}$$

The general t_Δ-dependence is given by $\prod_\Delta T_\Delta G$. For later convenience, let us also introduce the part of K responsible for fermion ultraviolet divergences:

$$K_2 = \sum_{l,\, j \ge i(\Delta)+1} \lambda \int S_{lj}(z) T_\Delta \Phi(z)\, \mathrm{d}z \tag{8}$$

and $K_1 = T_\Delta K - K_2$.

Perturbation Formulas

A perturbation step consists of:

$$G = T_\Delta G\Big|_{t_\Delta = 0} + \int_0^1 \frac{\mathrm{d}}{\mathrm{d}t_\Delta} T_\Delta G\, \mathrm{d}t_\Delta. \tag{9}$$

Each t_Δ-derivative generates a vertex of the Yukawa theory plus its counterterms;

we do explicitly the renormalization. We write a t_Δ-derivative as:

$$\int \omega(x_1, \ldots, x_N; y_1, \ldots, y_{N'}) \frac{\mathrm{d}}{\mathrm{d}t_\Delta} \det_{1 \le k, j \le N'} \left(\frac{1}{1+K}\right)(x_k, y_j)$$

$$\det_{\text{ren}}(1+K) \prod_{i=1}^{N'} \mathrm{d}x_i \, \mathrm{d}y_i = \int [\omega(x_1, \ldots, x_{N'}; y_1, \ldots, y_{N'})]$$

$$\times \left[F \det_{1 \le k, j \le N'} \left(\frac{1}{1+K}\right)(x_k, y_j)\right.$$

$$+ \int B(y_{N'+1}, x_{N'+1})$$

$$\times \det_{1 \le k, j \le N'+1}(x_k, y_j) \, \mathrm{d}x_{N'+1} \, \mathrm{d}y_{N'+1}$$

$$+ \int \sum_1^{N'} \omega(x_1, \ldots, x_{i-1}, z_i, x_{i+1},$$

$$\ldots, x_{N'}; y_1, \ldots, y_{N'}) A(z_i, x_i) \det_{1 \le k, j \le N'}$$

$$\left. \times \left(\frac{1}{1+K}\right)(x_k, y_j) \, \mathrm{d}z_i \right]$$

$$\times \det_{\text{ren}}(1+K) \prod_1^{N'} \mathrm{d}x_i \, \mathrm{d}y_i \tag{A}$$

with
$$F = \frac{\mathrm{d}}{\mathrm{d}t_\Delta} \left\{ -\frac{1}{2} :\mathrm{Tr}_{\text{ren}} K_2^2: - \frac{1}{4} :\mathrm{Tr}_{\text{ren}}^{(2)} K_2^4: - \frac{1}{4} :\mathrm{Tr}\, K_2^4: \right.$$

$$\left. - \frac{1}{6} \mathrm{Tr}\, K_2^6 + \frac{1}{6} \int \mathrm{Tr}\, K_2^6 \, \mathrm{d}\mu + \delta E_2 \right\}$$

$$A = \left(\sum_{i=0}^{5} (-1)^{i+1} K_2^i \right) \frac{\mathrm{d}K_2}{\mathrm{d}t_\Delta}$$

$$B = \left(-K_1 \sum_{i=0}^{5} (-1)^i K_2^i + K_2^6 \right) \frac{\mathrm{d}K_2}{\mathrm{d}t_\Delta} + \frac{\mathrm{d}K_1}{\mathrm{d}t_\Delta}$$

with the t_Δ-dependence of $:\mathrm{Tr}_{\text{ren}} K_2^2:$ and $:\mathrm{Tr}_{\text{ren}}^{(2)} K_2^4:$ defined through Formulas 2 and 6.

In A, the renormalization of pure fermionic divergences has been done; however, we still have to cancel explicitly the vacuum diagrams whose counterterms are the derivative of δE_2. Only diagrams with well-localized boson internal legs are divergent; to perform this last renormalization, we therefore have to contract the well-localized part of some boson legs.

Consider the well-localized part of a boson field Φ localized in cubes of a cover $\mathscr{D}_{i(\Delta)}$ (see Formula 4: $\sum_{l \ge i(\Delta)} \Phi^l(x)$). Typically, such a term appears as part of an ω-function (as they are defined implicitly in Formula A). Showing this explicit boson field dependence, the contraction formula is:

$$\int \Phi(z)\omega(z)(x_1, \ldots, x_{N'}; y_1, \ldots, y_{N'})\det_{1 \le k, j \le N'}\left(\frac{1}{1+K}\right)(x_k, y_j)$$

$$\det_{\text{ren}}(1+K)\, dz \prod_{i=1}^{N'} dx_i\, dy_i\, d\mu = \iint (C(v, z)$$

$$\times \left\{\left[\frac{\delta\omega(v)}{\delta\Phi(z)}(x_1, \ldots, x_{N'}; y_1, \ldots, y_{N'}) + \right.\right.$$

$$F'(z)\omega(v)(x_1, \ldots, x_{N'}; y_1, \ldots, y_{N'}) +$$

$$\left.\sum_1^{N'} \int \omega(v)(x_1, \ldots, x_{i-1}, w_i, x_{i+1}, \ldots, x_{N'}; y_1, \ldots, y_{N'})A'(w_i, x_i)\, dw_i\right]$$

$$\times \det_{1 \le k, j \le N'}\left(\frac{1}{1+K}\right)(x_k, y_j) + \int B'(y_{N'+1}, x_{N'+1})$$

$$\times\, \omega(v)(x_1, \ldots, x_{N'}; y_1, \ldots, y_{N'})\det_{1 \le k, j \le N'+1}\left(\frac{1}{1+K}\right)(x_k, y_j)$$

$$\left.\times\, dx_{N'+1}\, dy_{N'+1}\right\} dz\, dv \prod_{i=1}^{N'} dx_i\, dy_i \det_{\text{ren}}(1+K)\, d\mu, \qquad \textbf{(B)}$$

where $\int \Phi(z)\omega(z)(x_1, \ldots, x_N; y_1, \ldots, y_{N'})\, dz$ is a part of $\omega(x_1, \ldots, x_{N'}; y_1, \ldots, y_{N'})$ showing the explicit dependence on $\sum_{l \ge i(\Delta)} \Phi^l(x)$, and with

$$F'(z) = \frac{\delta}{\delta\Phi(z)}\left[-\frac{1}{2}:\mathrm{Tr}_{\text{ren}}\, K_2^2: - \frac{1}{4}:\mathrm{Tr}_{\text{ren}}^{(2)}\, K_2^4: - \frac{1}{4}:\mathrm{Tr}\, K_2^4:\right]$$

$$A' = \left[\sum_{i=0}^{4}(-1)^{i+1}K_2^i\right]\frac{\delta K_2}{\delta\Phi(z)}$$

$$B' = \left[-K_1\sum_{i=0}^{4}(-1)^i K_2^i - K_2^5\right]\frac{\delta K_2}{\delta\Phi(z)} + \frac{\delta K_1}{\delta\Phi(z)}.$$

We apply this contraction formula to all the well-localized boson legs of $F(S, \Phi)$; each vertex of F is localized in cubes of $\mathscr{D}_{i(\Delta)}$ (the vertexes of A, B, F', A', and B' are also localized in cubes of $\mathscr{D}_{i(\Delta)}$), canceling in this way the counterterms in δE_2.

A perturbation step relative to t_Δ consists first of applying Formula 9 to the result of the previous steps, computing the derivatives with respect to t_Δ by use of Formula A, and then contracting all the well-localized boson legs of F by use of Formula B; each new vertex is localized in cubes of $\mathscr{D}_{i(\Delta)}$.

Expansion

Using for Formula 9 the symbolic notation

$$G = I_\Delta G + P_\Delta G, \text{ with } I_\Delta G = G\Big|_{t_\Delta = 0},$$

and with G supposedly t_Δ-dependent, the expansion for the partition function or for the unnormalized Schwinger functions is given by

$$Z = \prod'_{\Delta \in \cup \mathscr{D}_i \cap \Lambda} (I_\Delta + P_\Delta)^{|\Delta|^{-\varepsilon_1}} Z, \tag{11}$$

where $\prod'$ means that we first apply Formula 9 in cubes of $\mathscr{D}_1$, then in cubes of $\mathscr{D}_2$, and so on. For a given cube Δ, the formula is applied $|\Delta|^{-\varepsilon_1}$ times† for some $\varepsilon_1 > 0$ to be fixed later.

Let $\mathscr{D} = \bigcup_{i=1}^{\xi} \mathscr{D}_i^{|\Delta|^{-\varepsilon_1}}$, then

$$Z = \sum_{c \subset \mathscr{D}} \left(\prod_{\Delta \in c} P_\Delta \right) Z$$

We will prove for a specific cutoff (nonpreserving Osterwalder–Schrader positivity) the following proposition.

Proposition 1. For $|\text{Arg}\, \lambda| < \pi/6$, $\varepsilon_0 > 0$, $\varepsilon_1 > 0$, and $\varepsilon_0 + \varepsilon_1 < \frac{1}{6}$, there exist $\varepsilon > 0$ that depend on ε_0 and ε_1 and $M_1 > 0$, $K_1 > 0$, and $K_2 > 0$ that depend on ε and λ such that

$$\left| \prod'_{\Delta \in c} P_\Delta Z \right| \le K_1 e^{K_2 |\Lambda|} \prod_{\Delta \in c} |\Delta|^{\varepsilon_0}.$$

As a corollary, one has[1]:

Theorem. Under the assumptions of Proposition 1, $\exists K_3 > 0$ such that

$$|Z| \le e^{K_3 |\Lambda|}.$$

We shall then argue that for $|\text{Arg}\, \lambda| < \pi/8$, the limit as $\zeta \to \infty$ is the same as for the theory defined with cutoff preserving the $O - S$ positivity (see Appendix 1).

Wick Bounds

$$\prod'_{\Delta \in c} P_\Delta Z = \sum_{\omega, N'} \int \omega(x_1, \ldots, x_{N'}; y_1, \ldots, y_{N'})$$
$$\times \det_{1 \le i, j \le N'} \left(\frac{1}{1 + K} \right) (x_i, y_j) \det_{\text{ren}}(1 + K).$$

The aim of this section is to bound each term of the sum.

Let $n(\Delta)$ be, for a given such term, the number of perturbation steps such that $t_\Delta \ne 0$ (i.e., the number of perturbation vertexes produced by t_Δ-derivatives). It follows a partition of $\Lambda : \Lambda = \bar{\mathscr{D}} \cup \bar{\bar{\mathscr{D}}}$, with

$$\bar{\mathscr{D}} = \{\Delta, \Delta \subset \Lambda \mid \xi - 3 > i(\Delta) > 1, n(\Delta) < |\Delta|^{-\varepsilon_1}$$

if $\Delta \subset \Delta'$, $\Delta' \in \mathscr{D}_{i(\Delta)-1}$ and

$$n(\Delta') = |\Delta'|^{-\varepsilon_1}\} \cup \{\Delta \in \mathscr{D}_1 \cap \Lambda, n(\Delta) = 0\},$$

$$\bar{\bar{\mathscr{D}}} = \{\Delta, \Delta \subset \Lambda \mid \Delta \in \mathscr{D}_{\xi - j}, j = 0, 1, 2, 3; \Delta \subset \Delta', \Delta' \in \mathscr{D}_{\xi - j - 1}, n(\Delta') = |\Delta'|^{-\varepsilon_1}\}. \tag{12}$$

† By $|\Delta|^{-\varepsilon_1}$, we mean the integer part of $|\Delta|^{-\varepsilon_1}$.

The main idea of this section is to use the "positivity" of the interaction to dominate the part of the boson legs that cannot be controlled by Gaussian integration (i.e., by Φ^2 positivity). For cubes in $\mathscr{D}$, the positivity comes from the effective potential (computed with K_2); for cubes of $\bar{\mathscr{D}}$, it comes from the mass counterterm. For $\Delta \in \mathscr{D}$, we reduce the problem to computing the effective potential by introducing periodic boundary conditions on Δ and remarking that in Δ (by Formula 12) Φ is constant; we write Φ_Δ for such a field.

With periodic boundary conditions on the Δ's of $\mathscr{D}$, one has

$$K_2^P := \sum_{\Delta \in \mathscr{D}} \sum_{l,\, j \geq i(\Delta)+1} \chi_\Delta U_i^P V_i^P \chi_\Delta \Phi_\Delta = \sum_{\Delta \in \mathscr{D}} \lambda S_\Delta^P \Phi_\Delta, \tag{13}$$

with

$$S_\Delta^P(x, y) = \sum_{\vec{n} \in \mathbb{Z}^3} \int e^{ip(x-y-n|\Delta|^{1/3})} \frac{-i\not{p} + \mathrm{M}}{p^2 + \mathrm{M}^2} [1 - \eta_{(i+1)}(p)]^2 \eta_{(\xi)}^2(p) \Gamma \chi_\Delta(x) \chi_\Delta(y)$$

and Fourier transform in $k = (k_o, k_1, k_2) k_i \in \mathbb{Z}|\Delta|^{-1/3}$

$$\tilde{S}_\Delta^P(k) = \frac{-i\not{k} + \mathrm{M}}{k^2 + \mathrm{M}^2} [1 - \eta_{(i+1)}(k)]^2 \eta_{(\xi)}^2(k).$$

One can prove that for $|\mathrm{Arg}\, \lambda| \leq \pi/2 - \varepsilon$, $\varepsilon > 0$ there exists $C(\varepsilon) \geq 1$ such that

$$\left\| \frac{1}{1 + K_2^P} \right\| \leq C(\varepsilon), \tag{14}$$

where $\| \ \|$ denotes the norm.

We write $K = K_1 + K_2 - K_2^P + K_2^P$ and remark that for $z \in \Delta$, $\Delta \in \bar{\mathscr{D}}$, $K(z) = K_1(z)$.

Now we write

$$1 + K = (1 + K_2^P)\left(1 + \frac{K_1}{1 + K_2^P} + \frac{K_2 - K_2^P}{1 + K_2^P}\right) \tag{15}$$

and explain how to factorize correspondingly the determinant.

The main lemma is, following the ideas of Seiler–Simon[15] and McBryan[8]:

Lemma 1.

$$\left\| \Lambda^{N'} \frac{1}{1 + K} \det(1 + K) \right\| \leq |\det(1 + K_2^P)|$$
$$\times \exp\left\{ N' + 2\,\mathrm{Tr}|K_1| + \left| \mathrm{Tr}\, \frac{(K_2 - K_2^P)}{1 + K_2^P} \right| + \mathrm{Tr}|K_2 - K_2^P| \right\}.$$

The first step in the proof of the lemma is the use of

Lemma 2.[15]

$$\left\| \Lambda^{N'} \left(\frac{1}{1 + P + Q} \right) \det(1 + P + Q) \right\|^2$$
$$\leq \left\| \Lambda^{N'} \left(\frac{1}{|1 + P| + |Q|} \right) \det(|1 + P| + |Q|) \right\|$$
$$\times \left\| \Lambda^{N'} \left(\frac{1}{|1 + P| + W|Q|W^{-1}} \right) \det(|1 + P| + W|Q|W^{-1}) \right\|,$$

where W is the unitary operator.

We take $1 + P = 1 + K_2$, $Q = K_1$, and rewrite

$$|1 + P| + |Q| = |1 + K_2^P| \left(\left| 1 + \frac{K_2 - K_2^P}{1 + K_2^P} \right| + \frac{|K_1|}{|1 + K_2^P|} \right), \tag{16}$$

and it is enough to develop further one of the terms (e.g., the first one) of the right-hand side of the inequality of Lemma 2.

Applying

Lemma 3.[15]

$$\left\| \Lambda^{N'} \frac{1}{R^*T} \det(R^*T) \right\| \leq \left\| \Lambda^{N'} \frac{1}{|R|} \det(|R|) \right\| \left\| \Lambda^{N'} \frac{1}{|T|} \det(|T|) \right\|$$

with argument R^*T, the right-hand side of Formula 16, it remains to bound

$$\Lambda^{N'} \frac{1}{|1 + A| + |B|} \det(|1 + A| + |B|),$$

with

$$A = \frac{K_2 - K_2^P}{1 + K_2^P}, \qquad B = \frac{K_1}{1 + K_2^P}.$$

For this purpose, we modify a lemma of Seiler–Simon[15] by use of a remark of McBryan:

Lemma 4.

$$\left\| \Lambda^{N'} \frac{1}{|1 + A| + |B|} \det(|1 + A| + |B|) \right\| \leq \exp\{N' + 2 \operatorname{Tr}|B| + \tfrac{1}{2} \operatorname{Tr}(A + A^* + AA^*)\}.$$

Proof. Let $C = |1 + A| - 1 = C_+ - C_- (\Rightarrow 0 \leq C_- \leq 1)$ and

$$D = \frac{1}{\sqrt{1 + C_+}} |B| \frac{1}{\sqrt{1 + C_+}} \qquad E = \frac{1}{\sqrt{1 + D}} C_- \frac{1}{\sqrt{1 + D}};$$

then,

$$\begin{aligned} X &= \left\| \Lambda^{N'} \frac{1}{|1 + A| + |B|} \det(|1 + A| + |B|) \right\| \\ &= \left\| \Lambda^{N'} \frac{1}{1 + C + |B|} \det(1 + C + |B|) \right\| \\ &\leq \det(1 + C_+) \left\| \Lambda^{N'} \frac{1}{1 + D - C_-} \det(1 + D - C_-) \right\| \\ &\leq \det(1 + C_+) \det(1 + D) \left\| \Lambda^{N'} \frac{1}{1 - E} \det(1 - E) \right\|. \end{aligned}$$

From

Lemma 5,[15] if $\operatorname{Tr}|T|^{n+1} < \infty$,

$$\left\| \Lambda^m \frac{1}{1 - T} \det(1 - T) e^{-\operatorname{Tr} T - \cdots - \operatorname{Tr} T^n/n} \right\| \leq e^{m[1 + 1/2 + \cdots + 1/n]}.$$

It follows that

$$X \le \det(1 + C_+)\det(1 + D)e^{N'}e^{-\mathrm{Tr}\,E}.$$

Now

$$\mathrm{Tr}\,E = \mathrm{Tr}(1 + D)^{-1}C_- \ge \mathrm{Tr}(1 - D)C_- \ge \mathrm{Tr}\,C_- - \mathrm{Tr}\,D\|C_-\| \ge \mathrm{Tr}\,C_- - \mathrm{Tr}|B|,$$

and

$$\det(1 + D) \le \det(1 + |B|) \le e^{\mathrm{Tr}|B|}.$$

Setting $O_A = A + A^* + AA^*$ and O_{A^+} its non-negative part,

$$1 + O_{A^+} = |1 + A|^2 P_+ + (1 - P_+) = (1 + C)^2 P_+ + (1 - P_+) = (1 + C_+)^2,$$

where P_+ is the projection associated with C_+; therefore,

$$\det(1 + C_+)^2 = \det(1 + O_{A^+}) \le e^{\mathrm{Tr}\,O_{A^+}}.$$

Thus,

$$X \le e^{N' + 2\,\mathrm{Tr}|B| - \mathrm{Tr}\,C_- + 1/2\,\mathrm{Tr}\,O_{A+}}.$$

Now we prove that $-\mathrm{Tr}\,C_- \le -\frac{1}{2}\mathrm{Tr}\,O_{A^-}$. As above, $1 - O_{A^-} = (1 - C_-)^2$ and $\sqrt{1 - O_{A^-}} = 1 - C_-$; thus, $\mathrm{Tr}\,C_- = \mathrm{Tr}(1 - \sqrt{1 - O_{A^-}}) \ge \frac{1}{2}\mathrm{Tr}\,O_{A^-}$ since $1 - \sqrt{1 - x} \ge x/2$ for $0 \le x \le 1$, and it follows that

$$X \le e^{N' + 2\,\mathrm{Tr}|B| + 1/2\,\mathrm{Tr}(A + A^* + AA^*)}.$$

Now with the bound of Formula 14 and $(\mathrm{Tr}|AB|^p)^{1/p} \le \|B\|(\mathrm{Tr}|A|^p)^{1/p}$, we have that

$$\mathrm{Tr}\left|\frac{K_1}{1 + K_2^p}\right| \le C\,\mathrm{Tr}|K_1| \qquad \mathrm{Tr}\left(\frac{|K_2 - K_2^p|}{|1 + K_2^p|}\right)^2 \le C^2\,\mathrm{Tr}|K_2 - K_2^p|^2,$$

and $\|\Lambda^{N'}(1/|1 + K_2^p|)\det(|1 + K_2^p|)\| \le |\det(1 + K_2^p)|$, which finishes the proof of the lemma.

Our next step is to bound the determinant with periodic boundary conditions; that is:

$$\left|\det(1 + K_2^p)\exp\left(1/2\lambda^2 \sum_{\Delta \in \bar{\mathscr{D}}} \Phi_\Delta^2 \sum \tilde{S}_\Delta^p(k)^2\right)\right| = \prod_{\Delta \in \bar{\mathscr{D}}} \exp \mathrm{Re}\,P_\Delta,$$

with

$$P_\Delta = \sum_{\Delta^{1/3}k \in \mathbb{Z}^3} \left(\log\left\{1 + \frac{\lambda^2\Phi_\Delta^2}{k^2 + \mathrm{M}^2}[1 - \eta_{(i+1)}(k)]^4\eta_{(\zeta)}^4\right\}\right.$$
$$\left. - \frac{\lambda^2\Phi_\Delta^2}{k^2 + \mathrm{M}^2}[1 - \eta_{(i+1)}(k)]^4\eta_{(\zeta)}(k)^4\right).$$

This is done by the following lemma, proved in Appendix 1.

Lemma 6. Let $\mathrm{Re}\,\lambda^3 > 0$; there then exist constants $0(1)$ and $\varepsilon > 0$, such that

$$e^{\mathrm{Re}\,P_\Delta} \le e^{-0(1)\mathrm{Re}\,\lambda^3|\Delta|\,|\Phi_\Delta|^3 + 0(1)|\Delta|^{-\varepsilon}} \qquad \text{for} \quad |\lambda|\,|\Phi_\Delta| < \mathrm{M}_\zeta$$

and

$$e^{\mathrm{Re}\,P_\Delta} \le e^{-0(1)\mathrm{Re}\,\lambda^2|\Delta|\mathrm{M}|\Phi_\Delta|^2} \qquad \text{for} \quad |\lambda|\,|\Phi_\Delta| \ge \mathrm{M}_\zeta.$$

The first part of the bound is the effective potential; the second part controls the error terms due to the cutoff and is controlled by the fact that in each $\Delta' \supset \Delta$, $\Delta \in \mathscr{D}$, one produces $|\Delta'|^{-\varepsilon_1}$ convergent vertexes, that is, a factor $0(1)^{-|\Delta'|^{-\varepsilon_1}}$ per cubes Δ'. This means that we can give $0(1)^{-1/2|\Delta'|^{-\varepsilon_1}|\Delta|/|\Delta'|}$ to each cube $\Delta \subset \Delta'$. Now, by use of the fact that $|\Delta'|/|\Delta| \leq 64|\Delta|^{-\varepsilon}$ and with $\varepsilon_1 > 3\varepsilon$, this gives a convergent factor $0(1)^{-|\Delta|^{-\varepsilon_2}}$ per cubes $\Delta \in \bar{\mathscr{D}}$ for some $\varepsilon_2 > 0$ big enough if ε_1 is chosen big enough in order that $|\Delta|^{-\varepsilon_2 2} \gg |\Delta|^{-\varepsilon}$ of Lemma 6.

We still have to control the correction terms to the introduction of periodic boundary conditions.

Lemma 7. For some $\varepsilon_3 > 0$ and related constants C_1, C_2, C_3, and C_4,

(i) $e^{\mathrm{Tr}|K_1|} \leq \prod_{\Delta \in \mathscr{D} \cup \bar{\mathscr{D}}} e^{C_1|\Delta|^{1/3-\varepsilon_3}|\Phi_\Delta|}$

(ii) $$e^{1/2(\mathrm{Tr}\,K2 - \mathrm{Tr}_{\mathrm{ren}}\,K2)+1/2 \sum_{\Delta, k} \lambda^2 \frac{1}{k^2+M^2}[1-\eta_{(i+1)}(k)]^4 \eta_{(\zeta)}^{(k)4}\Phi_\Delta^2} \leq \exp\left(C_2 \sum_{\Delta \in \bar{\mathscr{D}}} |\Delta|^{2/3}\Phi_\Delta^2\right)$$

(iii) $\mathrm{Tr}|K_2 - K_2^\xi|^2 \leq C_3 \sum_{\Delta \in \bar{\mathscr{D}}} |\Delta|^{2/3}\Phi_\Delta^2$

(iv) $$\left|\mathrm{Tr}\,\frac{K_2 - K_2^\xi}{1 + K_2^\xi}\right| \leq C_4 \sum_{\Delta \in \bar{\mathscr{D}}} (|\Delta|^{2/3}\Phi_\Delta^2 + |\Delta|^{1+\varepsilon_3}|\Phi_\Delta|^3).$$

Proof of the lemma. For $\Delta \in \bar{\bar{\mathscr{D}}}$, $K_1 = K$, and the cutoff is $|\Delta|^{-1/3-5\varepsilon}$; for $\Delta \in \bar{\mathscr{D}}$, the cutoff of K_1 is bounded by $\Delta^{-1/3-\varepsilon}$. Thus, in both cases,

$$\mathrm{Tr}|K_1(z)| \leq C_o|\Phi(z)|\,|\Delta|^{-2/3-0(1)\varepsilon}, \tag{17}$$

from which i follows.

Parts i and ii are almost the same. One decomposes K in K_1 and K_2. Because of the momentum cutoff in K_1, $\mathrm{Tr}\,K_1^2$, and $\mathrm{Tr}\,K_1K_2$ are bounded by $|\Delta|^{2/3+6\varepsilon}\Phi_\Delta^2$. Thus, we still have to bound $\mathrm{Tr}|K_2 - K_2^\xi|^2$, and if we prove it to be finite, this means the same thing for the corresponding counterterm.

Part iv reduces to a similar bound. In fact,

$$\mathrm{Tr}\,\frac{K_2 - K_2^\xi}{1+K_2^\xi} = \mathrm{Tr}(K_2 - K_2^\xi) - \mathrm{Tr}(K_2 - K_2^\xi)K_2^\xi + \mathrm{Tr}\,\frac{(K_2 - K_2^\xi)(K_2^\xi)^2}{1+K_2^\xi},$$

and $\mathrm{Tr}(K_2 - K_2^\xi) = 0$ (pseudoscalar theory)

$$\mathrm{Tr}(K_2 - K_2^\xi)K_2^\xi = \mathrm{Tr}(K_2 - K_2^\xi)K_2 - \mathrm{Tr}(K_2 - K_2^\xi)^2$$

$$\left|\mathrm{Tr}\,\frac{(K_2 - K_2^\xi)K_2^\xi K_2^\xi}{1+K_2^\xi}\right| \leq C(\varepsilon)(\mathrm{Tr}|K_2 - K_2^\xi|^2\,\mathrm{Tr}|(K_2^\xi)^2|^2)^{1/2}, \tag{18}$$

and by power counting $\mathrm{Tr}|(K_2^\xi)^2|^2 \leq \sum_{\Delta \in \bar{\mathscr{D}}} |\Delta|^{1+1/3-\varepsilon}\Phi_\Delta^4$ from the fact that the momenta in K_2 are bigger than $|\Delta|^{-1/3}$, thus proving the argument.

We now bound $\mathrm{Tr}|K_2 - K_2^p|^2$ and, in the same way, $\mathrm{Tr}(K_2 - K_2^\xi)K_2$. We remark that $S_\Delta^p(x)$ can be written as $\sum_{n \in |\Delta|^{1/3}\mathbb{Z}^3} S(x+n)$, and thus $(S - S_\Delta^p)(x) \times \chi_\Delta(x) = \sum_{n \neq 0} S(x+n)$. On the other hand, since the momenta of the fermion are, for $K_2(z)$, $z \in \Delta$, and $\Delta \in \bar{\mathscr{D}}$, bigger than $|\Delta|^{-1/3-\varepsilon/3}$, one can always sum over all the possible localizations $\Delta \in \bar{\mathscr{D}}$.

We then consider two cases:

1) that two of the localization cubes are not neighbors, in which case we use scaling to prove

$$\mathrm{Tr}\,|K_2 - K_2^\varrho|^2 \le C_3' \sum_{\Delta \in \bar{\mathscr{D}}} |\Delta|\,|\Delta|^{-1/3-\varepsilon}\Phi_\Delta^2, \tag{19}$$

2) that there are no two localization cubes that are not neighbors, in which case, with the remark above, they cannot all be equal.

A typical term is (with, e.g., $\Delta_1 \neq \Delta_2$):

$$\int \chi_{\Delta_1}(x_1)U(x_1, x_2)\chi_{\Delta_2}(x_2)V(x_2, x_3)\chi_{\Delta_3}(x_3)$$

$$\times\, U(x_3, x_4)\chi_{\Delta_4}(x_4)V(x_4, x_1)\, dx_1\, dx_2\, dx_3\, dx_4$$

$$= \int \tilde{\chi}_{\Delta_1}(l_1 - l_4)\tilde{U}(l_1)\chi_{\Delta_2}(l_2 - l_1)\tilde{V}(l_2)\tilde{\chi}_{\Delta_3}(l_3 - l_2)$$

$$\times\, \tilde{U}(l_3)\tilde{\chi}_{\Delta_4}(l_4 - l_3)\tilde{V}(l_4)\, dl_1\, dl_2\, dl_3\, dl_4. \tag{20}$$

We now expand $\tilde{U}(l_1)$ around l_2. Since $\int \tilde{\chi}_{\Delta_1}(l_1 - l_4)\tilde{\chi}_{\Delta_2}(l_2 - l_1)\, dl_1 = 0$, the first term is zero. By symmetry, the second one is also zero. The remaining integral is linearly convergent (the starting one was, in principle, linearly divergent). The price we pay in "gaining" two powers in l_2 is "losing" two powers in l_1. We finally obtain the bound

$$\sum_{\Delta \in \bar{\mathscr{D}}} |\Delta|\,|\Delta|^{-1/3-\varepsilon}\Phi_\Delta^2.$$

Collecting all these results gives the bounds iv and iii and ends the proof of the lemma.

We still have to bound, for the terms in $\Delta \in \bar{\bar{\mathscr{D}}}$, the mass counterterm: $\mathrm{Tr}\, K^2 - \mathrm{Tr}_{\mathrm{ren}}\, K^2$. By use of the fact that the cutoff for $\Delta \in \bar{\bar{\mathscr{D}}}$ is $|\Delta|^{-1/3-5\varepsilon}$, we obtain

$$[\mathrm{Tr}\, K^2 - \mathrm{Tr}_{\mathrm{ren}}\, K^2](z) \le -|\Delta|^{-1/3+\varepsilon}\Phi^2(z),$$

which dominates the bound of Formula 17 and gives $\prod_{\Delta \in \bar{\bar{\mathscr{D}}}} e^{0(1)|\Delta|^{-\varepsilon}}$ for some $\varepsilon > 0$.

By use of a part of the convergence given by the effective potential, one bounds in the same way the term $e^{\mathrm{Tr}\,|K_1|}$ from part i of Lemma 7.

To conclude this section, we have to bound

$$\int \omega(x_1, \ldots, x_{N'}; y_1, \ldots, y_{N'})\det{}_{1 \le i, j \le N'}\left(\frac{1}{1 + K}\right)(x_i, y_j) \prod_i dx_i\, dy_i. \tag{21}$$

In ω the variables x and y are either associated with $\hat{g}$ and $\hat{h}$ or belong to chain $B(x_i, z_1)A(z_1, z_2) \cdots A(z_p, y_j)$. Starting with B, or a similar chain with B' and A'. $B(x, y)$ is either of the form $K_1 K_2^i(d/dt)K_2$ or the form $K_2^6(dK_2/dt)$. We write it as $\int B^1(x, z)B^2(z, y)\, dz$. The splitting is chosen so that both Hilbert Schmidt norms of B^1 and B^2 exist. For the form with K_1, one splits the cutoff fermion line in K_1 into two equal parts that generate lines of power $(p^2 + m^2)^{-1/4}$ and an upper cutoff $\Delta^{-1/3-\varepsilon}$. For $K_2^6(dK_2/dt)$, the cut is made according to the figure

$$K_2^6 \frac{dK_2}{dt}$$

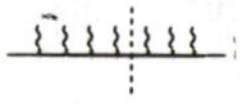

where for the cut line a power of 1/4 is given to the left and 3/4 to the right, giving rise to a finite Hilbert Schmidt norm (in particular, after boson integration, for the right one, to a finite six-order vacuum diagram). A similar splitting is made for B', but in this case one uses the fact that one boson field is derivated (this is the reason why one needs to develop only up to K_2^5).

With this splitting, $\omega(x_1, \ldots, x_{N'}; y_1, \ldots, y_{N'})$ can be written as:

$$\int \omega'(x_1, \ldots, x_{N'}; z_1, \ldots, z_N)\omega''(z_1, \ldots, z_N; y_1, \ldots, y_{N'})\, \mathrm{d}z_1, \ldots, \mathrm{d}z_N,$$

where $N' - N$ is the number of ψ fields (or $\bar{\psi}$ fields) in the case of a Schwinger function.

As a result, Formula 21 multiplied by $\det_{\text{ren}}(1 + K)$ is bounded by

$$\|\omega'\|_{L_2} \|\omega''\|_{L_2} \|\Lambda^{N'}(1/1 + K)\det_{\text{ren}}(1 + K)\|.$$

For shortage we note‡ $\|\omega'\|_{L^2} \|\omega''\|_{L^2} = \|\omega\|^2$.

Then, from Lemmas 1–7, one obtains:

$$\left|\int \omega(x_1, \ldots, x'_{N'}; y_1, \ldots, y_{N'})\det_{1 \le l, j \le N'} \left(\frac{1}{1 + K}\right)(x_l, y_j)\det_{\text{ren}}(1 + K) \prod_{i=1}^{N'} \mathrm{d}y_i\, \mathrm{d}x_i\right|$$

$$\le \|\omega\|^2 \prod_{\text{vert}} 0(1) \prod_{\Delta \in \mathscr{D} \cup \bar{\bar{\mathscr{D}}}} e^{0(1)|\Delta| - 0(1)\varepsilon} \prod_{\Delta \in \bar{\mathscr{D}}} e^{-0(1)|\Delta|\,|\Phi_\Delta|^3} \prod_{\Delta \in \bar{\bar{\mathscr{D}}}} e^{-0(1)|\Delta|^{-1/3+\varepsilon} \int_\Delta \Phi^2(z)}, \quad \textbf{(22)}$$

where $\prod_{\text{vert}} 0(1)$ is the product over all vertexes produced by the expansion steps, and $0(1)$ is the distribution of $e^{N'}$ of Lemma 4 from the fact that the order of the determinant is up to a fixed constant (the initial fermion fields), less than seven times the number of vertexes produced (P vertexes[9]).

Now we perform the boson integration by means of the method of combinatoric factors.

Gaussian integration consists of first contracting localized boson fields and then summing over the localizations. The combinatoric factor for the sum over all Δ', $|\Delta| \le |\Delta'|$ of a field localized in Δ contracting to a field localized in Δ': $\{\sup[1, \Delta^{-1/3}\, \mathrm{d}(\Delta, \Delta')]\}^4$ is controlled by the factor $[\mathrm{\tilde{d}}(\Delta, \Delta')]^{-q}0(q)$ $(q \ge 4)$ that one can extract from the propagator of the contraction line if M, its lower cutoff, is bigger than $|\Delta|^{-1/3}$, where $\mathrm{d}(\Delta, \Delta')$ is the Euclidean distance, and $\mathrm{\tilde{d}}(\Delta, \Delta') = \sup[1, \mathrm{M}\, \mathrm{d}(\Delta, \Delta')]$ is the scaled distance. In the case where $\mathrm{M} < |\Delta|^{-1/3}$, we cannot perform the Gaussian integration (the Φ^2 positivity is not enough), and we have to use domination by the last two factors of Formula 22, stronger positivity). This is the next section.

Domination of Badly Localized Legs

Consider a field Φ' localized in Δ and produced during a $t_{\Delta'}$ perturbation step. Let Φ be the field in Δ at the end of the expansion for the term that we are examining. The t_Δ-dependence of these two fields is not the same; however, by construction, $\Phi'(z) - \Phi(z)$ is well localized in Δ; that is, it is a field with components

‡ Note that this definition corrects an error in Reference 10.

Φ_i, only with $i \geq i(\Delta)$. This means that we write $\Phi'(z) = \Phi'(z) - \Phi(z) + \Phi(z)$, perform the Gaussian integration for $\Phi' - \Phi$, and dominate Φ as follows:

For $z \in \Delta$,

$$|\Phi(z)| = |\Phi_\Delta| \leq |\Delta|^{-\varepsilon-\varepsilon_1-1/3}\, \eth(\Delta, \Delta')^2 \exp\{|\Delta|^{2(\varepsilon+\varepsilon_1)+2/3}\, \eth(\Delta, \Delta')^{-4}\Phi_\Delta^2\}$$

and thus

$$\prod_{\Delta, \Delta'} |\Phi_\Delta| \leq \prod_{\text{propagators}} \eth(\Delta, \Delta')^2 \prod_{P \text{ vertexes}} |\Delta|^{-7(\varepsilon+\varepsilon_1)} \prod_{\text{by } \Phi_\Delta \text{ leg}} |\Delta|^{-1/3+\varepsilon} \exp[\Delta^{2(\varepsilon+\varepsilon_1)+2/3}\Phi_\Delta^2].$$

Now

$$\sum_{\substack{\Delta \in \mathscr{D}_{i(\Delta')} \\ t_\Delta \neq 0}} \Phi_\Delta^2 \leq \sum_{\Delta'' \in \bar{\mathscr{D}} \cup \bar{\bar{\mathscr{D}}}} \Phi_{\Delta''}^2$$

since $\bar{\bar{\mathscr{D}}} \cup \bar{\mathscr{D}}$ is obtained by subdivision of the cubes of $\mathscr{D}_{i(\Delta')}$. By use of

$$\sum_{i(\Delta)} \mathscr{L} \leq \sup |\Delta|^{-\varepsilon} \mathscr{L},$$

the fact that we create at most $|\Delta|^{-\varepsilon_1}$ P vertexes in Δ, and that

$$\int_\Delta \Phi^2(z)\, dz \geq |\Delta|\Phi_\Delta^2,$$

we can dominate all badly localized legs by

$$\exp\left[-\frac{0(1)}{2} \sum_{\Delta \in \bar{\mathscr{D}}} |\Delta|\,|\Phi_\Delta|^3 - \frac{0(1)}{2} \sum_{\Delta \in \bar{\bar{\mathscr{D}}}} |\Delta|^{-1/3}|\Delta|\Phi_\Delta^2 + \sum_{\Delta \in \bar{\mathscr{D}} \cup \bar{\bar{\mathscr{D}}}} |\Delta|^{2/3+\varepsilon}\Phi_\Delta^2\right] \leq \prod_{\Delta \in \bar{\mathscr{D}} \cup \bar{\bar{\mathscr{D}}}} 0(1). \quad \textbf{(23)}$$

With ω', the new ω obtained by suppressing badly localized legs or replacing them by well-localized legs, Formula 22 is now bounded by

$$\sup_{\omega'} \|\omega'\|^2 \prod_{P \text{ vertexes}} 0(1) \prod_{\Delta \in \bar{\mathscr{D}} \cup \bar{\bar{\mathscr{D}}}} 0(1) \prod_{P \text{ vertexes}} |\Delta|^{-7(\varepsilon+\varepsilon_1)} \prod_{\substack{\text{vertex without} \\ \text{boson}}} |\Delta|^{-1/3+\varepsilon} \prod_{\text{propagators}} \eth(\Delta, \Delta')^{0(1)}, \quad \textbf{(24)}$$

where we used $\eth(\Delta, \Delta') \leq 0(1)\, \eth(\Delta, \Delta_1') \cdots \eth(\Delta_5', \Delta')$ since there are at most six propagators between two bosons and since all cubes in such a chain of propagators are of the same size. The factor $\Delta^{-1/3+\varepsilon}$ per vertex without a boson is the power one will recover by not needing to integrate over the boson line.

We now perform the Gaussian integration of the left-hand side of Formula 22. Formula 24 reduces the bound we are looking for to a bound on

$$\int \|\omega'\|^2\, d\mu = \sum G, \quad \textbf{(25)}$$

where the sum extends over all contraction graphs.

In the next three sections, we give the combinatoric factors for the Gaussian integration, then for the terms created by the expansion steps, and finally the convergent factors that come from each graph (restricting ourselves to the case of the partition function; the Schwinger function case is as in Reference 10).

Gaussian Integration (of well-localized Φ)

Contractions are made starting from the smallest cubes. Let Φ be a field in Δ contracting with another one in Δ', $|\Delta| \leq |\Delta'|$. Each boson field in Δ is decomposed along the frequencies below $i(\Delta)$:

$$\Phi(x) = \sum_{l=0}^{i(\Delta)-1} \int \Phi(x-y)\tilde{\rho}_{(l)}(y)\,dy + \int \Phi(x-y)\left[\sum_{i(\Delta)}^{\infty} \tilde{\rho}_j(y)\right] dy = \sum_{l=0}^{i(\Delta)} \Phi^{(l)}(x).$$

In the contraction process, the choice of $i(\Delta')$ is made by $|\Delta'|^{-\varepsilon}$ since $\sum_{i(\Delta')} |\Delta'|^{\varepsilon} \leq 1$ for M_1 large enough. The choice of $\Phi^{(l)}$ is made also with $|\Delta|^{-\varepsilon}$ by use of $\sum_l 1 \leq |\Delta|^{-\varepsilon}$.

The propagator of $\Phi^{(l)}$ decreases like $[M_l\, d(\Delta, \Delta')]^{-4}$ when $d(\Delta, \Delta') \neq 0$, thus from

$$\sum_{\Delta' \in \mathscr{D}_{i(\Delta')}} \mathrm{d}(\Delta, \Delta')^{-4} \leq 0(1)|\Delta'|^{-1}M_l^{-3} \leq (|\Delta|^{-1}M_l^{-3})^{1/2}(|\Delta'|^{-1}M_l^{-3})^{1/2},$$

which holds because $|\Delta| \leq |\Delta'|$ and for $l = i(\Delta)$, because in $\mathscr{D}_{i(\Delta')}$ there are at most nine cubes such that $d(\Delta, \Delta') = 0$, the combinatoric factor for the choice of Δ' is:

$$\prod \mathrm{d}(\Delta, \Delta')^4 \prod_{\text{vertexes}} |\Delta|^{-1/2-\varepsilon}M_l^{-3/2}.$$

The choice of the field in Δ' is accomplished by remarking that such a field has been created by a perturbation in Δ'', with $i(\Delta') = i(\Delta'')$, so by use of the fact that fermions have lower cutoff $|\Delta|^{-1/3-\varepsilon}$, we take $\mathrm{d}(\Delta, \Delta')^4$.

Finally, the combinatoric factor for Gaussian integration is

$$\prod_{\text{propagators}} [\mathrm{d}(\Delta, \Delta')]^{0(1)} \prod_{P\text{ vertexes}} |\Delta|^{-7\varepsilon_1-14\varepsilon} \prod_{\text{vertexes}} |\Delta|^{-1/2+\varepsilon}M_l^{-3/2}. \qquad \mathbf{(25)}$$

Combinatoric Factors for the Expansion

These factors are the same as in Reference 2, so we will just give the result

$$\prod_{P\text{ vertexes}} |\Delta|^{-0(1)\varepsilon} \prod_{\text{propagators}} \mathrm{d}(\Delta, \Delta')^{0(1)}. \qquad \mathbf{(26)}$$

Estimates on Graphs and Proof of the Proposition

To bound a graph G of Formula 25, we follow Glimm and Jaffe.[5] We first extract the localization factors, then decompose each graph G in subgraphs, and finally estimate these subgraphs.

We first prove

Lemma 8. Let $N > 0$ be given (as large as we want); then for each G, there is a G' and a constant $0(1)$ such that

$$|G| \leq \prod_{\text{propagators}} \mathrm{d}(\Delta, \Delta')^{-N} \prod_{\text{vertexes}} 0(1) \prod_{\substack{\text{boson propagators}\\ \text{of frequency } M_l}} M_l(\Delta, \Delta')|G'|$$

and

$$M_l(\Delta, \Delta') = \begin{cases} |\Delta|^{1/3}|\Delta'|^{1/3}M_{l+1}^2 & \text{if } l < i(\Delta'), i(\Delta) \\ 1 & \text{otherwise} \end{cases}$$

Proof of the lemma. The graph G' is described during the proof. We first extract (following Glimm and Jaffe[5]), from each propagator that links two vertexes localized in Δ and Δ', a scaled factor $[\tilde{d}(\Delta, \Delta')]^{-N}$.

The contribution of a propagator comes from

$$\int \tilde{\chi}_\Delta(K + k)P(k)\tilde{\chi}_{\Delta'}(-k + K')\rho(k)\, dk,$$

where $P(k)$ is the Fourier transform of the propagator, and ρ is a cutoff with support in $|k| \geq \mathrm{M}$, and by the nature of the expansion $\mathrm{M} \leq \sup(|\Delta|^{-1/3}, |\Delta'|^{-1/3})$. Suppose $|\Delta| \leq |\Delta'|$. If $\mathrm{M}_l = \mathrm{M} > |\Delta'|^{-1/3}$, we subdivide Δ' in cubes of $\mathscr{D}_j := \mathscr{D}_{\inf[l, i(\Delta)]} : \Delta' = \bigcup_{\Delta'' \in \mathscr{D}_j} \Delta''$. We then translate all cubes to cubes centered at the origin:

$$\tilde{\chi}_\Delta(K + k)\tilde{\chi}_{\Delta''}(-k + K') = \tilde{\chi}_{\Delta_0}(K + k)\tilde{\chi}_{\Delta_0''}(-k + K')e^{irk + iKr_1 + iK'r_2},$$

where Δ_o, Δ_o'' are the translated cubes, and $r = r_2 - r_1$ is the vector translation between Δ and Δ''.

We now prove that

$$[\prod (r_2)^{N/2+2}]|G| \leq [\prod \mathrm{M}^{-(N+4)}]|G_1| \tag{27}$$

by use of

$$-r^2 = \sum_i \frac{\mathrm{d}^2}{\mathrm{d}k_i^2} e^{ikr}$$

and standard estimates[5,9,10] on derivations of characteristic functions $\tilde{\chi}$, propagators $P(k)$, and cutoff ρ. In G_1 the cutoff ρ is replaced by ρ' with support in $|k| \geq \mathrm{M}$ such that

$$\left|\prod_o^2 \left(\frac{\mathrm{d}}{\mathrm{d}k_i}\right)^{\alpha_i} \rho(k)\right| \leq C(\alpha)\tilde{\rho}(k)\mathrm{M}^{-\Sigma \alpha_i} \qquad \text{with} \qquad \sum \alpha_i \leq N + 4.$$

If $\mathrm{d}(\Delta, \Delta'') \neq 0$, one has from Formula 27

$$|G| \leq \sum_{\Delta''} \prod_{\text{propagators}} (\mathrm{M}|r|)^{-(N+4)} \prod_{\text{vertexes}} 0(1)|G_1|.$$

Since $|r| \geq \mathrm{d}(\Delta, \Delta'')$ and $\mathrm{M} \geq |\Delta''|^{-1/3}$, and since the number of Δ'' such that $\mathrm{d}(\Delta, \Delta'') = 0$ is bounded by 9, the sum over $|\Delta''|$ is controlled by

$$\sum_{\Delta''} (\mathrm{M}r)^{-4} \leq 0(1).$$

Thus,

$$|G| \leq \prod_{\text{propagators}} \tilde{d}_{(\Delta, \Delta')}^{-N} \prod_{\text{vertexes}} 0(1)|G_1|. \tag{28}$$

We now look at the factors for localization that come from the special type of cutoff on the boson fields.

Consider, therefore, a weakly well-localized field in $\Delta \in \mathscr{D}_i$ that contracts to a field localized in Δ', $|\Delta| \leq |\Delta'|$. By construction, this last field is either well localized (renormalization) or weakly well localized. The first case is treated as usual Gaussian contractions between well-localized fields. Thus, we suppose that the fields in Δ (and in Δ') are

$$\sum_l \sum_{k \geq i(\Delta)} \Phi_k^{(l)}(x) = \sum_l \left[\Phi^{(l)}(x) - \frac{1}{|\Delta|}\int_\Delta \Phi^{(l)}(x)\, \mathrm{d}x\right].$$

In the above formula, $\Phi(x) = \sum_l \Phi^{(l)}(x)$ is t-dependent, and equality results from the fact that for Φ produced during $\mathscr{D}_{i(\Delta)}$ steps, there is no t-dependence for modes $k \geq i(\Delta)$, and thus $\int_\Delta \Phi_k(x)\,\mathrm{d}x = 0$. The only case of interest is contractions with $l < i(\Delta')$ (since $l \geq i(\Delta')$ reduces to standard Gaussian contractions). Corresponding to fields of the above form (in Δ) is a vertex function whose Fourier transform is

$$\tilde{\chi}_\Delta(K + k) - \tilde{\chi}_\Delta(K)\frac{\tilde{\chi}_\Delta(k)}{|\Delta|} = \int_0^1 \frac{\mathrm{d}}{\mathrm{d}u}\left(\tilde{\chi}_\Delta(K + uk)\frac{\tilde{\chi}_\Delta[(1-u)k]}{|\Delta|}\right)\mathrm{d}u. \qquad \textbf{(29)}$$

By use of the fact[5] that

$$\left|\prod_{i=0}^{2}\left(\frac{\mathrm{d}}{\mathrm{d}k_i}\right)^{n_i}\tilde{\chi}_\Delta(k + K)\right| \leq C(n)|\Delta|^{1 + 1/3\,\Sigma n_i}F_\Delta(k + K),$$

where $F_\Delta(k)$ is $\prod_{i=1}^{3}\,[(\Delta^{-1/3}k_i)^2 + 1]^{-1/2}$, taking account of Formula 29 and of the fact that one has a similar factor for Δ', one obtains the bound

$$|\Delta|^{1/3}|k|\,|\Delta'|^{1/3}|k|\,|\Delta|\,\sup_u F_\Delta(K + uk)F_\Delta[(1-u)k].$$

Now by use of $|k| < \Delta'|^{-1/3}$, there exists $0(1)$ such that the supremum is bounded by $0(1)F_\Delta(K + k)$. By use also of $|k| \leq \mathrm{M}_{l+1}^2$, one has the required bound, that is, a factor $\mathrm{M}_l(\Delta, \Delta')$ per propagator.

The graph G' is the one obtained from G_1 after extraction of the localization factors $\mathrm{M}_l(\Delta, \Delta')$. This proves the lemma. We now decompose G' into subgraphs g' of the following types: an F term and the F' term it creates, a B^1 term or a B'^1 term, and a B^2 term (resp. B'^2 term) and all the A (resp. A') terms that form a chain with it: $B^2(x, z_1)A(z_1, z_2) \cdots A(z_{n-1}, y)$. By use of the Schwarz inequality,

$$|G'| = \left|\int\prod_{g'} g'(\{k_i\})\prod \mathrm{d}k_i\right| \leq \prod_{g'}\left(\int |g'(\{k_i\})|^2 \prod \mathrm{d}k_i\right)^{1/2} = \prod_{g'} \|g'\|^2_{\mathrm{H.S.}}\,.$$

Each $\|g'\|^2_{\mathrm{H.S.}}$ is a graph that contains one or several fermion loops. We again use the Schwarz inequality to decompose each loop of more than six fermion propagators in subgraphs g'' with loops of four or six fermion propagators. Finally, each $\|g''\|^2_{\mathrm{H.S.}}$ is decomposed in subgraphs g consisting of vacuum diagrams up to the sixth order with at least one K_1 vertex and where ~○~ and ⊓ are renormalized and of nonvacuum diagrams composed of fermion loops of eight or twelve propagators (fermions), of four propagators but with no internal boson line, or with a K_1 vertex, or with a boson internal line and renormalized, of six propagators but with at most two internal boson lines, and finally of ~○~○~ (ren, ren), of ~○~○~○~ (ren, ren, ren), or of ~○~ with a K_1 vertex.

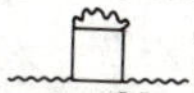

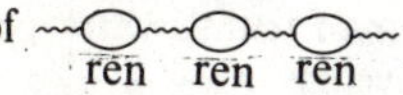

Estimates of $\|g\|^2_{\mathrm{H.S.}}$ are obtained by the power-counting argument (all integrations being convergent), following the method developed by Glimm and Jaffe,[5] where each vertex localized in $|\Delta|$ and with a well-localized boson leg gives a convergent factor $|\Delta|^{1/6-\varepsilon}$, and each vertex localized in $|\Delta|$ and without a boson leg (the boson field has been dominated) gives a $|\Delta|^{1/3-\varepsilon}$. However, we have to treat renormalization more explicitly. We will treat, for example, the case of the mass term. It is given by

k ~○~ k'
$q - l$

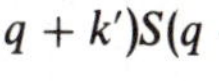

$$\mathrm{Tr}\int \tilde{\chi}_\Delta(k + q)\tilde{\chi}_{\Delta'}(-q + k')S(q - l)S(l)\rho(q)\rho(l)\,\mathrm{d}q\,\mathrm{d}l.$$

We suppose that $|\Delta| = |\Delta'|$, since if $|\Delta| < |\Delta'|$, we can subdivide Δ' in order to get equality; this is possible because l and $q - l$ are bigger than $|\Delta'|^{-1/3}$. The counterterm is

$$\mathrm{Tr} \int \tilde{\chi}_\Delta(k+q)\tilde{\chi}_{\Delta'}(-q+k')S(l)S(l)\rho(l)\rho(l)\, dl\, dq,$$

which is zero if $\Delta \neq \Delta'$. Thus,

$$\underset{\text{ren}}{\sim\!\bigcirc\!\sim} = \mathrm{Tr} \int \tilde{\chi}_\Delta(k+q)\tilde{\chi}_{\Delta'}(-q+k')$$
$$\times [S(q-l)\rho(q-l) - S(l)\rho(l)]S(l)\rho(l)\, dq\, dl.$$

The difference and its first derivative with respect to q are zero at $q \equiv 0$. It follows, expressing this difference as a second derivative, that it behaves as q^2. Now,

$$\left| \int \tilde{\chi}_\Delta(k+q)\tilde{\chi}_{\Delta'}(-q+k')q^2\, dq \right| \leq [F_\Delta(k+k')]^{1/3-\varepsilon},$$

which is the result expected from power counting.

Collecting the estimates and using Lemma 8, we get

Lemma 9. Let $N > 0$ be as large as we want; there exist $\varepsilon_4 > 0$ small enough and $0(1)$ depending on N and ε_4 such that

$$|G| \leq \prod_{\text{propagators}} d(\Delta, \Delta')^{-N} \prod_{\text{vertexes}} 0(1)|\Delta|^{-\varepsilon_4}$$
$$\times \prod_{\substack{\text{boson}\\\text{propagators}}} M_l(\Delta, \Delta') \prod_{\substack{\text{well-localized}\\\text{boson}}} |\Delta|^{1/6} \prod_{\substack{\text{dominated}\\\text{boson}}} |\Delta|^{1/3}.$$

Proposition 1 follows, then, from the above and Lemma 9.

Borel Summability

The infinite volume limit of the theory is obtained via a cluster expansion[5] as used for Φ_3^4 (Reference 9) and Y_2 (References 1 and 10). We need to prove the existence of this limit with λ in some domain of analyticity as well as bounds on the nth derivatives with respect to λ. In its present status, the cluster expansion requires the momentum cutoff to allow an exponential decrease of the propagators. We thus have to take cutoffs as in Feldman and Osterwalder[3] or Magnen and Seneor for the bosons; for the fermions, we use cutoffs inspired by the Roman team's work on Φ_3^4 (Reference 13). Defining functions of p, $\tilde{U}_\zeta(M)$ and $\tilde{V}(M)$ by

$$\tilde{V}(M) = [p^2 + M^2]^{-1/4} \qquad \tilde{U}_\zeta(M) = (i\not{p} + M)^{-1}[p^2 + M^2]^{1/4}\left(1 - \frac{i\not{p} + M}{i\not{p} + M_\zeta}\right),$$

we write

$$\tilde{V}(m) = \sum \tilde{V}_i$$

with

$$\tilde{V}_0 = \tilde{V}(m) - \tilde{V}(M_1), \qquad \tilde{V}_{i-1} = \tilde{V}(M_{i-1}) - \tilde{V}(M_i),$$
$$i < \zeta, \qquad \tilde{V}_\zeta = \tilde{V}(M_\zeta)$$

and

$$\tilde{U}_\zeta(m) = \sum \tilde{U}_i$$

with

$$\tilde{U}_0 = \tilde{U}(m) - \tilde{U}(\mathrm{M}_1), \qquad \tilde{U}_{i-1} = \tilde{U}(\mathrm{M}_{i-1}) - \tilde{U}(\mathrm{M}_i),$$
$$i < \zeta, \qquad \tilde{U}_\zeta = 0.$$

In this way, Formula 13 (p. 22) has to be interpreted with

$$\tilde{S}^P_\Delta(k) = (i\not{k} + \mathrm{M}_{i(\Delta)+1})^{-1} - (i\not{k} + \mathrm{M}_\zeta)^{-1}.$$

For the detailed expansion, we refer the reader to articles on Φ_3^4 and Y_2.[1,3,9,10,12]

We now explain briefly how to obtain bounds on the nth derivatives and the main points of the procedure. Derivatives are obtained in the following way[9]: we first perform a cluster expansion giving the infinite volume limit of the normalized Schwinger functions, with each unnormalized Schwinger function defined by a phase space expansion; we then derive with respect to the coupling constant λ (in the unit cubes). A λ-derivative can act on the exponent and then produce a new vertex D or act on already produced vertexes. With a factor of two per derivation, we decide which sort of derivative is to be taken.

First of all, let us treat the case of derivation acting on already produced vertexes. Old vertexes can be vertexes of the cluster expansion or vertexes of the phase space expansion. Possible derivations on vertexes of the cluster expansion are bounded by $0(1)^{|\Lambda|}$ (for an expression localized in Λ) since there are at most six bounds per unit cube. Vertexes created by the phase space expansion are of two types: P-vertexes or the vertexes that they generate through the renormalization procedure. If a derivated vertex is not a P-vertex, we decide to count this derivation to the generating P-vertex; the attribution is accomplished by use of the chain of propagators that connect these vertexes. If n_2 is the total number of derivations, we write

$$\left(\frac{\mathrm{d}}{\mathrm{d}\lambda}\right)^{n_2} = \sum_{\sum n_2(i) = n_2} \left(\frac{\mathrm{d}}{\mathrm{d}\lambda(i)}\right)^{n_2(i)} \frac{n_2!}{\pi n_2(i)!},$$

where $\mathrm{d}/\mathrm{d}\lambda(i)$ is a derivation acting on a P_Δ or on a vertex generated by a P_Δ-vertex with $\Delta \in \mathscr{D}_{i(\Delta)}$. The combinatoric factor for the above decomposition $(\sum_{\sum n_2(i)=n})$ is bounded by $|\Delta|^{-\varepsilon}$ per P_Δ-vertex. The problem with derivation of already produced vertexes is that they lose their λ-dependence; therefore, if the boson legs of the vertexes are not well localized, we cannot dominate them by use of the effective potential. Instead, we will dominate them by use of a part of the mass term by

$$(|\Delta|^{1/2}\Phi_\Delta)^n e^{-|\Delta|\Phi_\Delta 2} \leq (n!)^{1/2}. \tag{1}$$

In addition to the $(n!)^{1/2}$, the price we have to pay for such a domination is to be able to obtain a factor $|\Delta|^{1/6}$ per dominated vertex (since we already have $|\Delta|^{1/3-\varepsilon_4}$ for such vertexes). We need to do that to refine the renormalization procedure under *Perturbation Formulas*. In fact, if we create B terms with too long a fermion chain, and if all vertexes of these B terms are derivated with corresponding, badly localized boson fields, we may get into trouble. The solution is not to develop too much of a B term if the vertexes have badly localized legs. We treat this case in detail; the other ones are less singular and are left to the reader's understanding. Thus, suppose that we produce in Δ a P-vertex λ. We decompose the boson field in momentum range and consider two cases:

(1) The momentum ξ of the boson field is bigger than $|\Delta|^{-1/3}$. Then, we develop it twice more, generating a B term. Since the boson field is well localized, we also have to contract it (eliminating, by the way, the vacuum counterterm). By the above hypothesis, we have that the two remaining boson fields are badly localized and derivated. Let η be the momentum of the second fermion propagator (after a decomposition in momentum range). Consider now

(a) $\xi^N \geq \eta$ for some N large.

We now separate the B term in B_1 and B_2 according to; this gives

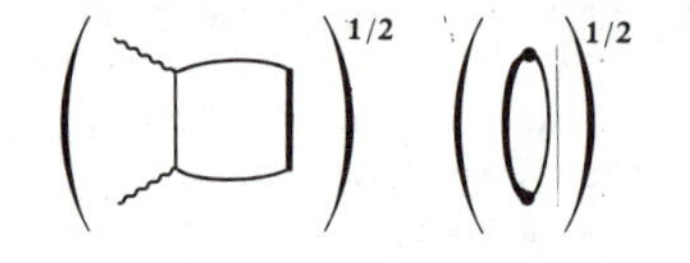

In the first diagram, the heavy-line propagator is in $(p^2 + M^2)^{-\varepsilon}$; in the second one, it is in $(p^2 + M^2)^{1-\varepsilon}$. The other ones are in $(p^2 + m^2)^{-1/2}$. The first fermion loop is convergent; the second diverges like η^{ε}. Since the boson line is now integrated, we can dominate the first term by its Hilbert Schmidt norm

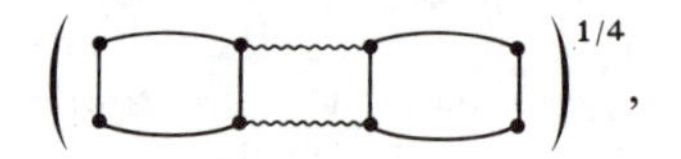

which behaves like $(|\Delta|\xi^{-1})^{1/4}$, being at least linearly convergent in the boson field. Thus, the B_1 and B_2 terms are bounded by

$$|\Delta|^{1/4}\xi^{-1/4}\eta^{\varepsilon/2} \leq |\Delta|^{1/4}\xi^{-1/4}\xi^{N\varepsilon/2} \leq |\Delta|^{1/4}$$

for ε small enough. We now have to dominate the badly localized boson field by use of Formula 1. If n_2^1 is the number§ of derivatives corresponding to this case, this means that we get

$$(|\Delta|^{-1/6-\varepsilon_4})^{2n_2^1/3}(n_2^1!)^{1/3}.$$

Addition of what we have from the first leg and its Gaussian integration gives

$$(|\Delta|^{-1/6-\varepsilon_4})^{2n_2^1/3}(|\Delta|^{1/6+2\varepsilon_4})^{n_2^1/3}(n_2^1!)^{1/3}(n_2^1!)^{1/6},$$

keeping $|\Delta|^{1/12-2\varepsilon_4}$ the convergence factor attributed to the first vertex. We obtain

$$(n_2^1!)^{1/2}(|\Delta|^{-1/18})^{n_2^1} = (n_2^1!)^{1/2}(|\Delta|^{-1/3})^{n_2^1/6} \leq (n_2^1!)^{1/2+1/6}e^{|\Delta|^{-1/3}}.$$

If now $n_2^1 \geq |\Delta|^{-1/3}$, with a factor $0(1)^{-1}$ per derivated vertex (compensated by a $0(1)$ per P_Δ-vertex), one dominates the exponential since

$$\pi 0(1)^{-1} = e^{-0(1)n_2^1} \leq e^{-0(1)|\Delta|^{-1/3}}.$$

If, on the other hand, $n_2^1 < |\Delta|^{-1/3}$, we can simply perform the Gaussian integration of the badly localized legs since the combinatoric factor for the Gaussian integration $(n_2^1)^{1/2}$ per leg is bounded by $|\Delta|^{-1/6}$, and we have this $|\Delta|^{1/6-\varepsilon_4}$ at our disposal (the missing $|\Delta|^{-\varepsilon_4}$ is taken from the first vertex).

§ Here, n_2^1 is, in fact, $n_2^1(i)$, or depends on the cover.

(b) $\xi^N < \eta$.

We then develop once more, producing a B term with four vertexes . We again suppose that the last one is derivated and has a badly localized boson leg. We split the B term into B_1 and B_2, cutting in the middle; this gives

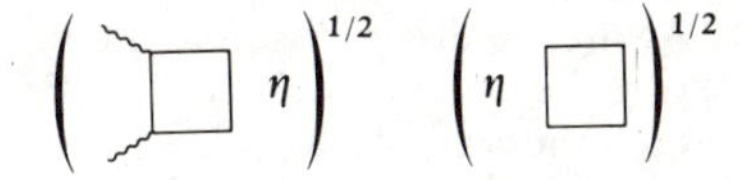

Both diagrams are convergent, giving an $\eta^{-\varepsilon}$, which, according to our hypothesis, is bounded by

$$|\Delta|^{N\varepsilon/3}.$$

Thus, with N such that $N\varepsilon/3 > \frac{3}{2}$, for instance, this gives a factor $|\Delta|^{1/2}$ for the other vertexes, allowing their Gaussian domination.

(2) The momentum ξ of the boson field is less than $|\Delta|^{-1/3}$. Again, we develop two times to form a B term, but this time we do not contract the first boson field (since it is badly localized). Let η again be the momentum of the second propagator; we distinguish two cases

(a) $\eta \leq |\Delta|^{-N/3}$.

We split the B term as in 1a. This gives

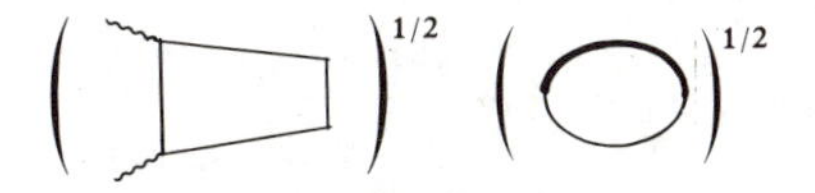

The worst divergence will be when the two boson fields of the first diagram contract. The first fermion loop is convergent; the second one diverges as $\eta^{\varepsilon/2}$. By use of translation invariance, we obtain

$$(|\Delta|\xi^{1-\varepsilon})^{1/2}(\eta^{\varepsilon}|\Delta|)^{1/2},$$

which is bounded by

$$|\Delta|^{5/6+\varepsilon/6}\eta^{\varepsilon/2}.$$

Now, by use of $\eta^{\varepsilon/2} \leq |\Delta|^{-\varepsilon N/6} \leq |\Delta|^{-\varepsilon'}$ with $\varepsilon' \geq \varepsilon N/6$, we have, finally, $|\Delta|^{5/6-\varepsilon''}$, with $\varepsilon'' = \varepsilon' - \varepsilon/6$. This factor is equivalent to $|\Delta|^{1/3-\varepsilon_4}$ for the last two vertexes and $|\Delta|^{1/6-\varepsilon_4}$ for the first one, as expected; the remaining $|\Delta|^{3\varepsilon_4-\varepsilon''}$ is kept for other convergence purposes if ε is small enough. We can now repeat the analysis of 1a, except that this time we have to dominate

$$(|\Delta|^{1/18-\varepsilon'''})^{n_2{}^2}$$

for some ε''', where n_2^2 is the number of derivated vertexes in the case under consideration. This domination is done by

$$(n_2^2!)^{1/6+3\varepsilon'''}e^{|\Delta|^{-1/3}},$$

and we proceed as above with $(n_2^2!)^{1/6+3\varepsilon'''}$ instead of $(n_2^2!)^{1/6}$.

(b) $\eta > |\Delta|^{-N/3}$.

We develop one step more. We split the B term into B_1 and B_2, giving a little convergence to the first diagram 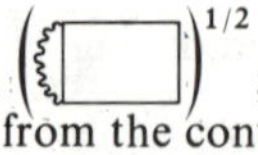 ; the second fermion loop is still convergent. Again, by use of the $\eta^{-\varepsilon}$ from the convergence, we can attribute a $|\Delta|^{1/2}$ for the Gaussian domination of the last three boson fields.

All these cases produce at most a factor $(n_2!)^{1/2+1/6+\varepsilon}$.

Let us now treat the case of derivations acting on the exponent, or those that produce new vertexes. Again, we decompose the number of such derivations n_1 as $\sum n_1(i)$, with the i indexing the cover associated with the cube in which the derivation acts. For each vertex produced, we also have to decide whether to develop it and, if so, how many times. These new vertexes can also be derivated. It is easy (as will be shown below) to say that in contrast to what happened in the above case, the worst situation is when the associated vertexes are not derived. Then, we need to dominate them by use of the effective potential which is $-|\Phi_\Delta^3||\Delta|$ or $-|\Delta|M_\zeta\Phi_\Delta^2$ according to whether Φ_Δ is bigger or less than M. Thus, from now on, the first vertex is with one λ less, and the other ones, if any, have their λ-dependence. We then consider different cases.

(1) Suppose that the cube Δ in which the derivative acts has a size less than $M_\zeta^{-3(1+\varepsilon)^{-4}}$. Then, we do not develop but contract the boson leg. Thus, splitting 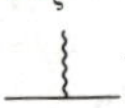 into 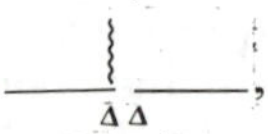, we bound them by (from the fact that the fermions have an upper cutoff M_ζ)

$$(|\Delta|M_\zeta^4)^{1/4}(|\Delta|M_\zeta^2)^{1/2} = |\Delta|^{3/4}M_\zeta^2 \leq |\Delta|^{\varepsilon'}$$

for some $\varepsilon' > 0$. The combinatoric factor is thus reduced to $(n_1^1!)^{1/2}$, the combinatoric factor for Gaussian contraction.

(2) Suppose now that the cube Δ has a size larger than $M^{-3(1+6)-4}$, that is, that the corresponding boson field is in $\bar{\mathscr{D}}$, and let n_1^2 be the number of derivatives of this type. We consider two cases.

(a) $n_1^2 \geq |\Delta|M_\zeta^3$.
We do not develop more 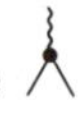 . We split it as in 1 and bound each term

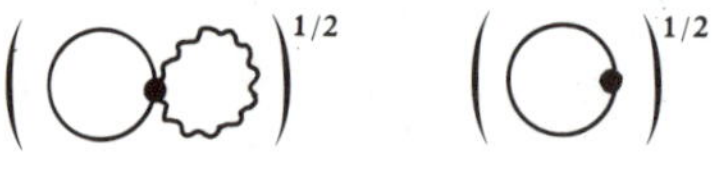

by

$$(|\Delta|\,|\Delta|^{-1/3}M_\zeta^2)^{1/2}(|\Delta|M_\zeta^2)^{1/2} = |\Delta|^{5/6}M_\zeta^2 = |\Delta|^{1/6}(|\Delta|^{1/3}M_\zeta)^2 \leq |\Delta|^{1/6}(n_1^2)^{2/3},$$

where we used the fact that we need a $|\Delta|^{-1/3}$ for the contraction of a Φ_Δ. This thus gives $(n_1^2!)^{1/2+2/3} = (n_1^2!)^{1+1/6}$ since we need a $(n_1^2!)^{1/2}$ for the contraction of the boson.

(b) $n_1^2 < |\Delta|M_\zeta^3$.
We will then need to expand the fermion two times or more, and since they present the worst cases, we suppose that all these new vertexes are not derivated. Therefore, we can dominate their corresponding boson fields by the effective potential, that is, $|\Delta|\Phi_\Delta^3$ or $-|\Delta|M_\zeta\Phi_\Delta^2$, according to the strength of the field.

We develop the fermion two times and decompose the second fermion propagator according to the momentum ranges. Let η be this momentum for one term. We then consider two cases.

(1) $\eta < (|\Delta|^{-1}n_1^2)^N$ for some N large, to be fixed later. We split the B term into B_1 and B_2 as [diagram with η]. A bound for B_1 and B_2 is (up to a constant in ε)

$$(|\Delta|\,|\Delta|^{-1/3})^{1/2}(|\Delta|\eta^{2\varepsilon})^{1/2} = |\Delta|^{5/6}\eta^{\varepsilon}.$$

We now dominate the two other boson fields by use of the effective potential. Let $2\alpha n_1^2$ be the number of those dominated by $|\Delta|\Phi_\Delta^3$ and $2(1-\alpha)n_1^2$ the number of those dominated by $|\Delta|M_\zeta\Phi_\Delta^2$ with $0 \le \alpha \le 1$. Thus, we need to control (after domination)

$$\begin{aligned}(|\Delta|^{5/6}\eta^{\varepsilon})^{n_1^2}(|\Delta|^{-1/3})^{2\alpha n_1^2}(|\Delta|^{-1/2}M_\zeta^{-1/2})^{2(1-\alpha)n_1^2}(n_1^2!)^{1/2+2\alpha/3+1-\alpha}\\ = (|\Delta|^{1/6}\eta^{\varepsilon})^{n_1^2}(|\Delta|^{-1/6}M_\zeta^{-1/2})^{2(1-\alpha)n_1^2}(n_1^2!)^{3/2-\alpha/3}\\ \le (|\Delta|^{1/6-\varepsilon N})^{n_1^2}(n_1^2)^{\varepsilon N n_1^2}(n_1^2!)^{1+1/6},\end{aligned}$$

where we have used $n_1^2 < |\Delta|M_\zeta^3$. Now, choosing ε small enough such that $\varepsilon N < \varepsilon'$ $\varepsilon' > 0$ very small given, the bound is controlled by $(n_1^2!)^{7/6+\varepsilon'}$, a convergent factor $|\Delta|^{\varepsilon''}$ being distributed to the vertexes ($\varepsilon'' = \frac{1}{18} - \varepsilon'/3$).

(2) $\eta \ge (|\Delta|^{-1}n_1^2)^N$.

We develop one time more, generating after splitting into B_1 and B_2 convergent fermion loops with lower cutoff of one propagator bigger than $(|\Delta|^{-1}n_1^2)^N$. With the same notation as above, the factor that we have to control after domination is

$$\begin{aligned}(|\Delta|^{5/6}\eta^{-\bar\varepsilon})^{n_1^2}(|\Delta|^{-1/3})^{3\alpha n_1^2}(|\Delta|^{-1/2}M_\zeta^{-1/2})^{3(1-\alpha)n_1^2}(n_1^2!)^{1/2+\alpha+3/2(1-\alpha)}\\ = (|\Delta|^{5/6-1}\eta^{-\bar\varepsilon})^{n_1^2}(|\Delta|^{-1/6}M_\zeta^{-1/2})^{3(1-\alpha)n_1^2}(n_1^2!)^{2-\alpha/2}\\ \le (|\Delta|^{5/6-1}|\Delta|^{\bar\varepsilon N})^{n_1^2}(n_1^2)^{-\bar\varepsilon N n_1^2}(n_1^2!)^{3/2}\end{aligned}$$

for some $\bar\varepsilon > 0$ given (lets say, $\bar\varepsilon > \frac{1}{16}$). Then, if we choose N such that $\bar\varepsilon N > \frac{1}{3}$, this factor is less than

$$(|\Delta|^{1/6})^{n_1^2}(n_1^2!)^{7/6},$$

which, as above, leads to a bound controlled by $(n_1^2!)^{7/6}$. Summing these results, we get in $n = n_1 + n_2$ a bound in $C^n(n!)^{7/6+\varepsilon}$; C is a constant blowing up when $\varepsilon \to 0$. This bound relative to derivative in unit cubes is transformed by the standard method[2,9] in an overall bound of the same type on the number of derivatives in λ. Then, following the argument of Reference 9, we proved the following theorem.

Theorem. Let $|\lambda|$ be small enough, m and M large enough, and Re $\lambda^3 > 0$. Then the normalized Schwinger functions S_Λ are such that

$$\left|\left(\frac{d}{d\lambda}\right)^n \lim_{\Lambda\to\infty} S_\Lambda\right| \le C'(n!)^{7/6+\varepsilon},$$

the constant depending on λ, m, M, and ε. Moreover, $\lim_{\Lambda\to\infty} S_\Lambda$ is analytic in Re $\lambda^3 > 0$, $|\lambda|$ small enough.

As a corollary, we have that $\lim_{q\to\infty} S_\Lambda$ are Borel summable at $\lambda = 0$.

Appendix 1: Lemma 6

We prove Lemma 6 in the case of cutoffs introduced at the beginning of Borel Summability, that is, with

$$\tilde{S}^{P}_{\Delta}(k) = (i\not{k} + \mathrm{M}_{i(\Delta)+1})^{-1} - (i\not{k} + \mathrm{M}_{\zeta})^{-1}.$$

This means that we have to estimate

$$\operatorname{Re} P_{\Delta} = \operatorname{Re} \sum_{\Delta^{1/3}k \in \mathbb{Z}^3} \log\{1 + \lambda[(i\not{k} + \mathrm{M}_{i(\Delta)+1})^{-1} - (i\not{k} + \mathrm{M}_{\zeta})^{-1}]\gamma_5 \Phi_{\Delta}\}$$
$$+ \frac{\operatorname{Re}}{2} \sum_{\Delta^{1/3}k \in \mathbb{Z}^3} \operatorname{Tr} \tilde{S}^{P}_{\Delta}(k)\gamma_5 \tilde{S}^{P}_{\Delta}(k)\gamma_5 \lambda^2 \Phi^2_{\Delta}$$

for $\operatorname{Re} \lambda^3 > 0$

To simplify what follows, we note $\mathrm{M}_{i(\Delta)+1} = m$, $\mathrm{M}_{\zeta} = \mathrm{M}$. Noticing that $(i\not{k} + m)^{-1} - (i\not{k} + \mathrm{M})^{-1} = (\mathrm{M} - m)[i\not{p}(\mathrm{M} + m) - p^2 + m\mathrm{M}]^{-1}$ and the symmetry $(p, \gamma_5) \to (-p, -\gamma_5)$, one has

$$\operatorname{Re} P_{\Delta} = \frac{\operatorname{Re}}{2} \sum_{\Delta^{1/3}k \in \mathbb{Z}^3} \left\{\log\left[1 + \frac{(\mathrm{M} - m)^2\lambda^2\Phi^2_{\Delta}}{(k^2 + m^2)(k^2 + \mathrm{M}^2)}\right] - \frac{(\mathrm{M} - m)^2\lambda^2\Phi^2_{\Delta}}{(k^2 + m^2)(k^2 + \mathrm{M}^2)}\right\}$$
$$= \frac{\operatorname{Re}}{2} \sum_{\Delta^{1/3}k \in \mathbb{Z}^3} \left\{\frac{1}{2} \log\left[\left(1 + \frac{\operatorname{Re} \lambda^2(\mathrm{M} - m)^2\Phi^2_{\Delta}}{(k^2 + m^2)(k^2 + \mathrm{M}^2)}\right)^2\right.\right.$$
$$\left.\left. + \left(\frac{\operatorname{Im} \lambda^2(\mathrm{M} - m)^2\Phi^2_{\Delta}}{(k^2 + m^2)(k^2 + \mathrm{M}^2)}\right)^2\right] - \frac{\operatorname{Re} \lambda^2(\mathrm{M} - m)^2\Phi^2_{\Delta}}{(k^2 + m^2)(k^2 + \mathrm{M}^2)}\right\}.$$

Setting $X = \operatorname{Re} \lambda^2(\mathrm{M} - m)^2\Phi^2_{\Delta}$ and $Y = \operatorname{Im} \lambda^2(\mathrm{M} - m)^2\Phi^2_{\Delta}$, one writes

$$\operatorname{Re} P_{\Delta} = \frac{1}{2} \sum \log\left(1 + \frac{Y^2}{[(k^2 + m^2)(k^2 + \mathrm{M}^2)]^2}\right)$$
$$+ \sum \left\{\log\left(1 + \frac{X}{(k^2 + m^2)(k^2 + \mathrm{M}^2)}\right) - \frac{X}{(k^2 + m^2)(k^2 + \mathrm{M}^2)}\right\}.$$

We now estimate the two terms by continuous integrals by use of the fact that

$$\sum_{p \in \Delta^{1/3}\mathbb{Z}} G(p) \leq |\Delta|^{1/3} \int G(p)\, \mathrm{d}p + G(0)$$

for G monotone decreasing in p^2 and that

$$\sum_{p \in \Delta^{1/3}\mathbb{Z}} G(p) \leq |\Delta|^{1/3} \int G(p)\, \mathrm{d}p$$

for G monotone increasing in p^2. We obtain

$$\operatorname{Re} P_{\Delta} \leq \frac{1}{2} |\Delta| \operatorname{Re} \int \left[\log\left(1 + \frac{X + iY}{(k^2 + m^2)(k^2 + \mathrm{M}^2)}\right) - \frac{X + iY}{(k^2 + m^2)(k^2 + \mathrm{M}^2)}\right] \mathrm{d}^3k$$
$$+ \frac{3}{2} |\Delta|^{2/3} \int \log\left(1 + \frac{Y^2}{[(k^2 + m^2)(k^2 + \mathrm{M}^2)]^2}\right) \mathrm{d}^2k$$
$$+ \frac{3}{2} |\Delta| \int \log\left(1 + \frac{Y^2}{[(k^2 + m^2)(k^2 + \mathrm{M}^2)]^2}\right) \mathrm{d}k + \log\left(1 + \frac{Y^2}{m^4\mathrm{M}^4}\right).$$

By use of the fact that $X \geq 0$ (Re $\lambda^2 \geq 0$) and doing explicitly the integrations, one obtains

$$\operatorname{Re} P_\Delta \leq \operatorname{Re}\Big\{|\Delta|\left(\frac{2\pi^2}{3}(\mathrm{M}^3 + m^3 - a_+^3 - a_-^3) - \pi^2\frac{(X+iY)}{m+\mathrm{M}}\right)$$

$$+ |\Delta|^{2/3}\frac{3\pi}{2}\Bigg(-\frac{m^2+\mathrm{M}^2}{2}$$

$$\times \log\left(1 + \frac{Y^2}{m^4\mathrm{M}^4}\right) - 2\left(\frac{(\mathrm{M}^2 - m^2)^2}{4} - i|Y|\right)^{1/2}$$

$$\times \log\frac{\mathrm{M}^2 + m^2 + [(\mathrm{M}^2 - m^2)^2 - 4i|Y|]^{1/2}}{\mathrm{M}^2 + m^2 - [(\mathrm{M}^2 - m^2)^2 - 4i|Y|]^{1/2}}$$

$$+ \frac{1}{2}(\mathrm{M}^2 - m^2)\log\frac{\mathrm{M}^2}{m^2}\Bigg) + 2|\Delta|^{1/3}\pi\left\{\left(\left\{\frac{\mathrm{M}^2+m^2}{2} + \frac{1}{2}[(\mathrm{M}^2 - m^2)^2 - 4i|Y|]^{1/2}\right\}^{1/2}\right.\right.$$

$$\left.\left. + \left\{\frac{\mathrm{M}^2+m^2}{2} - \frac{1}{2}[(\mathrm{M}^2 - m^2)^2 - 4i|Y|]^{1/2}\right\}^{1/2} - m - \mathrm{M}\right) + \frac{1}{2}\log\left(1 + \frac{Y^2}{m^4\mathrm{M}^4}\right)\right\}\Big\},$$

where

$$a_t^2 = \frac{\mathrm{M}^2 + m^2}{2} \pm \frac{1}{2}\{(\mathrm{M}^2 - m^2)^2 - 4(X+iY)\}^{1/2}$$

$$= \frac{\mathrm{M}^2}{2}\left[1 + \rho^2 \pm (1-\rho^2)\left\{1 - \frac{4}{(1+\rho)^2}\lambda^2\frac{\Phi_\Delta^2}{\mathrm{M}^2}\right\}^{1/2}\right], \qquad \rho = \frac{m}{\mathrm{M}}.$$

Imposing Re $\lambda^3 > 0$, that is, $\sigma|\operatorname{Re}\lambda| = |\operatorname{Im}\lambda|$ for $0 \leq \sigma \leq 1/\sqrt{3}$ with $X = (\operatorname{Re}\lambda)^2 \times (1-\sigma^2)$, $Y = 2(\operatorname{Re}\lambda)^2\sigma$ and from the fact that $m = |\Delta|^{-1/3(1+\varepsilon)}$ and that $m/\mathrm{M} \leq m^{-\varepsilon}$, the coefficient of $|\Delta|$ is bounded by

$$-0(1)|\lambda|^3|\Phi_\Delta|^3 + |\Delta|^{-\varepsilon'-1} \quad \text{for} \quad |\lambda\Phi_\Delta| \leq \mathrm{M} \qquad (*)$$

$$-0(1)\mathrm{M}|\lambda|^2|\Phi_\Delta|^2 \quad \text{for} \quad |\lambda\Phi_\Delta| \geq \mathrm{M}$$

In fact, one can show that it behaves like

$$0\left(\frac{\lambda^4\Phi_\Delta^4}{m}\right) \qquad \text{for } |\lambda\Phi_\Delta| \ll m,$$

that is, like

$$|\lambda|^3|\Phi_\Delta|^3 + |\lambda|^3|\Phi_\Delta|^3 + 0\left(\frac{\lambda^4\Phi_\Delta^4}{m}\right) \simeq -|\lambda|^3|\Phi_\Delta|^3 + 0(|\Delta|^{-1-3\varepsilon})$$

like

$$\operatorname{Re}\frac{m^2}{3}\left\{m^3\left[1 - \left(1 + \frac{1-\rho}{1+\rho}\lambda^2\frac{\Phi^2}{m^2}\right)^{3/2}\right]\right\} + 0\left(\frac{\lambda^4\Phi_\Delta^4}{\mathrm{M}}\right) \qquad \text{for} \quad m \ll |\lambda\Phi_\Delta| \ll \mathrm{M},$$

that is, like

$$-\frac{\mathrm{M}^2}{4}\left(\frac{1-\rho}{1+\rho}\right)^{3/2}|\lambda|^3\frac{1-3\sigma^2}{(1+\sigma^2)^{3/2}}|\Phi_\Delta|^3 + 0\left(\frac{\lambda^4\Phi_\Delta^4}{\mathrm{M}}\right) \sim -|\lambda|^3|\Phi_\Delta|^3,$$

where use is made of the condition on λ, and like

$$0(M^{3/2}|\lambda|^{3/2}|\Phi_\Delta|^{3/2}) - \pi^2 M \frac{(1-\rho)^2}{1+\rho}|\lambda|^2 \frac{1-\sigma^2}{1+\sigma^2}\Phi_\Delta^2 \sim -M|\lambda|^2\Phi_\Delta^2$$

for $|\Phi_\Delta| \gg M$. The coefficient of $|\Delta|^{2/3}$ is bounded by

$$0(1)|\lambda\Phi_\Delta|^2 \log|\lambda\Phi_\Delta| \qquad \text{for} \quad |\lambda\Phi_\Delta| \le M$$

and by

$$0(1)M|\lambda\Phi_\Delta| - 0(1)M^2 \log\frac{|\lambda\Phi_\Delta|^2}{M^2} \qquad \text{for} \quad |\lambda\Phi_\Delta| \ge M.$$

The coefficient of $|\Delta|^{1/3}$ is bounded by $0(\Phi_\Delta)$. Finally, the last term is bounded by $\log|\Phi_\Delta|/m$. It is now a matter of careful analysis to show that including the correction terms, the bounds $*$ are just modified by a change in their coefficients.

Remark. If we take for the fermion fields, standard cutoffs $\eta_i(k) = \eta(k/M_{i+1}) - \eta(k/M_i)$ with $\eta(k)$ a C_0^∞ function such that $\eta(k) = 0$ if $|k| \ge 3/2$ and $\eta(k) = 1$ for $|k| \le 1/2$ defining

$$U_i(X, Z) = \int e^{ik(X-Z)} \frac{-i\not{k} + M}{(k^2 + M^2)^{3/4}} \eta_i(k)\eta(k/M_\zeta)\, d^3k$$

$$V_j(Z, Y) = \int e^{ik(Z-Y)} \frac{1}{(k^2 + M^2)^{1/4}} \eta_j(k)\eta(k/M_\zeta)\, d^3k.$$

One would have to estimate

$$\mathrm{Re} \sum_{k \in (\Delta^{-1/3}\mathbb{Z})^3} \left(\log\left\{1 + \frac{\lambda^2\Phi_\Delta^2}{k^2 + M^2}[1 - \eta(k/M_{i(\Delta)+1})]^2 \eta(k/M_\zeta)^2\right\} - \frac{\lambda^2\Phi_\Delta^2}{k^2 + M^2}[1 - \eta(k/M_{i(\Delta)+1})]^2\eta(k/M_\zeta)^2 \right) = \mathrm{Re} \sum_k F_\eta(k)$$

(here M is the fermion mass). Introducing $\chi_{(i,\zeta)}(k)$, the characteristic function of the k's such that $[1 - \eta(k/M_{i(\Delta)+1})]^2\eta(k/M_\zeta)^2 = 1$, it is easy to check that

$$\mathrm{Re}\, F_\eta(k) \le \mathrm{Re}\, F_\chi(k)$$

for $\mathrm{Re}\,\lambda^4 \ge 0$, where $F_\chi(k)$ is $F_\eta(k)$ with $(1 - \eta/M_{i(\Delta)+1})^2\, \eta(k/M_\zeta)^2$ replaced by $\chi_{(i,\zeta)}(k)$.

It is then possible to use monotonicity arguments as above to get a bound with continuous integrals and finally to obtain Lemma 6.

The results in PHASE SPACE EXPANSION show the existence in a finite volume of theories defined with both types of fermion cutoffs. They are both analytic real in λ and obviously coincide for $\mathrm{Re}\,\lambda^4 \ge 0$, defining therefore the same theory. The cutoffs of the second type allow one to connect the starting expression to a cutoff Hamiltonian formulation; the first ones are necessary for the proof of the convergence of the cluster expansion, that is, to perform the infinite volume limit.

APPENDIX 2: ON PRINCIPLES

We present here a succession of arguments that together build a skeleton for any constructions involving Y_3 or similar theories. They will clear up some of the steps that are presented in this paper.

(1) The model is a Yukawa (Y_3) model given by the Matthews–Salam formula, that is, after the fermions have been integrated out. As a result, we have to study a purely bosonic theory with a nonlocal interaction; the nonlocality is given by the fermion propagator. To define the theory, we choose to put a momentum cutoff M_ζ on this fermion propagator.

(2) The theory is superrenormalizable; that is, each vertex $\curlywedge$ is convergent with a power counting $-1/2$, and a finite number of diagrams are primitively divergent.

(3) The cutoff action is bounded from below.

$$\frac{\exp(-\int_\Lambda \bar{\psi}(i\not{p} + m + \Gamma\Phi)\psi)\, d\bar{\psi}\, d\psi}{\exp(-\int \bar{\psi}(i\not{p} + m)\psi)\, d\bar{\psi}\, d\psi} = \det(1 + K),$$

Λ is the space cutoff, and $K \simeq \Gamma\Phi/(i\not{p} + m)$, then

$$|\det(1 + K)| \leq e^{\mathrm{Tr}\,|K|} \leq e^{M_\zeta{}^2 \int |\Phi(X)|}.$$

Adding the mass counterterm $\frac{1}{2}\,\mathrm{Tr}_\infty\, K^2 \sim -M_\zeta \int \Phi^2(X)$, one has

$$|\det(1 + K)e^{-\mathrm{Tr}\,K + 1/2\,\mathrm{Tr}_\infty\, K^2}| \leq e^{|\Lambda| M_\zeta{}^3},$$

so that $\mathscr{L} \geq |\Lambda| M_\zeta^3$.

(4) The effective cutoff action $\mathscr{L}_{\mathrm{eff}}$ is positive. Take $\Phi = C^t$; then,

$$e^{-\mathscr{L}_{\mathrm{eff}}} = \det(1 + K)e^{-\mathrm{Tr}\,K + 1/2\,\mathrm{Tr}_\infty\, K^2} \sim e^{-0(1)|\Phi^3|\,|\Lambda|} \quad \text{or} \quad e^{-0(1)|\Lambda| M_\zeta \Phi^2},$$

according to whether one looks at the behavior for M_ζ very large or Φ very large.

The arguments now presented are directly related to the phase space cell expansion for Y_3.

(5) The expansion is a perturbative expansion in which each step produces a vertex P_Δ $\curlywedge$ localized in cubes Δ with a leg of high momentum k, $|k| > M_{\mathrm{low}}$ ensuring that $|P_\Delta| \approx M_{\mathrm{low}}^{-1/2}$. The perturbation is in cubes belonging to lattices $\mathscr{D}_0$, $\mathscr{D}_1, \ldots, \mathscr{D}_\zeta$ such that $\Delta_i \in \mathscr{D}_i$ implies $|\Delta_i| \simeq M_i^{-3}$, $i = 1, \ldots, \zeta$. Each lattice $\mathscr{D}_i$ is a refinement of the preceding one $\mathscr{D}_{i-1}$.

(6) For each vertex P_Δ produced, one has to perform a finite perturbation expansion to renormalize the divergent diagrams that contain P_Δ as a vertex. This produces C-vertexes (attached to the P_Δ-vertex).

(7) At the end of the expansion, we have to integrate (or dominate) each boson of the C-vertexes. We have to compute combinatoric factors (see Reference 1), taking account of all the possible C-vertexes. We replace them by combinatoric factors on the P_Δ-vertexes that are associated with them. For that, we require that all the propagators between a given C-vertex localized in $\Delta_i \in \mathscr{D}_i$ are such that

$$\sum_{\Delta_i' \in \mathscr{D}_i} (\text{propagators between } \Delta_i \text{ and } \Delta_i')^{0(1)^{-1}} < \infty,$$

where $0(1)$ is the number of associated vertexes of a given P-vertex. Such a condition is satisfied if we require that the momenta of these propagators be bigger than $|\Delta_i|^{-1/3} = M_i$. Then,

$$(\text{propagators between } \Delta_i \text{ and } \Delta_i') \sim e^{-|\Delta_i|^{-1/3}\,\mathrm{dist}(\Delta_i,\, \Delta_i')}.$$

(8) To use argument 4 to dominate Φ localized in Δ, this Φ has to be constant

in Δ. We are therefore led to perturb relatively the variation of Φ in Δ. We write for $X \in \Delta$

$$\Phi(X) = \Phi_\Delta + \delta\Phi_\Delta(X) \qquad \Phi_\Delta = \frac{1}{|\Delta|}\int_\Delta \Phi(X)\,dX$$

$$\delta\Phi_\Delta(X) = \Phi(X) - \Phi_\Delta .$$

Then, roughly speaking, Φ_Δ is with momentum cutoff $|\Delta|^{-1/3}$, and $\delta\Phi_\Delta$ is with *low*-momentum cutoff $|\Delta|^{-1/3}$. Perturbation relative to $\delta\Phi_\Delta$ produces vertexes with a $\delta\Phi_\Delta$ boson leg, that is, with a high-momentum leg (see argument 5). This leads to a decomposition of $\Phi(X) = \sum \Phi_i(X)$, where $\Phi_i(X)$ is constant in cubes of $\mathscr{D}_{i+1}$ and of mean zero in cubes of $\mathscr{D}_i$. Roughly speaking, their momenta are between M_i and M_{i+1}. The field Φ_i is said to be weakly well localized.

(9) The Gaussian integration of the weakly well-localized boson fields of the P_Δ-vertexes. These bosons (in Δ), roughly speaking, have momenta bigger than $|\Delta|^{-1/3}$. This gives for each P_Δ-vertex a convergent factor $|\Delta|^{1/6}$ (see arguments 2 and 5). In fact, the condition of argument 7 is in some sense fulfilled

$$(\text{propagator between } \Delta \text{ and } \Delta' \text{ for } \Phi_\Delta \text{ fields}) \sim [|\Delta|^{-1/3}\,\mathrm{dist}(\Delta, \Delta')]^{-3-\varepsilon}.$$

(10) The above behavior is not enough to fulfill condition 7 for the propagators generated by the renormalization procedure. We thus require sharper cutoff on those propagators. The corresponding fields are well localized.

(11) To integrate boson fields localized in Δ but not well localized, one needs an extra factor $|\Delta|^{-1/2}$ since

$$\sum_{\Delta'} (\text{propagator between } \Delta' \text{ and } \Delta) \sim |\Delta'|^{-1} \le |\Delta'|^{-1/2}|\Delta|^{-1/2},$$

where the sum extends over Δ' with size $|\Delta'| \ge |\Delta|$.

(12) A C_Δ-vertex is, according to argument 2, small like $|\Delta|^{1/6}$; thus, we cannot integrate the boson of a C_Δ-vertex (see argument 11). Instead, we dominate it by use of

$$|\Delta|^{1/3}|\Phi_\Delta| e^{-|\Delta|\int|\Phi(X)|^3\,dX} \le C^t$$

or

$$|\Delta|^{1/3}|\Phi_\Delta| e^{-|\Delta|\mathrm{M}_\xi\int|\Phi(X)|^2\,dX} \le C^t$$

since $|\Delta|^{-1/3} \le \mathrm{M}_\zeta$ (equality is for the cube of the smallest cover).

(13) One uses argument 4 in a cube Δ by introducing periodic boundary conditions on $\partial\Delta$. This approximation is controlled if the momenta in Δ are big relative to $|\Delta|^{-1/3}$. We therefore apply periodic boundary conditions when the fermion momenta are bigger than $|\Delta|^{-1/3(1+\varepsilon)}$ and the bound of argument 3 on the determinant for the low-momenta part.

(14) To be able to obtain enough small factors to compensate a bound

$$|\det_{\mathrm{ren}}(1 + K)|_{\text{restricted to } \Delta} \le e^{|\Delta|^{-2\varepsilon}}$$

coming from argument 3 applied for momenta below $|\Delta|^{-1/3(1+\varepsilon)}$ (see argument 13), one is led to produce at least $|\Delta|^{-3\varepsilon}$ factors for each Δ.

Appendix 3: The Scalar Case

In this case, the theory has no more of the symmetry that eliminates odd terms in Φ; thus, one has to add linear counterterms in Φ: $-\mathrm{Tr}\,K$ and the divergent part of $-\frac{1}{3}[\mathrm{Tr}\,K^3 - :\mathrm{Tr}\,K^3:]$ and the corresponding vacuum diagrams.

The expansion in PHASE SPACE EXPANSION is then similar to the one of the pseudoscalar case. However, there is a difference due to the fact that $1 + K_2^p$ can now have zero as an eigenvalue (see *Wick Bounds*).

To solve this problem, we proceed as follows. Suppose we are in a cube Δ. Then, if for $|\text{Arg } \lambda| \leq \pi/6$

(1) $|\lambda||\Phi_\Delta| \geq 2\text{M}$,

one obtains

$$\left\| \frac{1}{1 + K_2^p} \right\| \leq \text{Const.}$$

(2) $|\lambda|\,|\Phi_\Delta| \leq 2\text{M}$.

One introduces a new $\tilde{K}_2^p$, corresponding to fermion mass 4M instead of M. For this new $\tilde{K}_2^p$,

$$\left\| \frac{1}{1 + \tilde{K}_2^p} \right\| \leq \text{Const.}$$

Instead of estimating $\det(1 + K_2^p)$, we will have to estimate $\det(1 + \tilde{K}_2^p)$. We use

$$1 + K_2^p = (1 + \tilde{K}_2^p)\left[1 + K_2^p - \tilde{K}_2^p - \frac{(K_2^p - \tilde{K}_2^p)\tilde{K}_2^p}{1 + \tilde{K}_2^p}\right]$$

and the fact that $\text{Tr } K_2^p - \tilde{K}_2^p$, $\text{Tr } |K_2^p - \tilde{K}_2^p|^2$, $\text{Tr}(K_2^p - \tilde{K}_2^p)\tilde{K}_2^p$, and $\text{Tr } |(K_2^p - \tilde{K}_2^p)(K_2^p)^2|$ are finite to control the change of K_2^p into $\tilde{K}_2^p$ in the determinant and in the renormalization counterterms. Estimates that use the finiteness of $\text{Tr } |(K_2^p - \tilde{K}_2^p) \times (K_2^p)^2|$ come from

$$\left| \text{Tr } \frac{(K_2^p - \tilde{K}_2^p)\tilde{K}_2^p}{1 + \tilde{K}_2^p} - \text{Tr}(K_2^p - \tilde{K}_2^p)\tilde{K}_2^p \right| = \left| \text{Tr } \frac{(K_2^p - \tilde{K}_2^p)(\tilde{K}_2^p)^2}{1 + \tilde{K}_2^p} \right|$$

$$\leq \left\| \frac{1}{1 + \tilde{K}_2^p} \right\| \text{Tr } |(K_2^p - \tilde{K}_2^p)(\tilde{K}_2^p)^2|,$$

where we use the fact that we can control any powers of Φ_Δ since we are in the case $|\lambda|\,|\Phi_\Delta| \leq 2\text{M}$.

REFERENCES

1. COOPER, A. & L. ROSEN. 1977. Trans. Am. Math. Soc. **234:** 1–88.
2. ECKMANN, J. P., J. MAGNEN & R. SENEOR. 1975. Commun. Math. Phys. **39:** 251–271.
3. FELDMAN, J. & K. OSTERWALDER. 1976. Ann. Phys. **97:** 80–135.
4. FRÖHLICH, J. & E. SEILER. 1976. Helv. Phys. Acta **49:** 889–924.
5. GLIMM, J. & A. JAFFE. 1973. Fortschr. Phys. **21:** 327–376.
6. GLIMM, J., A. JAFFE & T. SPENCER. 1973. Springer Lect. Notes Phys. **25:** 199–242.
7. MCBRYAN, O. 1975. Commun. Math. Phys. **44:** 237–243.
8. MCBRYAN, O. 1975. Personal communication.
9. MAGNEN, J. & R. SENEOR. 1977. Commun. Math. Phys. **56:** 237–276.
10. MAGNEN, J. & R. SENEOR. 1976. Commun. Math. Phys. **51:** 297–313.
11. PARISI, G. Phys. Lett. **66B**(4): 382–384.
12. RENOUARD, P. 1978. Ann. Inst. Henri Poincaré **27:** 237–277.
13. X, Y, ... 1978. Commun. Math. Phys. **49:** 143–166.
14. SEILER, E. 1975. Commun. Math. Phys. **42:** 163–182.
15. SEILER, E. & B. SIMON. 1975. J. Math. Phys. **16:** 2289–2293.

DEFORMED SHAPES OF HADRONS*

John Parmentola

Center for Theoretical Physics
Laboratory for Nuclear Science and Department of Physics
Massachusetts Institute of Technology
Cambridge, Massachusetts 02139

In the past several years, the M.I.T. bag model[1] has contributed a great deal to our understanding of the properties of elementary particles. The basis and motivation for the model have been of phenomenological origin. In a recent paper, Johnson and Thorn[2] extended the application of the bag model to Regge phenomenology. They consider the possibility that deformed states of the bag, in particular stringlike states, result in a rotational energy spectrum that accounts for linear rising Regge trajectories. They assume that the hadronic states that maximize angular momentum for a fixed mass are long in shape and that the deformed bag rotates about its center of mass with angular frequency ω. The ends of this configuration are assumed to move at the speed of light, with the quarks contributing concentrations of color at the ends. The colored gluons terminate on the quarks and provide an internal pressure that prevents the bag from collapsing. The length of the bag is a variational parameter. Subsequent to variation, they find that the stringlike states with minimum mass for a fixed angular momentum correspond to an asymptotically linear Regge trajectory, with a slope determined by the fundamental colored-quark-gluon coupling constant, α_c, and the universal bag constant, B. The agreement with the experimentally determined value for the slope is remarkable, and the model accounts for its universality.

There have been several attempts in the literature to derive the bag model from first principles.[3]† In a recent paper, Huang and Stump[4] considered a variational approach to quark confinement in a quantum-field theory of quarks coupled linearly to a real scalar field. Their trial functions are chosen to be eigenstates of linear and angular momentum and are such that the quarks find themselves in a spherically symmetric well. Their hope is that such a well simulates some aspects of confinement and that the problem of confinement will emerge from quantum chromodynamics while retaining some characteristics of the model. They find that the total energy, including the contributions from the negative-energy Dirac sea, possesses a minimum for nonzero well radius and that it is finite in the limit where the renormalization cutoff is taken to infinity. They show that the contributions of the negative-energy Dirac sea result in a minimum of the total energy for which the depth and height of the confining well are unequal. In fact, they find that the well height is infinite as the renormalization cutoff goes to infinity. Upon choosing the depth to be zero, they obtain a model that is similar to the M.I.T. bag model for a spherical cavity. By choosing two remaining constants phenomenologically and assuming that hadrons are color singlets, they apply the model to hadron spectroscopy.

* Supported in part through funds provided by the U.S. Department of Energy under contract EY-76-C-02-3069.

† For a discussion of baglike models in QCD, see, for example, Lee[3] and references therein.

The purpose of this paper is to present a variational calculation that demonstrates that the ground-state configuration of a system of quarks and scalar particles can be nonspherical. Our model is similar to that of Huang and Stump; however, there are important differences. Our trial states are more general in that they include states in which quarks find themselves in a deformed potential well. However, we neglect the contributions of the negative-energy Dirac sea to the total energy. Instead, we judiciously choose the well depth to be zero and the well height to be infinite.

The plan of this paper is as follows. In the next section, we describe our model and the particular trial functions of interest. Under THE TOTAL ENERGY, we simplify the expression for the energy and determine the renormalized scalar particle mass. Under SEMICLASSICAL LIMIT AND BOSON ENERGY, we specialize to an ellipsoidal well of minor axis, b, and major axis, a, and take the limit when the renormalized scalar particle mass goes to infinity. We calculate the expectation value of the scalar particle Hamiltonian, which takes on a semiclassical form in this limit. Under SEMICLASSICAL LIMIT AND QUARK ENERGY, we determine the energy eigenvalues of a quark in an ellipsoidal well by an interpolation approximation. Justification for this approximation is given via perturbation theory. In the final section, we define the hadronic mass and carry out the final variations to determine the minor and major axes of the ellipsoidal well. The model is then applied to hadronic spectroscopy after assuming that hadrons are color singlet states.

THE MODEL AND TRIAL WAVE FUNCTIONS

For our model, we choose a simple field theory of quarks coupled linearly to a real scalar field for which the Hamiltonian is given by

$$\mathscr{H} = \mathscr{H}_B + :\mathscr{H}_Q:,$$

where

$$\mathscr{H}_B = \int d^3r(\tfrac{1}{2}\pi^2(\vec{r}) + \tfrac{1}{2}|\vec{\nabla}\phi(\vec{r})|^2 + P(\phi(\vec{r}))\}, \tag{1}$$

with

$$P(\phi(\vec{r})) = \sum_{n=1}^{4} \frac{C_n}{n!}\phi^n(\vec{r})$$

and

$$\mathscr{H}_Q = \int d^3r\psi^+(\vec{r})\left(\frac{1}{i}\vec{\alpha}\cdot\vec{\nabla} + g\beta\phi(\vec{r})\right)\psi(\vec{r}),$$

with the colored quark indices suppressed. The scalar and quark fields obey the usual commutation and anticommutation relations, respectively, with $\vec{\alpha}$ and β the conventional Dirac matrices.[5] The C_n are bare cutoff-dependent coupling constants, which, subsequent to renormalization, are to be taken finite in the limit when the renormalization cutoff goes to infinity and the quark-scalar particle-coupling constant, g, may also be cutoff dependent. We have also taken the bare mass of the quarks to be zero.

It is important to note our introduction of the normal ordering prescription of the quark Hamiltonian. This will systematically ignore the contributions of the negative-energy Dirac sea to the total energy.

We will carry out this calculation in a representation in which the scalar field is diagonal, as in Reference 4. In this representation, our choice of trial state is given by[6]

$$|j, m, \vec{p} = 0\rangle = \sum_{\Lambda} A_{\Lambda}^{*} \int d^3R \, d\Omega D_{m,\Lambda}^{*j}(\Omega)U(\Omega)T(\vec{R})G(\phi - \Phi)|Q\rangle, \qquad (2)$$

where $U(\Omega)$ is the rotation operator expressed in terms of the usual Euler angles and $D_{m,\Lambda}^{j}(\Omega)$,[7] the Wigner functions, are eigenfunctions of the symmetric top. $T(\vec{R})$ is the translation operator. The A_{Λ}'s are variational parameters that will eventually drop out of our calculation. The quantity $G(\phi - \Phi)$ is a Gaussian functional in $\phi(\vec{x})$[8] given by

$$G(\phi - \Phi) = \exp\left(-\frac{1}{2}\int d^3x \, d^3y(\phi(\vec{x}) - \Phi(\vec{x})f(|\vec{x} - \vec{y}|)(\phi(\vec{y}) - \Phi(\vec{y}))\right), \qquad (3)$$

where $\Phi(\vec{x})$ is a classical field about which the quantum field $\phi(\vec{x})$ fluctuates and corresponds to the average field produced by the quarks.

The function $f(\vec{x})$‡ corresponds to the width function given by

$$f(\vec{x}) = \int \frac{d^3k}{(2\pi)^3} \exp(i\vec{k} \cdot \vec{x})\omega(k). \qquad (4)$$

Both $\Phi(\vec{x})$ and $f(\vec{x})$ are variational functions. For the time being, the state vector $|Q\rangle$ represents an arbitrary number of quarks and/or antiquarks that will be chosen more specifically later in the calculation.

The state vector in Equation 2 can be shown to have the following properties:

$$J^2|j, m, \vec{p} = 0\rangle = j(j + 1)|j, m, \vec{p} = 0\rangle$$
$$J_z|j, m, \vec{p} = 0\rangle = m|j, m, \vec{p} = 0\rangle \qquad (5)$$
$$\vec{P}|j, m, \vec{p} = 0\rangle = 0$$

It therefore represents a system of quarks and scalar particles with definite total angular momentum and zero total linear momentum. For a more detailed discussion of this model and the interpretation and limitations of this trial function for the special case when $\Phi(\vec{x})$ is spherically symmetric, the reader is referred to Reference 4.

To attempt a general variation of the total energy with respect to $f(\vec{x})$ and $\Phi(\vec{x})$ is a formidable task. Instead, we consider a specific form for $\Phi(\vec{x})$, namely,

$$\Phi(\vec{r}) = \phi_0\theta(S - 1),$$

where

$$S = \frac{x^2 + y^2}{b^2} + \frac{z^2}{a^2}, \qquad (6)$$

so that $S = 1$ corresponds to the surface of an ellipsoid of major axis, a, and minor axis, b. The parameters a, b, and ϕ_0 are to be determined variationally.

‡ We have assumed that the width function depends on $|\vec{x} - \vec{y}|$, which, in general, is not true. As an example of a variational calculation when the width function is of the form $f(\vec{x}, \vec{y})$, see Cornwall *et al.*[9]

The Total Energy

With these specifications, we can proceed to determine the expectation value of our Hamiltonian. If we require

$$J_z|Q\rangle = K|Q\rangle, \tag{7}$$

so that K is the component of angular momentum of the quarks along the symmetry axis, as a result of rotational and translational invariance, we obtain from Equations 1 and 2 the result

$$E(j, K) = \frac{\int d^3R\, d\Omega_\beta \mathscr{D}\phi\, d^j_{KK}(\beta)\langle\Psi|T(\vec{R})^{-1}T(\beta)^{-1}\mathscr{H}|\Psi\rangle}{\int d^3R\, d\Omega_\beta\, \mathscr{D}\phi\, d^j_{KK}(\beta)\langle\Psi|T(\vec{R})^{-1}T(\beta)^{-1}|\Psi\rangle}, \tag{8}$$

where $T(\beta)$ is given by

$$T(\beta) = \exp(-i\beta J_y), \tag{9}$$

which corresponds to a rotation about the y axis through an angle β and the $d^j_{KK}(\beta)$ are related to the $D^j_{K,K}(\Omega)$ by the relation

$$D^j_{KK}(\Omega) = \exp(-iK\alpha)\, d^j_{KK}(\beta)\exp(-iK\gamma). \tag{10}$$

Note that the energy depends on the total z component of angular momentum through the relation

$$j \geq |m|$$

and an additional dependence on K given by

$$j \geq |K|. \tag{11}$$

To simplify the expression for the energy, we must perform the functional integration with respect to the variable $\phi(\vec{x})$. Making the following change of variable[4,8]

$$\phi(\vec{x}) = \phi(\vec{x}) + \Phi(\vec{x})$$

and making use of the following identities:

$$\int \mathscr{D}\phi G^2(\phi)\phi^n(\vec{r}) = 0 \qquad n = \text{odd integer}$$

$$\frac{\int \mathscr{D}\phi G^2(\phi)\phi(\vec{r}_1)\phi(\vec{r}_2)}{\int \mathscr{D}\phi G^2(\phi)} = \frac{1}{2}g(|\vec{r}_1 - \vec{r}_2|)$$

$$\frac{\int \mathscr{D}\phi G^2(\phi)\phi^{2n}(\vec{r})}{\int \mathscr{D}\phi G^2(\phi)} = (2n-1)!!\left[\frac{1}{2}g(0)\right]^n,$$

where $g(\vec{r})$ is the functional inverse of $f(\vec{r})$ given by

$$g(\vec{r}) = \int \frac{d^3k}{(2\pi)^3}\exp(i\vec{k}\cdot\vec{r})\frac{1}{\omega(k)},$$

we obtain

$$E(j, K) = E_b(j, K) + E_Q(j, K), \tag{12}$$

where

$$E_b(j, K) = E_0' - \frac{1}{2}\int d^3r\langle\Phi^T(\vec{r})\nabla^2\Phi(\vec{r})\rangle$$

$$+ \frac{\mu^2}{4}\int d^3r\Phi(\vec{r})\langle\Phi^T(\vec{r}) - \Phi(\vec{r})\rangle + \int d^3r\langle\tilde{P}(\bar{\Phi}(\vec{r}))\rangle \qquad \textbf{(12a)}$$

and

$$E_Q(j, K) = \int \frac{d^3R\, d\Omega_\beta\, d^j_{KK}(\beta)G^2(\Delta/2)\langle Q|T(\vec{R})^{-1}T(\beta)^{-1}:\mathscr{H}_Q:|Q\rangle}{\chi(j, K)}. \qquad \textbf{(12b)}$$

The above symbols have the following meaning:

$$\Phi^T(\vec{r}) = \Phi(\vec{r}\,' - \vec{R}),$$

with

$$\vec{r}\,' = R_y(\beta)\vec{r},$$

where $R_y(\beta)$ is a rotational matrix about the y axis through an angle β.

The symbol $\bar{\Phi}$ corresponds to the average classical field, namely,

$$\bar{\Phi}(\vec{r}) = \frac{\Phi(\vec{r}) + \Phi^T(\vec{r})}{2},$$

while Δ corresponds to the difference given by

$$\Delta(\vec{r}) = \Phi(\vec{r}) - \Phi^T(\vec{r}).$$

The symbol $\langle\ \rangle$ refers to an average of the quantity in brackets, for example,

$$\langle F\rangle = \int \frac{d^3R\, d\Omega_\beta\, d^j_{KK}(\beta)G^2(\Delta/2)F(\vec{R}, \beta, \vec{r})}{\chi(j, K)},$$

where

$$\chi(j, K) = \int d^3R\, d\Omega_\beta\ d^j_{KK}(\beta)G^2\left(\frac{\Delta}{2}\right)\langle Q|T(\vec{R})^{-1}T(\beta)^{-1}|Q\rangle,$$

with $G^2(\Delta/2)$, the weight function, given by

$$G^2\left(\frac{\Delta}{2}\right) = \exp\left(-\frac{1}{4}\int d^3x\, d^3y\ \Delta(\vec{x})f(|\vec{x} - \vec{y}|)\,\Delta(\vec{y})\right). \qquad \textbf{(13)}$$

The remaining quantities contributing to E_b are

$$\tilde{P}(z) = \sum_{n=1}^{4} \frac{\tilde{C}_n}{n!} z^n,$$

where the renormalized coupling constants, $\tilde{C}_n$, are given by

$$\begin{aligned}
\tilde{C}_1 &= C_1 + C_3 A \\
\tilde{C}_2 &= C_2 + C_4 A \\
\tilde{C}_3 &= C_3 \\
\tilde{C}_4 &= C_4,
\end{aligned} \qquad \textbf{(14)}$$

with

$$4A = \frac{1}{2\pi^2}\int_0^\Lambda \frac{dk\, k^2}{\omega(k)}$$

and

$$E_0' = \frac{\Omega}{4}\left(\int \frac{d^3k}{(2\pi)^3}\frac{\omega^2(k) + k^2}{\omega(k)} + C_2 + \frac{C_4 A}{2}\right),$$

where Ω is the volume of space.

For the contribution E_Q, we have

$$\mathscr{H}_Q = \int d^3r\psi^+(\vec{r})\left(\frac{1}{i}\vec{\alpha}\cdot\vec{\nabla} + g\beta\bar{\Phi}(\vec{r})\right)\psi(\vec{r}). \tag{15}$$

A detailed evaluation of E_Q will be presented under SEMICLASSICAL LIMIT AND QUARK ENERGY.

We can now proceed to determine the variational function $\omega(k)$ and hence $f(\vec{x})$ in Equation 4. Since the terms proportional to Ω make the largest contributions to the energy, we collect these terms and vary E_b with respect to $\omega(k)$ to obtain

$$\frac{\delta E_b}{\delta\omega(k)} = \frac{\Omega}{4}\left(1 - \frac{1}{\omega^2(k)}(k^2 + \tilde{C}_2 + \tilde{C}_3\phi_0 + \frac{1}{2}\tilde{C}_4\phi_0^2)\right) = 0, \tag{16}$$

which implies that

$$\omega(k) = (k^2 + \mu^2)^{1/2}, \tag{17}$$

where

$$\mu^2 = \tilde{C}_2 + \tilde{C}_3\phi_0 + \tfrac{1}{2}\tilde{C}_4\phi_0^2$$

is the renormalized scalar particle mass squared, which is to be taken finite in the limit when the cutoff $\Lambda \to \infty$. Since the sign of the second derivative of Equation 16 is positive at the point given by Equation 17, we, in fact, have a minimum.

SEMICLASSICAL LIMIT AND BOSON ENERGY

Consider for the moment the weight function given by

$$G^2\left(\frac{\Delta}{2}\right) = \exp(-K(\beta, \vec{R}, \eta)), \tag{18}$$

where

$$K(\beta, \vec{R}, \eta) = \frac{1}{4}\int d^3k\,\omega(k)\left|\int \frac{d^3r}{(2\pi)^{3/2}}\exp(i\vec{k}\cdot\vec{r})(\Phi(\vec{r}) - \Phi^T(\vec{r}))\right|^2, \tag{19}$$

and $\Phi(\vec{r})$ is given by Equation 6 with

$$\eta = \frac{b^2}{a^2} - 1. \tag{20}$$

The function $K(\beta, \vec{R}, \eta)$ has the following property:

$$K(\beta, \vec{R}, \eta) \geq 0,$$

and it takes on its minimum value under three circumstances

$$
\begin{aligned}
&1)\ \vec{R}=0; \qquad \beta=0 \\
&2)\ \vec{R}=0; \qquad \beta=\pi \\
&3)\ \vec{R}=0; \qquad \eta=0.
\end{aligned} \tag{21}
$$

It is easy to show that for the forms of $\omega(k)$ and $\Phi(\vec{r})$ given by Equations 17 and 6, respectively, that $K(\beta, \vec{R}, \eta)$ essentially reduces to the volume of intersection of two ellipsoids in the limit when $\mu \to \infty$. Since the weight function is steeply sloped in this limit, it is sufficient to expand $K(\beta, \vec{R}, \eta)$ about the points given in Equations 21 to obtain the leading contributions to the total energy. The relevant forms are given by

$$
\begin{aligned}
&1)\ G_1^2\left(\frac{\Delta}{2}\right) \cong \exp(-\tau\vec{R}\cdot\vec{d})\exp(-\varepsilon|\eta|\beta) \\
&2)\ G_2^2\left(\frac{\Delta}{2}\right) \cong \exp(-\tau\vec{R}\cdot\vec{d})\exp(-\varepsilon|\eta|(\pi-\beta)) \\
&3)\ G_3^2\left(\frac{\Delta}{2}\right) \cong \exp\left(-\frac{\tau' R}{b}\right)\exp(-\varepsilon'|\eta|\sin\beta),
\end{aligned} \tag{22}
$$

where we have

$$
\begin{aligned}
\tau &= \frac{\pi\mu\phi_0^2 b^2 a}{2} \qquad & \varepsilon &= \frac{2\mu\phi_0^2 b a^2}{3} \\
\tau' &= \frac{\pi\mu\phi_0^2 b^3}{2} \qquad & \varepsilon' &= \frac{2\mu\phi_0^2 b^3}{3},
\end{aligned} \tag{23}
$$

and the vector $\vec{d}$ is given by

$$\vec{d} = \frac{\hat{i}}{b} + \frac{\hat{j}}{b} + \frac{\hat{k}}{a}.$$

The above expressions are easily obtained upon evaluating the following general approximate form derived in Appendix A:

$$G^2\left(\frac{\Delta}{2}\right) \cong \exp\left(-\frac{\mu\phi_0^2}{4}\int d^3k\,\Delta S\ \delta(1-S)\right),$$

where ΔS is the change in the surface about the points in Equations 21. For the moment, we will assume that the parameters τ, ε, τ', and ε' are all much greater than one, and our variational solutions to ϕ_0, a, and b will be shown to be consistent with these assumptions. The approximate forms of the weight function given by Equation 22 essentially define two cases in the limit $\mu \to \infty$, namely, 1) and 2) contribute to the total energy when $\eta \neq 0$, and 3) contributes when $\eta \approx 0$. As a typical leading contribution to the boson energy, E_b, for the case when $\eta \neq 0$, consider the kinetic energy term given by

$$\text{Kinetic Energy} = -\frac{1}{2}\int d^3r\langle\Phi^T(\vec{r})\nabla^2\Phi(\vec{r})\rangle$$

$$\cong \frac{\phi_0^2}{2\chi(j,K)}\int d^3r\,\delta(1-S)|\vec{\nabla}S|^2\left(\left(\int d^3R\int_0^{\pi/2} d\Omega_\beta\, d^j_{KK}(\beta)G_1^2\left(\frac{\Delta}{2}\right)\right.\right.$$

$$\left.\left.+\int d^3R\int_{\pi/2}^{\pi} d\Omega_\beta\, d^j_{KK}(\beta)G_2^2\left(\frac{\Delta}{2}\right)\right)\delta(\Delta S)\langle Q|T(\vec{R})^{-1}T(\beta)^{-1}|Q\rangle\right),$$

where

$$\Delta S = \frac{2xz\beta\eta}{b^2} - 2\vec{r} \cdot \vec{R}'$$

and the vector $\vec{R}'$ is given by

$$\vec{R}' = \frac{R_x}{b^2}\hat{\imath} + \frac{R_y}{b^2}\hat{\jmath} + \frac{R_z}{a^2}\hat{k}.$$

Noting the following approximations:

$$T(\vec{R})^{-1} \cong 1$$

$$d^j_{KK}(\beta) \cong 1 - \frac{\beta^2}{4}(j(j+1) - K^2)$$

$$\langle Q | T(\beta)^{-1} | Q \rangle \cong 1 - \frac{\beta^2}{2}\langle J_y^2 \rangle$$

we have to leading order in μ

$$\text{Kinetic Energy} \cong \frac{\mu\pi a^2 b^2 \phi_0^4 (I_0 + \eta I_1)}{2},$$

where

$$I_0 = \int_0^{\pi/2} \mathrm{d}\phi \int_0^1 \mathrm{d}(\cos\theta) \frac{(1 + (9\pi/8)\sin 2\theta \cos\phi)}{(1 + (3\pi/8)\sin 2\theta \cos\phi)^3}$$

$$I_1 = \int_0^{\pi/2} \mathrm{d}\phi \int_0^1 \mathrm{d}(\cos\theta) \frac{(1 + (9\pi/8)\sin 2\theta \cos\phi)\cos^2\theta}{(1 + (3\pi/8)\sin 2\theta \cos\phi)^3}.$$

A numerical integration of I_0 and I_1 reveals

$$I_0 \cong 1.185$$

$$I_1 \cong 0.367.$$

A similar analysis of the remaining terms in Equation 12a gives, to leading order in μ, the result

$$E_b \cong E_0' - \frac{5\mu}{2} + (\Omega - V_0)\tilde{P}(\phi_0) + \frac{\mu\pi\phi_0^4 a^2 b^2}{2}(I_0 + \eta I_1), \tag{24}$$

where

$$V_0 = \frac{4\pi b^2 a}{3}.$$

The corresponding expression for E_b when $\eta = 0$ is given by

$$E_b \cong E_0' - \frac{3\mu}{2} + (\Omega - V)\tilde{P}(\phi_0) + \frac{\mu\pi^2\phi_0^4 b^4}{4}. \tag{25}$$

At this point, a brief description of the various terms which contribute to E_b is in order. For example, when $\eta \neq 0$, the expression $-5\mu/2$ arises from a $(1/\mu)$ contribution to the term proportional to μ^2 in Equation 12a. The next term, $(\Omega - V_0)\tilde{P}(\phi_0)$, is the classical energy density, modulo renormalization corrections of $\tilde{P}(\phi_0)$, multiplied by the volume for which the ellipsoidal well is nonzero. The form of the kinetic energy term results from the sharp edge of our potential well. Roughly speaking, it is equal to the intensity of the scalar field, multiplied by the surface area of the ellipsoid, divided by the skin thickness, and modified by shape dependent factors. It is a purely quantum mechanical contribution.

The most striking contrast between Equations 24 and 25 results from a comparison of the kinetic energy terms as $a \to b$. In fact, strictly speaking, the boson energy is discontinuous in the neighborhood of $a = b$ in the limit $\mu \to \infty$. Looking at the variationally dependent terms, we conclude that the contribution of the kinetic energy for a spherical well is greater than that for the ellipsoidal well by a factor $\pi/(2I_0)$ when $a \to b$. In fact, the spherical configuration is sitting on a cusp.

To complete the specification of our model, we make the following judicious choice of constants,[4] namely:

$$K_0 = \frac{\mu\pi}{2} z_0^4 \left(\frac{\Lambda}{g}\right)^4 > 0$$

$$U_0 = -\frac{4\pi}{3}\tilde{P}(\phi_0) = -\frac{1}{3\pi}(\Lambda z_0)^4 \sum_{n=1}^{4} b_n,$$

where

$$z_0 = \frac{m_0}{\Lambda} \gg 1$$

with

$$m_0 = g\phi_0$$

and

$$b_n = \frac{4\pi^2 \tilde{C}_n (\Lambda z_0)^{n-1}}{n!\, g^n} \gg 1.$$

The quark-scalar particle-coupling constant is taken such that

$$g = (2\pi^2 b_4)^{1/6} K_0^{-1/3} (z_0 \Lambda)^{5/3},$$

and the renormalized coupling constants, $\tilde{C}_n$, are chosen according to the prescription

$$\tilde{C}_1 = 0$$

$$\tilde{C}_2 = \frac{b_4 z_0^2 g^2 \Lambda^2}{2\pi^2}$$

$$\tilde{C}_3 = \frac{3 b_4 z_0 g^2 \Lambda}{\pi^2}$$

$$\tilde{C}_4 = \frac{6 b_4 g^4}{\pi^2}.$$

For this choice of constants, the renormalized scalar particle mass becomes

$$\mu = \left(\frac{b_4}{2\pi^2}\right)^{1/2} g z_0 \Lambda.$$

To determine the value of ϕ_0 that minimizes the energy, it is sufficient to minimize $\tilde{P}(\phi_0)$, namely,

$$\frac{\partial \tilde{P}(\phi_0)}{\partial \phi_0} = 0,$$

since the largest contributions that are dependent on ϕ_0 are proportional to Ω. The result is given by

$$\phi_0 = (2\pi^2 b_4)^{-1/6} K_0^{1/3} (z_0 \Lambda)^{-2/3}.$$

With these specifications, we will proceed in the next section to evaluate the quark contribution to the total energy given by Equation 12b. Since we have taken $z_0 \gg 1$, as $\Lambda \to \infty$ in the renormalized limit, the quarks will be confined by an ellipsoidal well with zero mass inside the well and infinite mass outside.

Semiclassical Limit and Quark Energy

We will concern ourselves with the case when $\eta \neq 0$, since $\eta = 0$ has been discussed elsewhere.[4] Upon substituting the appropriate form of the weight function into the expression for the quark energy in Equation 12b, we have to leading order in μ

$$E_Q \cong \langle Q | : \tilde{H}_F : | Q \rangle, \tag{26}$$

where

$$\tilde{H}_F = \int d^3 r \psi^+(\vec{r}) \left(\frac{1}{i} \vec{\alpha} \cdot \vec{\nabla} + g\beta\Phi(\vec{r}) \right) \psi(\vec{r}).$$

Note that we have taken the normalization of the quark state vectors to be one. We will now choose the quark state vectors, $|Q\rangle$, such that they diagonalize $:\tilde{H}_F:$ and hence minimize the quark contribution to the total energy. If we expand the quark field, $\psi(\vec{r})$, in creation and annihilation operators such that

$$\psi(\vec{r}) = \sum_\lambda A_\lambda U_\lambda + B_\lambda^+ V_\lambda$$

and define the vacuum state by the relations

$$B_\lambda | \text{vac} \rangle = 0$$
$$A_\lambda | \text{vac} \rangle = 0,$$

Equation 26 becomes

$$E_Q \cong \sum_\lambda E_\lambda (n_\lambda + \bar{n}_\lambda), \tag{26a}$$

where the E_λ's are determined by the equations

$$H_F U_\lambda = E_\lambda U_\lambda$$
$$H_F V_\lambda = -E_\lambda V_\lambda \tag{27}$$

with

$$H_F = \frac{1}{i} \vec{\alpha} \cdot \vec{\nabla} + \beta g \Phi(\vec{r}).$$

The quantities U_λ and V_λ are Dirac wave functions satisfying the usual orthonormality relations.[5] The index λ summarizes the quantum numbers of the quarks with n_λ and $\bar{n}_\lambda$, which represent the quark and antiquark occupation numbers, respectively.

It can be shown that the problem of solving the eigenvalue Equations 27 for an infinite well of arbitrary shape is equivalent to finding a solution to the following equation inside the well,[1] namely:

$$(\nabla^2 + E_\lambda^2)\begin{pmatrix} u_\lambda \\ v_\lambda \end{pmatrix}_{\text{inside well}} = 0, \tag{28}$$

where u_λ and v_λ are two-component spinors, and satisfying the following boundary condition on the surface:

$$\vec{\sigma} \cdot \hat{n} u_\lambda = -\frac{1}{E_\lambda} \vec{\sigma} \cdot \vec{\nabla} u_\lambda, \tag{29}$$

where $\hat{n}$ is the unit vector normal to the surface. To obtain analytic solutions to the above equations for an ellipsoidal surface is a formidable task. As an alternative, we consider three separate limiting cases and interpolate between them. We consider the problems of a quark between infinite parallel plates, inside a sphere, and inside an infinite cylinder. Based on dimensional arguments, the energy eigenvalues of a quark in an infinite ellipsoidal well of minor axis b and major axis a are approximately given by

$$\begin{aligned} E_K &\cong \frac{\nu_K^{(\pm)}}{b} && \text{when } a \gg b \\ E_K &\cong \frac{\lambda_K^{(\pm)}}{b} && \text{when } a = b \\ E_K &\cong \frac{\gamma_K^{(\pm)}}{a} && \text{when } a \ll b, \end{aligned} \tag{30}$$

where $(\pm)$ refers to the parity of the state. The solutions to Equations 28 and 29 and the eigenvalue conditions that determine $\nu_K^{(\pm)}$, $\lambda_K^{(\pm)}$ and $\gamma_K^{(\pm)}$ are presented in Appendix B. The results for the lowest eigenvalues are given by

$$\begin{aligned} \nu_{1/2}^{(+)} &= 1.44 \\ \lambda_{1/2}^{(+)} &= 2.04 \\ \gamma_{1/2}^{(+)} &= \pi/4. \end{aligned} \tag{31}$$

We choose to connect the three limiting cases in Equations 30 by a curve of the form

$$E_K \cong \sqrt{\frac{C_{1K}}{b^2} + \frac{C_{2K}}{ab} + \frac{C_{3K}}{a^2}}, \tag{32}$$

where

$$\begin{aligned} C_{1K} &= \nu_K^{(\pm)2} \\ C_{2K} &= \lambda_K^{(\pm)2} - \nu_K^{(\pm)2} - \gamma_K^{(\pm)2} \\ C_{3K} &= \gamma_K^{(\pm)2}. \end{aligned}$$

If we consider ordinary lowest-order perturbation theory as presented in Appendix C and perturb about the spherical solution, we obtain for the lowest eigenvalue

$$E^*_{K=1/2} \cong \frac{2.04}{b} + \frac{0.34}{b}(R^2 - 1), \tag{33}$$

where

$$R = \frac{b}{a}.$$

The ratio of the perturbative result in Equation 33 to the interpolated result in Equation 32 is presented as a function of R with b held fixed in FIGURE 1. The

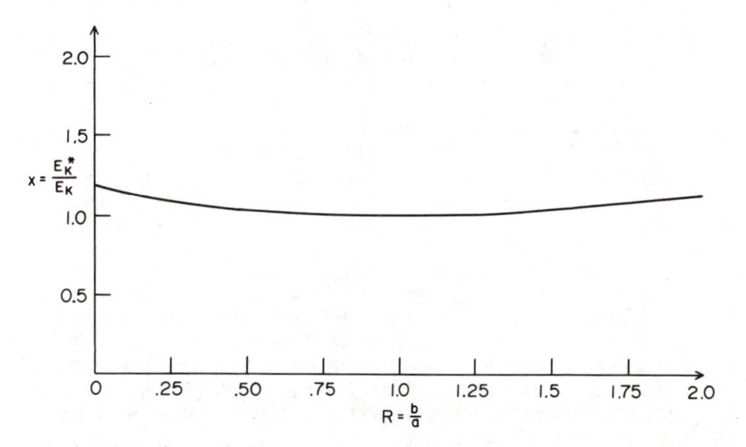

FIGURE 1. A comparison of the ground-state energy of a quark in an ellipsoidal well obtained from lowest-order perturbation theory, $E^*_{K=1/2}$, to our interpolated expression, $E_{K=1/2}$, in Equation 32 as a function of $R = b/a$ with the minor axis, b, held fixed.

agreement is excellent for R in the range

$$0.9 \leq R \leq 1.1,$$

so that one would expect that our interpolation is a good approximation for small deformations.

VARIATIONAL DETERMINATION OF THE HADRONIC MASS

Upon neglecting all terms that are independent of the variational parameters a and b in Equations 24 and 26a, we define the hadron mass by the relation

$$M = K_0 a^2 b^2 (I_0 + \eta I_1) + U_0 b^2 a + \sum_\lambda E_\lambda (n_\lambda + \bar{n}_\lambda), \tag{34}$$

where E_λ is given by Equation 32, and we choose U_0 to be greater than zero.

We will now proceed to apply this model to hadronic spectroscopy by considering the low-lying (mass < 2 GeV) nonstrange baryons with $j \leq \frac{3}{2}$. The quark occupation numbers summarized by the index λ in Equation 34 are now chosen to conform to the desired quantum numbers of the hadronic states of interest, with a minimum number of required quarks.[1] The vacuum state is defined as the state of the system in the absence of quarks. Inspection of Equation 34 indicates that the minimum vacuum energy is zero with $a = b = 0$. In the presence of quarks, the mean field, $\Phi(\vec{r})$, is determined by minimizing the energy with respect to the minor axis, b, and the major axis, a. This is carried out numerically for the following values of K_0 and U_0

$$U_0^{1/4} = 172.17 \text{ MeV}$$

$$K_0^4 = 0.41 \cdot U_0^5,$$

which are the values taken by Huang and Stump.[4] The results are presented in TABLE 1. As TABLE 1 indicates, the degenerate energy of the nucleon and Δ is

TABLE 1
LOW-LYING NONSTRANGE BARYONS WITH $j \leq \frac{3}{2}$*

$N(I = \frac{1}{2})$	M^*	M	R	$\Delta(I = \frac{3}{2})$
$N(\frac{1}{2})+$	1036	917	1.048	$\Delta(\frac{3}{2})+$
$2N(\frac{1}{2}-)_1\, N(\frac{3}{2}-)$	1286	1141	1.142	$\Delta(\frac{1}{2}-)_1\, \Delta(\frac{3}{2}-)$
$2N(\frac{1}{2}+)_1\, N(\frac{3}{2}+)$	1495	1322	1.222	$\Delta(\frac{1}{2}+)_1\, \Delta(\frac{3}{2}+)$
$2N(\frac{1}{2}+)_1\, N(\frac{3}{2}+)$	1527	1453	1.217	$\Delta(\frac{1}{2}+)_1\, \Delta(\frac{3}{2}+)$
$4N(\frac{1}{2}-)_1 2N(\frac{3}{2}-)$	1730	1524	1.284	$2\,\Delta(\frac{1}{2}-)_1\, \Delta(\frac{3}{2}-)$
$N(\frac{1}{2}-)$	1760	1555	1.279	$\Delta(\frac{3}{2}-)$

* A comparison of the mass values M^*, obtained in Reference 4 with a spherical well to those values, M, for an ellipsoidal well of minor axis, b, and major axis, a, with $R = b/a$ the equilibrium ratio. The symbol $N(j^P)$ indicates there are N states of spin j and parity P, differing from one another by internal quantum numbers. For an explanation of the spectroscopic notation, see Reference 1.

lowered by 59 MeV with respect to the case of a spherical well. The degree of deformation is rather small and corresponds to $R = b/a = 1.048$. This value of R indicates that our interpolation formula in Equation 32 for the energy of a quark in an ellipsoidal well is essentially exact to three decimal places. This also justifies our assumption that the quantities in Equations 23 are much greater than one. Hence, we conclude that for this model and our choice of trial functions, the ground-state energy of a system of quarks and scalar particles is lowered by deformation.

SUMMARY

In this paper, we have considered the possibility that the shape of the ground-state configuration of a system of quarks interacting with a real scalar field is nonspherical. We have explored this possibility through application of the variational principle to a simple model field theory of quarks coupled linearly to a real scalar field. Our trial functions have been, by construction, eigenstates of linear and angular momentum and have corresponded to quarks confined by an ellipsoidal well of major axis a and minor axis b with zero depth and height $g\phi_0$, where g is the quark-scalar particle coupling constant. The parameters a, b, and ϕ_0 were determined such that their values minimize the total energy of the system. We carried out these variations in the limit when the renormalized scalar particle mass, μ, is taken to infinity and found that the energy takes on a semiclassical form. Furthermore, in the renormalized limit (i.e., when the renormalization cutoff goes to infinity), we found that the energy depends on two finite arbitrary constants and that there exists a minimum of the energy for which the eccentricity of the confining well is nonzero. In fact, the spherical configuration is unstable and of higher energy. The model is applied to hadronic spectroscopy for the low-lying (mass < 2 GeV) nonstrange baryons with $j \leq \frac{3}{2}$.

ACKNOWLEDGMENT

The author thanks Professor Kerson Huang for suggesting this problem and contributing his ideas and critical comments throughout its development.

APPENDIX A

In the limit $\mu \to \infty$, Equation 19 becomes

$$K(\vec{R}, \beta, \eta) = \frac{\mu\phi_0^2}{2}\left(V_0 - \int d^3r\theta(1 - S)\theta(1 - S^T)\right), \tag{A.1}$$

where

$$V_0 = \frac{4\pi b^2 a}{3}$$

and

$$S^T = S + \Delta S,$$

with $\vec{r}' = R_y(\beta)\vec{r}$ and $\eta = b^2/a^2 - 1$. The volume of intersection of two ellipsoids is given by

$$Y = \int d^3r\theta(1 - S)\theta(1 - S^T),$$

which can be written in the equivalent form

$$Y = \int d^3r\theta(\Delta S)\theta(1 - S^T) + \int d^3r\theta(-\Delta S)\theta(1 - S). \tag{A.2}$$

If we consider the approximation where $|\Delta S/S| \ll 1$, we have

$$\theta(1 - S^T) \cong \theta(1 - S) - \Delta S\, \delta(1 - S)$$

and

$$\theta(1 - S) \cong \theta(1 - S^T) + \Delta S\, \delta(1 - S^T).$$

Substituting these approximate forms into Equation A.2, we obtain two approximate forms for Y, namely,

$$Y \cong V_0 - \int d^3r\theta(\Delta S)\, \Delta S\, \delta(1 - S)$$

and

$$Y \cong V_0 + \int d^3 r \theta(-\Delta S) \Delta S\, \delta(1 - S).$$

Adding these expressions and substituting in Equation A.1, we obtain the desired result.

Appendix B

The purpose of this appendix is to obtain solutions to Equations 28 and 29 for a quark in an infinitely long cylindrical well, a spherical well, and infinite parallel plates. This will then determine the quantities $\nu_K^{(\pm)}$, $\lambda_K^{(\pm)}$, and $\gamma_K^{(\pm)}$ in Equations 30.

Cylindrical Well

We will consider solutions for which $K \geq \frac{1}{2}$ since the corresponding solutions with $K \leq -\frac{1}{2}$ have the same energy. It is sufficient to discuss the upper spinor component, u, of the Dirac wave function since the lower component, v, is related by the relation

$$v = \frac{1}{iE} \vec{\sigma} \cdot \vec{\nabla} u. \tag{B.1}$$

Since the z component of the total angular momentum, $J_z = \sigma_z/2 + L_z$, commutes with the Dirac Hamiltonian in Equation 27, we take

$$u = A'_0 \begin{pmatrix} A(\rho, z)\exp(i(K - \frac{1}{2})\phi) \\ B(\rho, z)\exp(i(K + \frac{1}{2})\phi) \end{pmatrix}, \tag{B.2}$$

where

$$A(\rho, z) = J_{K-(1/2)}(\omega\rho)(A_1 \cos(k_z z) + \sin(k_z z))$$
$$B(\rho, z) = J_{K+(1/2)}(\omega\rho) \cdot B_0(B_1 \cos(k_z z) + \sin(k_z z))$$

with k_z the z component of momentum, $\omega = (E^2 - k_z^2)^{1/2}$, and A'_0, A_1, B_0, and B_1 are constants. With these specifications, the boundary condition, Equation 29, for $\rho = b$ becomes

$$\frac{\partial A}{\partial z} + \frac{\partial B}{\partial \rho} + \frac{(K + \frac{1}{2})B}{b} = -E \cdot B$$
$$-\frac{\partial B}{\partial z} + \frac{\partial A}{\partial \rho} - \frac{(K - \frac{1}{2})A}{b} = -E \cdot A. \tag{B.3}$$

Substituting the explicit forms for $A(\rho, z)$ and $B(\rho, z)$ in Equations B.3, we obtain two conditions

$$\sin(k_z z)(B_0 \cdot \omega - k_z \cdot A_1 + E \cdot B_0 \cdot \delta) + \cos(k_z z)(k_z + B_0 \cdot B_1 \omega + B_0 \cdot B_1 \cdot E \cdot \delta) = 0$$

$$\sin(k_z z)(k_z \cdot B_0 \cdot B_1 - \omega + E/\delta) + \cos(k_z z)(-k_z \cdot B_0 - \omega \cdot A_1 + E \cdot A_1/\delta) = 0, \tag{B.4}$$

where

$$\delta = J_{K+(1/2)}(\omega b)/J_{K-(1/2)}(\omega b).$$

Equations B.4 imply that $\delta = \pm 1$. If we define our zero of energy such that $k_z = 0$, which is equivalent to the lowest energy eigenvalues, the solutions to Equations B.4 are given by $\delta = +1$ with $B_0 \cdot B_1 = 0$ and A_1 arbitrary and $\delta = -1$ with $A_1 = 0$ and $B_0 \cdot B_1$ arbitrary. A solution to Equation B.2 that satisfies these requirements is

$$u = A_0 \begin{pmatrix} C \cdot (1 + \delta) J_{K-(1/2)}(E\rho) \exp(i(K - \frac{1}{2})\phi) \\ (1 - \delta) J_{K+(1/2)}(E\rho) \exp(i(K + \frac{1}{2})\phi) \end{pmatrix}$$

with A_0 the normalization and C an arbitrary constant. For $\delta = +1$ and $(K - \frac{1}{2})$ an even integer, the states have $(+)$ parity, whereas for $\delta = -1$, they have $(-)$ parity. In the case when $(K - \frac{1}{2})$ is an odd integer, the reverse is true. The quantity $\nu_K^{(\pm)}$ is determined by the conditions $\delta = \pm 1$. For the applications of interest in this paper, the following values of $\nu_K^{(\pm)}$ are to be considered

$$\nu_{1/2}^{(+)}(1) = 1.44$$

$$\nu_{1/2}^{(-)}(1) = 3.11$$

$$\nu_{1/2}^{(+)}(2) = 4.68,$$

where (1) refers to the first radial excitation and (2) the second.

Infinite Parallel Plates

We now address ourselves to the situation of a quark between infinite parallel plates located at $z = \pm a$. Since this problem possesses cylindrical symmetry, the solutions in Equations B.2 apply when the radius, b, is taken to infinity. The boundary conditions are given by

$$\frac{\partial A}{\partial z} + \frac{\partial B}{\partial \rho} + \frac{(K + \frac{1}{2})B}{b} = \begin{cases} -E \cdot A & z = a \\ +E \cdot A & z = -a \end{cases}$$

$$\frac{\partial B}{\partial z} - \frac{\partial A}{\partial \rho} + \frac{(K + \frac{1}{2})A}{b} = \begin{cases} -E \cdot B & z = a \\ +E \cdot B & z = -a \end{cases}.$$

Upon substituting the solutions, Equation B.2, we obtain the conditions

$$B_0^2 = \frac{A_1}{B_1},$$

where

$$\frac{A_1}{B_0} = \frac{-E \sin(k_z a) - k_z \cos(k_z a)}{\omega \cos(k_z a)}$$

and

$$\frac{A_1}{B_0} = \frac{\omega \sin(k_z a)}{-E \cos(k_z a) + k_z \sin(k_z a)},$$

from which we obtain the eigenvalue condition

$$\cos(2k_z a) = 0.$$

Taking the zero of energy such that $\omega = 0$, we find for the lowest-energy eigenvalue $\gamma_{1/2}^{(+)} = \pi/4$. The eigenfunctions follow trivially and will not be presented.

Spherical Well

In this section, we will develop the solutions for a quark in a spherical well in some detail since they will be of use in Appendix C. By use of the general relation in Equation B.1, the solution to Equation 28 is given by

$$U = \begin{pmatrix} \dfrac{iG_j^{(s)}}{r}\phi_{jK}^{(s)} \\ \dfrac{F_j^{(s)}}{r}\dfrac{\vec{\sigma}\cdot\vec{r}}{r}\phi_{jK}^{(s)} \end{pmatrix} \tag{B.6}$$

Where $G_j^{(s)}$ and $F_j^{(s)}$ satisfy

$$-EG_j^{(s)} = \left(\frac{\mathrm{d}}{\mathrm{d}r} + \frac{1 - I_s}{r}\right)F_j^{(s)}$$

$$EF_j^{(s)} = \left(\frac{\mathrm{d}}{\mathrm{d}r} + \frac{1 + I_s}{r}\right)G_j^{(s)}, \tag{B.7}$$

which are the recursion relations of the spherical Bessel functions and $I_s = -s(j + \frac{1}{2})$ with $s = \pm 1$. The positive sign corresponds to $j = l + \frac{1}{2}$ and the negative sign $j = l - \frac{1}{2}$. The functions $\phi_{jK}^{(s)}$ are given by

$$\phi_{jK}^{(s)} = \begin{pmatrix} \sqrt{\dfrac{l + \frac{1}{2} + S \cdot K}{2l + 1}}\, Y_{l,\, K-(1/2)} \\ s \cdot \sqrt{\dfrac{l + \frac{1}{2} + S \cdot K}{2l + 1}}\, Y_{l,\, K+(1/2)} \end{pmatrix}$$

with the states in Equation B.6 having parity $(-)^l$. Consider the case when $s = +1$, then the upper spinor component is given by

$$u = n_{I+}(Er)\phi_{jK}^{(+)},$$

where n_{I+} is a spherical Neumann function. If we further specialize by taking $j = K \geq \frac{1}{2}$, the boundary condition, Equation 29, gives

$$E\left(\frac{\mathrm{d}n_{I+}}{\mathrm{d}r} + n_{I+}\right) - \frac{(K - \frac{1}{2})}{b}n_{I+} = 0,$$

which implies

$$j_{K+(1/2)}(Eb) = j_{K-(1/2)}(Eb). \tag{B.8}$$

For the case when $s = -1$, we find

$$j_{K+(1/2)}(Eb) = -j_{K-(1/2)}(Eb), \tag{B.9}$$

which determines $\lambda_K^{(\pm)}$. The values of interest for this investigation are

$$\lambda_{1/2}^{(+)}(1) = 2.04$$
$$\lambda_{1/2}^{(-)}(1) = 3.84$$
$$\lambda_{1/2}^{(+)}(2) = 5.39.$$

Appendix C

The purpose of this appendix is to obtain an expansion of the ellipsoidal well in the parameter η and then apply ordinary first-order perturbation theory to obtain the first-order correction in η to the energy of a quark in a spherical well. The essential point of the calculation is that to determine the above correction to the energy in the limit when $M_0 \to \infty$, one must determine the $1/M_0$ corrections to the wave function.

For $\eta \ll 1$ and the minor axis, b, held fixed, our Hamiltonian in Equation 27 is approximately given by

$$H \cong H_0 + H',$$

where

$$H_0 = -i\vec{\alpha} \cdot \vec{\nabla} + \beta M_0\left(1 - \theta\left(1 - \frac{r^2}{b^2}\right)\right) \tag{C.1}$$

is the unperturbed Hamiltonian and

$$H' = \xi M_0 b\, \delta(b - r)\left(Y_{00}(\Omega) + \frac{2}{\sqrt{5}} Y_{20}(\Omega)\right)\beta$$

the perturbation with

$$\xi = \frac{\sqrt{\pi}}{3} \cdot \eta.$$

First-order perturbation theory gives

$$E^* = E^{(0)} + \langle j, K | H' | j, K\rangle,$$

where $E^{(0)}$ is determined by the eigenvalue equation

$$H_0 U_\lambda = E_\lambda^{(0)} U_\lambda,$$

and U_λ is of the general form given by Equation B.6, for which a direct substitution gives

$$E^{(1)} = \langle j, K | H' | j, K\rangle = \xi M_0 b^3(|G_j^{(s)}|^2 - |F_j^{(s)}|^2)_{r=b} \cdot R_{(j,K)}^{(s)} \tag{C.2}$$

with

$$R^{(s)}(j, K) = \int d\Omega \phi_{j,K}^{+(s)}(\Omega)\phi_{j,K}^{(s)}(\Omega)\left(Y_{00}(\Omega) + \frac{2}{\sqrt{5}} Y_{20}(\Omega)\right)$$

$$= \frac{1}{\sqrt{4\pi}}\left(1 - 2 \cdot \frac{3(K^2 + \frac{1}{4}) - l(l+1) - s \cdot 6K^2/(2l+1)}{(2l-1)(2l+3)}\right)$$

The radial wave functions, G and F, inside the spherical well, $r < b$, satisfy[5]

$$\frac{dG_{\text{in}}^{(s)}}{dr} + \frac{(1 + I_s)}{r} G_{\text{in}}^{(s)} = E^{(0)} F_{\text{in}}^{(s)}$$

$$\frac{dF_{\text{in}}^{(s)}}{dr} + \frac{(1 - I_s)}{r} F_{\text{in}}^{(s)} = -E^{(0)} G_{\text{in}}^{(s)}, \quad \textbf{(C.3)}$$

and outside, $r > b$, we have

$$\frac{dG_{\text{out}}^{(s)}}{dr} + \frac{(1 + I_s)}{r} G_{\text{out}}^{(s)} = (M_0 + E^{(0)}) F_{\text{out}}^{(s)}$$

$$-\frac{dF_{\text{out}}^{(s)}}{dr} + \frac{(I_s - 1)}{r} F_{\text{out}}^{(s)} = (E^{(0)} - M_0) G_{\text{out}}^{(s)} \quad \textbf{(C.4)}$$

with the boundary conditions at the surface, $r = b$, given by

$$G_{\text{in}}^{(s)} = G_{\text{out}}^{(s)}$$

$$F_{\text{in}}^{(s)} = F_{\text{out}}^{(s)}. \quad \textbf{(C.5)}$$

For the case when $I_+ = -(j + \frac{1}{2})$, Equations C.3 and C.4 have the solutions

$$G_{j\,\text{in}}^{(+)} = -A_\sigma^{(+)} j_{\sigma-1}(x) \qquad F_{j\,\text{in}}^{(+)} = A_\sigma^{(+)} j_\sigma(x)$$

$$G_{j\,\text{out}}^{(+)} = -B_\sigma^{(+)}(M_0 + E^{(0)})^{1/2} k_{\sigma-1}(z) \qquad F_{j\,\text{out}}^{(+)} = B_\sigma^{(+)}(M_0 - E^{(0)})^{1/2} k_\sigma(z), \quad \textbf{(C.6)}$$

where $j_\sigma(x)$ and $k_\sigma(z)$ are the regular and modified spherical Bessel functions, respectively, with $\sigma = j + \frac{1}{2}$, $x = E^{(0)} r$, $z = r \cdot (M_0^2 - E)^{1/2}$. $A_\sigma^{(+)}$ and $B_\sigma^{(+)}$ are constants. Substituting these solutions into Equation C.2 and using the boundary conditions in Equations C.5, we obtain

$$E^{(1)} = \xi \cdot M_0 \cdot b^3 |B_\sigma^{(+)}|^2 \cdot \frac{\pi}{2b(M_0^2 - E^{(0)2})^{1/2}} ((M_0 + E^{(0)}) \cdot K_{\sigma+(1/2)}^2(z)$$

$$- (M_0 - E^{(0)}) \cdot K_{\sigma+(1/2)}^2(z))_{r=b} \cdot R^{(+)}(j, K), \quad \textbf{(C.7)}$$

where $K_\sigma(z)$ is the regular modified Bessel function. As $M_0 \to \infty$, we have the expansions

$$E^{(0)} \cong \frac{\lambda_{j,K}^{(+)}}{b} + 0\left(\frac{1}{M_0}\right)$$

$$K_\sigma(z_0) \cong \sqrt{\frac{\pi}{2z_0}} \cdot \exp(-z_0)\left(1 + \frac{a-1}{8z_0} + \cdots\right), \quad \textbf{(C.8)}$$

where $\lambda_{j,K}^{(+)}$ is determined by Equation B.8, $z_0 = M_0 b$, and $a = 4\sigma^2$. To complete the determination of $E_1^{(+)}$, we must determine $|B_\sigma^{(+)}|^2$ in the limit when $M_0 \to \infty$. To this end, we use Equations C.5, C.6, and C.8 and the normalization condition to obtain

$$|B_\sigma^{(+)}|^2 \cong \frac{8 \cdot M_0}{\pi^2 b \exp(-2bM_0)\left(2 - \frac{J_{\sigma-(3/2)}(x')}{J_{\sigma+(1/2)}(x')} - \frac{J_{\sigma+(3/2)}(x')}{J_{\sigma-(1/2)}(x')}\right)},$$

where $x' = \lambda_{j,K}^{(+)}$. If we further specialize to the case when $j = K \geq \frac{1}{2}$, Equation C.7

becomes

$$E^* \cong \frac{\lambda^{(+)}_{K,K}}{b} + \frac{2\eta}{3b} \frac{\left(\lambda^{(+)}_{K,K} - \left(K + \frac{1}{2}\right)\right) \cdot \left(1 - \frac{2K^2 - 3K + 1}{2(K^2 - 1)}\right)}{2 - \frac{J_{K-1}(x')}{J_{K+1}(x')} - \frac{J_{K+2}(x')}{J_K(x')}}. \tag{C.9}$$

For the case when $I_s = K + \frac{1}{2}$ and $K \geq \frac{1}{2}$, we have

$$E^* \cong \frac{\lambda^{(-)}_{K,K}}{b} + \frac{2\eta}{3b} \frac{\left(\lambda^{(-)}_{K,K} - \left(K + \frac{1}{2}\right)\right)\left(1 - \frac{2(K^2 - 1) + 3K}{2(K + 2)(K + 1)}\right)}{2 - \frac{J_{K-1}(x'')}{J_{K+1}(x'')} - \frac{J_{K+2}(x'')}{J_K(x'')}},$$

where $x'' = \lambda^{(-)}_{K,K}$. Equation 33 follows from Equation C.9 when $K = \frac{1}{2}$.

References

1. Chodos, A., R. L. Jaffe, K. Johnson, C. B. Thorn & V. F. Weisskopf. 1974. Phys. Rev. D **9:** 3471; Chodos, A., R. L. Jaffe, K. Johnson & C. B. Thorn. 1974. Phys. Rev. D **10:** 2599; Jaffe, R. L. 1975. Phys. Rev. D **11:** 1953; DeGrand, T., R. L. Jaffe, K. Johnson & J. Kiskis. 1975. Phys. Rev. D **12:** 2060.
2. Johnson, K. & C. B. Thorn. 1976. Phys. Rev. D **13:** 1934.
3. Creutz, M. 1974. Phys. Rev. D **10:** 1749; Creutz, M. & K. S. Soh. 1975. Phys. Rev. D **12:** 443; Bardeen, W. A., M. S. Chanowitz, S. D. Drell, M. Weinstein & T.-M. Yan. 1975. Phys. Rev. D **11:** 1094; Lee, T. D. 1979. Phys. Rev. D **19:** 1802.
4. Huang, K. & D. R. Stump. 1976. Phys. Rev. D **14:** 223.
5. Bjorken, J. D. & S. D. Drell. 1964. Relativistic Quantum Mechanics. McGraw-Hill Book Company. New York, N.Y.
6. Peierls, R. E. & D. J. Thouless. 1962. Nucl. Phys. **38:** 154.
7. Gottfried, K. 1966. Quantum Mechanics. Vol. I. Fundamentals. W. A. Benjamin. New York, N.Y.
8. Rosen, G. 1967. Phys. Rev. **160:** 1278.
9. Cornwall, J. M., R. Jackiw & E. Tomboulis. 1974. Phys. Rev. D **10:** 2428.

SOME QUESTIONS OF S-MATRIX STRUCTURE

Yu. A. Gol'fand

Moscow, U.S.S.R.

In this paper, some general properties of the S-matrix will be considered. The important point of the approach is construction of quasilocal structure. The concrete form of the quasilocal structure depends on a given set of fields and their interactions. The S-matrix will be represented in the form of expansion into series on a quasilocal structure. The coefficient functions of this expansion are amplitudes of the S-matrix.

The amplitudes are analytic functions of coordinates. This situation can be proved in simple cases, and, naturally, it seems that it remains correct in general.

The main result of this investigation is that the unitarity and causality of the S-matrix turn out to be the consequence of simple analytic properties of the amplitudes in x-space.

We shall consider the simplest model of a self-interacting scalar field, although the results can be generalized to other cases without any difficulty. We shall then obtain the explicit form of quasilocal structure on the basis of Lagrangian formalism. In this case, amplitudes may be represented by means of T-products of local Heisenberg operators. At this stage, T-products have a purely formal role. Their meaning will be determined later. It should be noted that application of the Lagrangian method has only a heuristic character.

In the third section, we shall obtain a modified representation of the Wightman functions. In addition to the usual set of Wightman axioms, the specific Quantum Field Theory representation of operators by means of expansion on normal products of the free fields is taken into account. As a result, the n-point Wightman function is expressed through the function of $n(n-1)/2$ 4-vectors.

We shall next impose some restrictions on the general form of the Wightman functions. In this way, we arrive at the simplified form of the Wightman functions. In this form, different properties of the Wightman functions can be ascertained easily; in particular, a convenient expression for the Wightman functions that correspond to permutation of the field operators is obtained as a boundary value of the same analytic function. We thus are able to construct the expression for vacuum expectation of the T-products of the operators and also for more complicated forms of ordered products, which are needed for representation of the amplitude. We then use this representation to prove unitarity and Bogolyubov's causality of the S-matrix.

Functional Representation of the S-Matrix and Related Quantities

For every operator A of Quantum Field Theory, a corresponding functional $A[\lambda]$, which depends on the numerical function $\lambda(x)$, can be constructed. By definition, the functional $A[\lambda]$ has the same coefficient functions as does the expansion of operator A through normal products of the free fields. In essence, the functional $A[\lambda]$ contains as much physical information as does operator A.

The functional $A[\lambda]$ can be expressed through the operator A by means of the relation

$$A[\lambda] = \langle A(\varphi + \lambda)\rangle_o. \quad (2.1)$$

The brackets $\langle\ \rangle_o$ denote vacuum expectation. The formula is true irrespective of the mode in which operator A is represented. By means of the Wick theorem, it is easy to ascertain that Equation 2.1 corresponds to the definition of functional $A[\lambda]$.

It is convenient here to express the S-matrix in terms of the Lagrangian. The usual expression for the S-matrix is

$$S = T \exp i \int \mathscr{L}(\varphi)\, \mathrm{d}x. \quad (2.2)$$

The interaction Lagrangian $\mathscr{L}(\varphi)$ may be an arbitrary function of the scalar field $\varphi(x)$. At this stage, the question of divergence is not of interest. It should be noted that the treatment of Lagrangian formalism has only a heuristic role. The final expression has an application outside the framework of the Lagrangian method.

According to Equation 2.1, the functional $S[\lambda]$ may be expressed in the form

$$S[\lambda] = \langle S(\varphi + \lambda)\rangle_o, \quad (2.3)$$

where

$$S(\varphi + \lambda) = T \exp i \int \mathscr{L}(\varphi + \lambda)\, \mathrm{d}x. \quad (2.4)$$

However, it is more convenient to express the functional $S[\lambda]$ in a somewhat different form:

$$S[\lambda] = \langle R(\varphi, \lambda)\rangle_o, \quad (2.5)$$

where

$$R(\varphi, \lambda) = S^+(\varphi)S(\varphi + \lambda). \quad (2.6)$$

If the vacuum state is stable (we conjecture that it is), Equation 2.5 is equivalent to Equation 2.3. To obtain the final expression for $S[\lambda]$, the expansion

$$\mathscr{L}(\varphi + \lambda) = \sum_{n=0}^{\infty} \frac{\lambda^n}{n!} \mathscr{L}^{(n)}(\varphi) \quad (2.7)$$

is used.

It is convenient here to introduce the following notation. The quantity

$$\mathscr{L}^{(n)}(\varphi) \equiv \mathrm{d}^n \mathscr{L}(\varphi)/\mathrm{d}\varphi^n = j_n^0(x) \quad (2.8)$$

will be called current of n-s order (in interaction representation). These currents form a "vector" (in the general case, infinite dimensional)

$$j^0(x) = (j_1^0(x), j_2^0(x), \ldots).$$

By analogy, the quantities

$$\Lambda_n(x) = \lambda^n(x)/n! \quad (2.9)$$

form the "vector" $\Lambda(x)$. This "vector" will be called a quasilocal structure. In the case under consideration, the quasilocal structure is of simple form.

In these notations, Equation 2.7 can be rewritten in the form

$$\mathscr{L}(\varphi + \lambda) = \mathscr{L}(\varphi) + j^0(x)\Lambda(x). \quad \textbf{(2.10)}$$

The second term in Equation 2.10 means "scalar product" of "vectors" $j^0(x)$ and $\Lambda(x)$. By use of Equation 2.10, the quantity 2.6 can be transformed into the form

$$R(\varphi, \lambda) = S^+ T \exp\left\{S \exp i \int j^0(x)\Lambda(x)\, dx\right\}$$

$$= T \exp i \int j(x)\Lambda(x)\, dx, \quad \textbf{(2.11)}$$

where $j(x) = (j_1(x), j_2(x), \ldots)$ is the "vector" of Heisenberg currents corresponding to quantity 2.8. The functional Equation 2.5 now takes the form

$$S[\lambda] = \left\langle T \exp i \int j(x)\Lambda(x)\, dx \right\rangle_0. \quad \textbf{(2.12)}$$

The following remarks should be made regarding this formula:

To define the S-matrix in this form, it is necessary to have a set of local Heisenberg operators $j_n(x)$. This set of operators enables one to use non-Lagrangian formalism for Quantum Field Theory.

Coefficient functions of the functional $S[\lambda]$ are represented as vacuum expectations of T-products of any number of Heisenberg operators $j_n(x)$. At this stage, these T-products are considered in a quite formal way. In the next two sections, however, precise definitions will be given for these quantities.

The functional $S[\lambda]$ explicitly depends on the set of functions $\Lambda_n(x)$. According to Equation 2.9, this means the isolation of quasilocal terms in functional $S[\lambda]$, which correspond to emission and absorption of any number of particles at the same point x.

The analogous expression for amplitudes of the S-matrix has been obtained in Reference 1.

To consider the properties of unitarity and causality, it is convenient to introduce the functional in a more general form that depends on two numerical functions $\lambda(x)$ and $\mu(x)$.

$$S[\lambda, \mu] = \langle S^+(\varphi + \lambda)S(\varphi + \mu)\rangle_0. \quad \textbf{(2.13)}$$

By use of Equations 2.6 and 2.11, it is easy to represent this functional in the form

$$S[\lambda, \mu] = \langle R^+(\varphi, \lambda)R(\varphi, \mu)\rangle_0$$

$$= \left\langle \tilde{T} \exp\left(-i \int j(x)\Lambda(x)\, dx\right) T \exp i \int j(x)M(x)\, dx \right\rangle_0. \quad \textbf{(2.14)}$$

The symbol $\tilde{T}$ denotes the antichronological product. The functional $S[\lambda, \mu]$ depends on two quasilocal structures, Λ and M. The amplitudes of the functional Equation 2.14 are vacuum expectations of the "mixed" products of the currents

$$\langle \tilde{T}(j_{n_1}(x_1) \cdots j_{n_k}(x_u)) T(j_{m_1}(y_1) \cdots j_{m_l}(y_l))\rangle_0. \quad \textbf{(2.15)}$$

It is convenient to represent the functional Equation 2.14 in the exponential form

$$S[\lambda, \mu] = e^{i\Delta[\Lambda, M]}. \quad \textbf{(2.16)}$$

The amplitudes of functional $\Delta[\Lambda, M]$ are of the form given by Equation 2.15, but in this case it is necessary to take into account only the contributions of connected diagrams.

Generalized Wightman Function

The amplitudes of the functionals in the previous section are formally determined by Equation 2.15. The next task is to give them exact mathematical meaning, which will be accomplished by representing the amplitudes as boundary values of analytic functions.

In this section, some generalizations of the Wightman theorem[2] of the vacuum expectation of the product of local field operators will be considered. In addition to the usual postulates of the Wightman theorem (e.g., see References 3 and 4), the specific property of field theoretical operators is taken into account. This additional property is that the local operator $j_n(x)$ may be represented in the form of an expansion by normal products of free field operators.

Consider the Wightman function

$$W(x_1, \ldots, x_k) = \langle j_{n_1}(x_1) \cdots j_{n_k}(x_k)\rangle_0 . \tag{3.1}$$

The calculation of vacuum expectation in Equation 3.1 is performed by forming all pairs of the free fields that belong to different factors. This procedure can be described by means of generalized matrix elements

$$\begin{aligned} M_l &= \langle p_{1l}, \ldots, p_{l-1,l} | j_{n_l}(x_l) | p_{l,l+1}, \ldots, p_{lk}\rangle \\ &= \langle p_{1l}, \ldots, p_{l-1,l} | j_{n_l}(0) | p_{l,l+1}, \ldots, p_{lk}\rangle \exp\left[ix_l\left(\sum_{k}^{l-1} p_{sl} - \sum_{l+1}^{k} p_{ls}\right)\right] \end{aligned} \tag{3.2}$$

Here, $p_{r,s}$ $(r < s)$ denotes momentum transfer from $j_{n_r}(x_r)$ to $j_{n_s}(x_s)$. For the sake of brevity, all internal quantum numbers are omitted in Equation 3.2. The Wightman function of Equation 3.1 may be represented in the form

$$\begin{aligned} W(x_1, \ldots, x_k) &= \int \prod_{r<s} dp_{rs} \sum \prod_{l=1}^{k} M_l \\ &= \int \prod_{r<s} dp_{rs}\, f(p_{rs}) \exp\left[-i \sum_{r<s} p_{rs}(x_r - x_s)\right], \end{aligned} \tag{3.3}$$

where summation in the integrand spreads over all internal quantum numbers. The integration goes over $k(k-1)/2$ independent momenta p_{rs} $(1 \le r < s \le k)$.

The function $F(Z_{rs})$ of $k(k-1)/2$ independent complex 4-vectors Z_{rs} $(1 \le r < s \le k)$ can be defined by the expression:

$$F(Z_{rs}) = \int \prod_{r<s} dp_{rs}\, f(p_{rs}) \exp\left(-i \sum_{r<s} p_{rs} Z_{rs}\right). \tag{3.4}$$

It follows from the spectral condition that the function $F(Z_{rs})$ is analytic in the tube $\mathrm{Im}\, Z_{rs} \in V_+$. The Wightman function of Equation 3.3 is the boundary value of the analytic function of Equation 3.4 defined by the relation

$$W(x_1, \ldots, x_k) = \lim_{\substack{\gamma_{rs} \to 0 \\ (\gamma_{rs} \in V_+)}} F(x_r - x_s - i\gamma_{rs}). \tag{3.5}$$

These considerations are not strict enough from a mathematical point of view. However, they reveal sufficiently clearly that the k-point Wightman function can be expressed as the boundary value of an analytic function that depends on $k(k-1)/2$ complex 4-vectors.

Representation of the Amplitudes

To represent the amplitudes, the following limitation will be imposed on the general form of the function in Equation 3.4: the function $F(Z_{rs})$ depends only on squares of vectors Z_{rs}

$$F(Z_{rs}) = G(Z_{rs}^2). \tag{4.1}$$

The condition in Equation 4.1 is strong, and, in fact, it is possible that it is not fulfilled identically in the whole space but only in some of its subspace. For subsequent considerations, it is necessary to give the condition about analytic properties of the function $G(Z_{rs}^2)$. Of course, this condition should not contradict the analytic properties of function $F(Z_{rs})$ stated in the previous section.

The condition on analytic properties is that the function is analytic in certain complex environments of the physical domain, except on a set of cuts along the positive semiaxes $z_{rs}^2 > 0$.

The physical domain is a manifold of values Z_{rs} of the form

$$Z_{rs}^2 = (x_r - x_s), \tag{4.2}$$

where x_r $(1 \leq r \leq k)$ is the set of k-independent real 4-vectors.

At this stage, the conjecture in Equation 4.1 and the analyticity condition have a model character. It would be interesting to see whether they can be obtained from general considerations.

Under the condition in Equation 4.1, the Wightman function in Equation 3.5 can be expressed in the form

$$W(x_1, \ldots, x_k) = G(x_{rs}^2 - i\varepsilon x_{rs}^0). \tag{4.3}$$

Here, $x_{rs} = x_r - x_s$ and vectors γ_{rs} are taken in the form $\gamma_{rs} = (\varepsilon/2, 0, 0, 0)$, where $\varepsilon > 0$ is infinitesimal.

Consider now the Wightman function $W_\pi(x_1, \ldots, x_k)$, where π denotes any permutation of operators $j_n(x)$ in the right-hand side of Equation 3.1. As was shown in the previous section, calculation of the Wightman function is performed by means of the Wick theorem, and, as a result, the Wightman function W_π will be a boundary value of the same analytic function:

$$W_\pi(x_1, \ldots, x_k) = G(x_{rs}^2 \pm i\varepsilon x_{rs}^0). \tag{4.4}$$

The rule of signs of the imaginary part is simple: a minus sign is used if in Equation 3.1 the operator $j_{n_r}(x_r)$ is on the left of operator $j_{n_s}(x_s)$; a plus sign is used if $j_{n_r}(x_r)$ is on the right of $j_{n_s}(x_s)$.

This simple rule of signs allows one to construct representations for the amplitudes of the type 2.15:

$$\Delta(x_1, \ldots, x_p | x_{p+1}, \ldots, x_k) = \langle \tilde{T}(j_{n_1}(x_1) \cdots j_{n_p}(x_p)) T(j_{n_{p+1}}(x_{p+1}) \cdots j_{n_k}(x_k)) \rangle_0. \tag{4.5}$$

For the sake of brevity, the indexes n_s in this notation are dropped. The symbols for the chronological and antichronological products in the right-hand side of Equation 4.5 define the order of operators $j_n(x)$. By means of the rule of signs, the function Δ can be represented in the form:

$$\Delta(x_1, \ldots, x_p | x_{p+1}, \ldots, x_k) = G(x_{rs}^2 + i\varepsilon\zeta_{rs}), \tag{4.6}$$

where

$$\zeta_{rs} = \begin{cases} 1 & \text{for} \quad 1 \leq r, s \leq p \\ -x_{rs}^0 & \text{for} \quad 1 \leq r \leq p;\ p+1 \leq s \leq k. \\ -1 & \text{for} \quad p+1 \leq r, s \leq k \end{cases} \tag{4.7}$$

The conditions in Equation 4.7 are an immediate consequence of the rule of signs. Consider, for example, the case where $1 \leq r, s \leq p$. The corresponding operators are antichronologically ordered, and that is why the condition $x_{rs}^0 < 0$ means that operator $j_{n_r}(x_r)$ is located at the left of operator $j_{n_s}(x_s)$.

Representation 4.6 is relativistically invariant, because the function G is analytic by $x_{rs}^2 < 0$, and therefore, in this case, the sign of the imaginary part is not of importance.

Equation 4.6 is the definition of vacuum expectations of the mixed ordered operator product. All singularities that can arise by coincidence of two points are taken into account by selection of the analytic function G.

Some relations for the function Δ will now be proved. These relations are useful for demonstrating the unitarity and causality conditions of the S-matrix.

Let A denote the set of points $\{x_1, x_2, \ldots x_k\}$ and indexes $\{n_1, n_2, \ldots n_k\}$. Consider subdividing A into two nonintersecting sets P and Q: $A = P + Q$. In particular, either P or Q may be empty. Let p denote the number of elements in set P. The function of the form of Equation 4.6, for which the first p arguments belong to set P and the rest to set Q, will be denoted by $\Delta(P|Q)$. The function $\Delta(P|Q)$ obeys the relation:

$$\sum_{P+Q=A} (-1)^p \, \Delta(P|Q) = 0. \tag{4.8}$$

The summation in Equation 4.8 is spread over all subdivisions of set A.

The proof follows. Among the points that belong to set A there is always a point x_{s_0} that fulfills the conditions

$$x_{s_0} \gtrsim x_r \qquad (1 \leq r \leq k), \tag{4.9}$$

where the relation $x \gtrsim y$ means that either point x is later than y or the interval between points x and y is spacelike. To every set P that does not contain the point x_{s_0}, there exists a corresponding set $P' = P + (x_{s_0})$. For such sets, P and P', the equality

$$\Delta(P|Q) = \Delta(P'|Q') \tag{4.10}$$

is fulfilled. Since these quantities enter in the summation in Equation 4.8 with opposite signs, Equation 4.8 follows from Equation 4.10. To prove Equation 4.10, it is necessary to calculate the sign factors ζ_{rs_0} (the other sign factors are the same). By means of Equations 4.6 and 4.7, it is easy to see that

for $\quad \Delta(P|Q) \qquad \zeta_{rs_0} = \begin{cases} -x_{rs_0}^0 & (x_r \in P) \\ -1 & (x_r \in Q) \end{cases}$

and for $\quad \Delta(P'|Q') \qquad \zeta_{rs} = \begin{cases} 1 & (x_r \in P) \\ -x_{s_0 r}^0 & (x_r \in Q) \end{cases}$.

If $x_{s_0} > x_r$, according to the condition in Equation 4.9, $x_{s_0 r}^0 = -x_{r s_0}^0 > 0$, and the sign factors coincide for both cases. If $x_{s_0} \sim x_r$, the function G is analytic at the point $x_{s_0 r}^2$, and the sign of the imaginary part is unimportant. Equation 4.10 is thus proved.

Now consider that point x and index n do not belong to set A. Define the retarded function $W^R(x\,|\,A)$ by the expression:

$$W^R(x\,|\,A) = \sum_{P+Q=A} (-1)^p\, \Delta(P\,|\,(x) + Q). \tag{4.11}$$

This function has the following property: if the condition

$$x_r \gtrsim x \tag{4.12}$$

is fulfilled for at least one point $x_r \in A$,

$$W^R(x\,|\,A) = 0. \tag{4.13}$$

The proof is based on the remark that under the condition in Equation 4.12 there exists a point $x_{s_0} \in A$ for which

$$x_{s_0} \gtrsim x;\; x_{s_0} \gtrsim x_r \quad (1 \le r \le k). \tag{4.14}$$

By use of 4.14, the condition in Equation 4.13 can be proved by the same considerations that have been used for the proof of Equation 4.8.

Unitarity and Causality of the S-Matrix

In terms of functional Equation 2.13, the condition of unitarity is

$$S[\lambda, \lambda] = 1,$$

or, using Equation 2.16,

$$\Delta[\Lambda, \Lambda] = 0. \tag{5.1}$$

The coefficient functions of the expansion of the left-hand side of Equation 5.1 have the form given by Equation 4.8. As has been proved in previous sections, these coefficient functions are equal to zero, and the condition in Equation 5.1 is therefore fulfilled.

Bogolyubov's causality condition[5] in terms of the functional in Equation 2.13 can be expressed in the form:

for $x \lesssim y$
$$\frac{\delta}{\delta\lambda(y)}\left\{\frac{\delta S[\lambda, \mu]}{\delta\mu(x)}\right\}_{\mu=\lambda} = 0. \tag{5.2}$$

According to Equation 2.9,

$$\delta/\delta\lambda(x) = \sum_{n=1}^{\infty} \Lambda_{n-1}(x)\delta/\delta\Lambda_n(x),$$

and therefore it is sufficient to prove that for $x \lesssim y$, the relation

$$\frac{\delta}{\delta\Lambda_m(y)}\left\{\frac{\delta S[\lambda, \mu]}{\delta M_n(x)}\right\}_{\mu=\lambda} = 0 \tag{5.3}$$

is fulfilled for every value of indexes m and n. By use of Equation 2.16 and the unitarity condition in Equation 5.1, Equation 5.3 can be transformed into the form

$$\frac{\delta}{\delta\Lambda_m(y)}\left\{\frac{\delta\Delta[\Lambda, M]}{\delta M_n(x)}\right\}_{M=\Lambda} = 0 \qquad \text{for } x \lesssim y. \tag{5.4}$$

The coefficient functions of the expansion of the functional $\{\delta\Delta[\Lambda, M]/\delta M_n(x)\}_{M=\Lambda}$ are the functions $W^R(x|A)$ defined by Equation 4.11. Therefore, the causality condition in Equation 5.4 is equivalent to Equation 4.13, which had been proved in a previous section.

Concluding Remarks

To prove the properties of unitarity and causality, only the representation in Equation 4.6 for amplitudes and analytic properties of functions G has been used. The amplitudes of different types were considered to be independent. Any relations between different amplitudes have not been used in the proof of unitarity and causality of the S-matrix.

The subsequent development of this scheme from my point of view should relate to restrictions on the representation in Equation 4.6 for amplitudes. These restrictions will be expressed by relations between different functions G in Equation 4.6. Such relations can arise from the condition of positive definiteness, which plays an important part in Wightman theory. Further restrictions on the choice of functions G may be obtained by considering concrete variants of Quantum Field Theory and its symmetry properties.

References

1. Gol'fand, Yu. A. 1968. Nucl. Phys. (USSR) **8:** 600.
2. Wightman, A. S. 1956. Phys. Rev. **101:** 860.
3. Streater, F. R. & A. S. Wightman. 1964. PCT, Spin and Statistics and All That. New York and Amsterdam.
4. Jost, R. 1965. The General Theory of Quantized Fields. American Mathematical Society. Providence, R.I.
5. Bogolyubov, N. N. & D. V. Shirkov. Introduction to the Theory of Quantized Fields.

FORMAL METHODS FOR CLASSIFICATION OF COMPLEX SITUATIONS

Alexander Ya. Lerner

Weizmann Institute of Science
*Rehovot, Israel**

Formulation of the Task

By classification of complex situations, we shall understand the procedure of discriminating to what category a given situation belongs. By formal methods, we understand such methods, which can be realized by computers. If the symptoms used for classification correspond to those on which the classification of situation is based, the actual classification simply consists of determining the truth (or falsity) of a statement that can be represented as a logical (Boolean) function, and the task of discrimination becomes trivial. Much more complicated are cases in which classificatory symptoms cannot be used (if, for instance, they are not yet established, or if they have not yet made themselves known), and such a classification must be made on the basis of indirect (i.e., nonclassificatory) symptoms. Tasks of this kind appear rather frequently in engineering, medicine, economics, sociology, and in many other areas of contemporary life.

Until recently, such tasks were either solved intuitively, on the basis of the experience of experts, or remained unsolved. But even when the opinion of experts made it possible to assign a given situation to a given category, the reliability of the decision still left much to be desired. The number of wrong classifications used to be (one may say, still is) too great, with serious consequences: people died, engineering systems failed, and resources or efforts were wasted.

In recent years, however, outlines of formalized classification procedures have begun to appear that promise to bring a more effective solution to the problem of classification of complex situations than was possible in the past. In many cases, such formalized procedures enable us to considerably enhance the reliability of classification and also to classify situations to which intuitive classification was not applicable. The frequency of such tasks in engineering is rather high. They occur in the classification of phenomena associated with an engineering project, that is, performance pattern, proper or defective operation, quality of design, and so on. The tasks of engineering diagnostics include, for instance, checking the performance of an engine by its sound, checking a computer by testing, and establishing the presence of mineral ores on the basis of a set of geophysical and geologic data. The tasks of medical diagnostics (which is also a classification problem) consist of determining the state of an animal or human organism on the basis of accumulated anamnestic and clinical data. The important point is to establish the presence of a disease, type of disease, its stage, pathologic changes in relation to the normal situation, pregnancy, sex of the fetus, and so on. The need for the classification of situations appears in many other fields as well. Although the tasks of classification of complex situations on the basis of the complete set of symptoms are difficult,

* Temporary address: USSR, B-333, Dm. Uliahova 4, Korp. 2, Dept. 322.

they are extremely important, and I believe that the use of formal methods for these purposes might be highly beneficial and effective. By formal classification methods, I mean procedures that are basically suited for computer processing.

The task of formalization therefore consists of elaborating such formal procedures that are accessible not only to man but also to the computer and that make it possible, on the basis of raw data relating to all existing symptoms, to make decisions (with a degree of certainty) in regard to the predetermined category to which a given situation belongs.

Difficulties Encountered in the Solution of Classification Tasks

The complexity of situation classification depends mainly on the number of symptoms involved in the choice of the classes, on the nature of interdependence between all symptoms involved, and on whether the classification is performed by man or machine. Thus, for instance, the separation of numbers into odd and even ones can be performed with equal ease by man and machine, since the classification is based on a single, easily perceived symptom; the separation of men from women according to external appearance is comparatively easy for man, based as it is on a few informal symptoms, but it presents considerable difficulties to the machine. Man, on the other hand, finds it rather difficult to establish the presence of mineral ores on the basis of geophysical and geologic data, whereas the computer readily solves the task. If we leave aside trivial tasks of classification in which only a small number of symptoms is involved in determining whether a situation belongs to a given category and concentrate on complicated tasks, which are, naturally, of major interest, we shall soon be able to cite a long series of causes that render the intuitive solution of classification tasks extremely intricate. One of the principal difficulties is that extensive data are needed for separation. Let us evaluate the number of data used in classification by the length of a questionnaire in which every question can be answered only by "yes" or "no." Experiments have shown that in cases that are of any practical importance, such questionnaires contain hundreds of questions. But not all of them can be regarded as equally important. The actual role of each question in the solution of the task at hand will depend on its importance and code number. Since we usually have no presumptive knowledge of the relation between the symptoms involved and the category to which the situation belongs, or of the importance of each symptom, one will readily appreciate the complicated nature of tasks of this kind. The classification procedure apparently involves singling out vital symptoms and, on the basis of personal experience as a kind of yardstick, gauging these symptoms consciously or unconsciously, combining the findings, and making appropriate decisions. It is hardly feasible that a man would be capable of performing these operations sufficiently accurately if large numbers of symptoms were involved. Hence, it is no wonder that in cases that go beyond the trivial, classification is art rather than science.

Formalization of Classification Tasks

If formal procedures are to be used in the solution of classification tasks, the initial step should consist of formulating the task in the appropriate mathematical language, followed by formal elaborating procedures that would enable us, relying on available data, to assign the situation under analysis to one or another category. For this purpose, we shall make use of a space of symptoms in which it becomes possible to make a geometric interpretation of situations from the point of view of situation categories. The space of symptoms will be such a space, in which the value of one of the indexes used for classification is plotted on each axis. If the set of symptoms is such that each symptom can have only two values, namely, its presence or absence, every coordinate of the entire set of symptoms can have only one of two values: 1 (symptom present) or 0 (symptom absent). Such an arrangement of starting data is in most cases practically feasible. All possible situations can be shown as angular points of an n-dimensional cube.

Direct Nomination

The trivial solution of the classification problem would consist of recording all possible situations, while noting the category to which each belongs. However, bearing in mind that in cases of any degree of complexity the number of symptoms is quite large (as great as dozens and hundreds), it must be recognized that a procedure of this kind, although possible in principle, is practically unfeasible, since the number n of probable combinations of n symptoms (N), and consequently the number of angular points of the n-dimensional cube, would be extremely high: $N = 2^n$. If $n = 100$, for example, which is a rather common number in practical cases, the number of situations N becomes so high that even the fastest modern computer would take many thousands of years to perform categorization. It follows from this example that effective methods of classification cannot be based on such a principle. Fortunately, however, it has been found that the solution of a classification problem does not necessarily require the information about the categories of all possible situations. Theoretical investigations, confirmed by personal observations, have demonstrated that information about only a small number of all possible situations is adequate for classification purposes, provided that this information is sufficiently representative of the whole. The number of situations about log N, which is incomparably smaller than the total number of situations involved, suffices to solve such tasks.[1] Thus, in cases of practical importance, the number of situations required for classification is not too high, and the process can be performed by modern computers.

Logical Classification

In some cases, the category to which a particular situation belongs can be determined after calculating the value of a Boolean function of the coordinates of the point that represents the situation in the space of symptoms. This function manifests the logic of classification of the situations under consideration. A value of 1 for that function may symbolize that the situation belongs to Category A and a value of 0 that it belongs to Category B.

The partition, of course, may be prolonged, and categories A and B can be divided into subcategories. A1, A2, · · · and B1, B2, · · · and so on, as is shown by the following tree of Boolean functions:

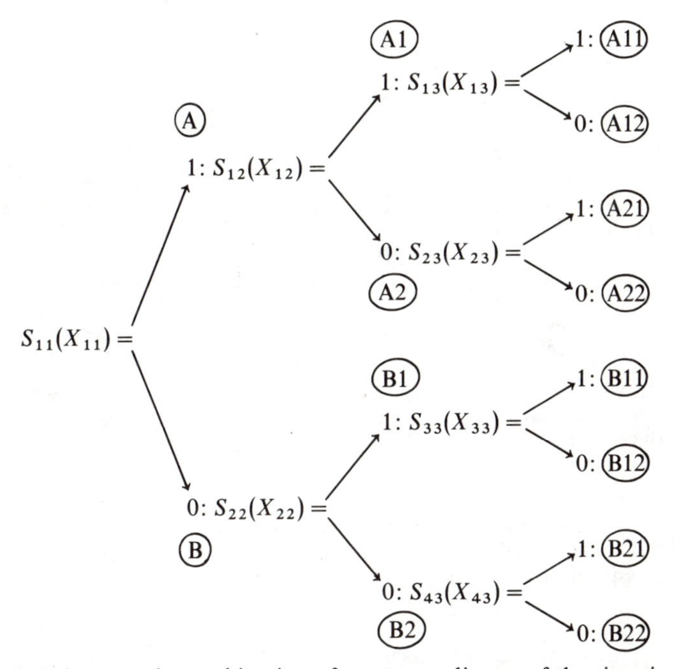

where each X is a certain combination of some coordinates of the situation.

Such a hierarchic system of classification can be understood as a process, every step of which gives more and more detailed characterization of a given situation. For instance, if such a classification system is used for medical diagnostic purposes, the first step may be to determine to what category of diseases the case belongs (somatic or mental), and the second step would be to ascertain whether the case belongs to the group of cardiac or noncardiac somatic diseases and organic or functional mental disorders, and so on.

Sometimes, function S can be established by theoretical treatment of the subject. However, in most practical cases, the theory fails to do so, and it is necessary to restore the structure of function B from empirical data. The set of empirical data can contain, for instance, information about the specific categories to which several situations belong (FIGURE 1).

It now seems as if the problem of classification is reduced to the problem of restoration of Boolean functions, a classic problem of relay circuits synthesis. However, this is not the case, because we need to restore the function by providing only some information about specific categories under which the situations fall. This problem does not have a single solution. In the case shown in FIGURE 1, the vertexes designated by a cross can be separated from those designated by a circle by different B functions, for instance:

$$S(X) = (x_1 \wedge \bar{x}_2 \wedge \bar{x}_3) \vee (x_1 \wedge x_2 \wedge x_3)$$

or

$$S(X) = x_1.$$

To solve the classification problem properly, it is necessary to choose from

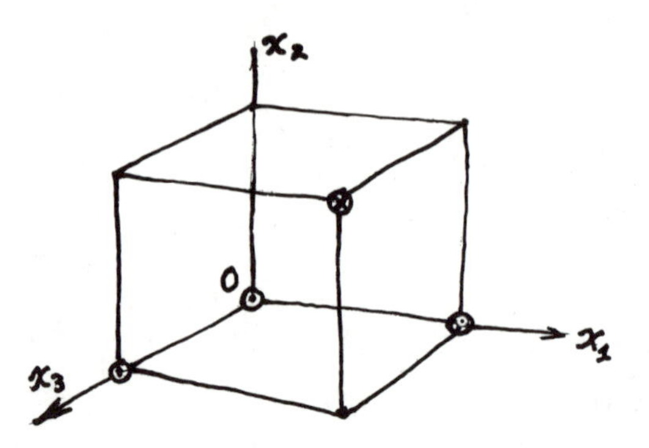

FIGURE 1. A simple example of binary state space: each variable x_i can take the value 0 or 1. Partial information is recorded concerning whether a given vertex belongs to classification ⊗ or classification ⊙.

among the many functions that satisfy empirical data the one that gives the best extrapolation, that is, the function that minimizes the number of mistakes when used for classification of new situations.

Classification by Separating the Space of Symptoms

Let us assume, therefore, that some of the angular points of the *n*-dimensional cube represent the situation of Category A and others the situation of Category B. In that case, the task of classification would consist of finding a space boundary (assuming that it exists) in the field of symptoms that separates the angular points of the *n*-dimensional cube that belong to Category A from all other angular points of that cube. When this boundary has been found, the category to which a situation belongs can be determined by simply establishing on which side of the boundary lies the point representing the situation in question. To make this formalized procedure possible, it is essential that all symptoms available should be sufficient for classification. Unless this is so, we should get angular points on our *n*-dimensional cube that belong simultaneously to two, or even more, different categories. It is obvious that, in this case, categories would get mixed up, and it would be impossible to define a method to discriminate them.

To avoid this dilemma, we should extend the space of symptoms by adding some new symptoms to the ones that have already been used, until classification becomes possible. An opposite situation is likewise feasible, namely, that some symptoms used for the classification are redundant, since the classification problem lends itself readily to solution in the absence of these symptoms. Here, the boundary between categories can be established, but the classification procedure is rendered more complicated by redundant symptoms that necessitate superfluous changes in the space of symptoms.

Computer Training for Situation-Classification Procedures

Formalized procedures are now used in computer training for the classification of situations into different categories.

The process of training consists of the trainer (teacher) implanting the knowledge or the patterns of solution tasks. These two methods differ substantially from each other. The first method consists of providing the trainee with the algorithm for the solution of the task.

The second method comprises training by means of examples. Thus, both people and machines are trained to perform arithmetical or logical operations by explaining the algorithm for the execution of the program, by means of which this algorithm is realized. However, procedures for the solution of tasks such as, for instance, the reading (identification) of letters or numerals in various types of print or written by hand are not transferred by the teacher to the learner by explaining the structure of these symbols or the mechanism of recognition (with which, in addition, the teacher himself is not familiar) but, rather, by teaching through examples. In this process, it is important that the learner, after he has been trained to recognize a certain number of objects, should be capable of identifying correctly new objects with which he did not come into contact in the process of training.

Training through examples is of extreme importance for the existential activity of many kinds of living organisms. When animals teach their young to find food and avoid danger, they do not transfer to them the algorithms for the solution of these tasks but merely make use of training procedures based on examples. A large proportion of habits and examples enabling human beings to resolve various tasks of importance for their existence are also acquired through observation and by analogy, without resorting to any algorithms.

Until recently, machines, even those as advanced as digital computers, obtained their solutions according to programs containing a completely clear algorithm for every task. It can be considered that man, designing programs for digital computers, teaches the computer how to solve tasks of a particular type, using in the process the first method of training. However, this method proved useless in tasks, since there were solutions for which no algorithms were known, although we could resolve them by intuition. For instance, man can be taught easily to distinguish between a cat and a dog, to recognize his acquaintances, or to catch a ball in mid-air, but he does not know how to design a program for a computer that would perform all these actions.

The desire to extend computer capabilities has forced scientists and engineers to attempt to simulate in computers the ability to learn solutions of a variety of tasks by means of examples. The first successful attempt to teach computers through examples was the design of machines capable of identifying visual patterns, such as geometric figures, letters, numerals, and other symbols.

The task of recognition of patterns is as follows. We have a cluster D, containing a large number of various objects

$$D = d_1, d_2, \cdots d_N,$$

and the objects belong to a relatively limited number of known categories D_1, D_2, ..., D_n. We shall suppose that the machine resolves the recognition task if it always (or at least with a certain degree of reliability) classifies each object d into a particular category-pattern D_i.

If, for instance, cluster D represents all possible graphic symbols for numerals, the recognition task consists of classifying each graphic symbol into one of the 10 categories: 0, 1, 2, ..., 9, each of which represents the pattern for the corresponding numeral. Cluster D can contain various geometric figures, among which the patterns should be distinguished as belonging to the triangle category, square category, circle category, and so on.

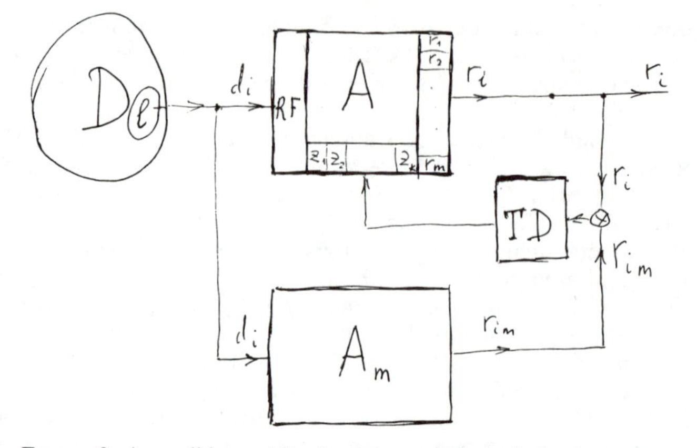

FIGURE 2. A possible machine for the recognition of visual patterns.

The machine (FIGURE 2,A) expected to recognize visual patterns should obviously contain an input device for receiving information about the object to be recognized. This input device is called the receptor field (RF). It can be represented, for instance, by a punch card or by a mosaic of photoelements onto which the pattern to be recognized is projected. If the receptor field consists of the number of cells p, each of which is capable of being in one of the two states (activated or unactivated), the number of all possible configurations in the input device would be $N = 2^p$. Even in cases where p is relatively small, the number of all possible configurations N is of the order of billions. Therefore, it is practically impossible to store data on every configuration in the computer memory.

The output device of the recognition machine must have m outputs. Depending on which of the outputs has been activated, it is possible to establish to which category the object in question has been assigned by the machine.

The feasibility of training the machine necessitates an adequate number of possible internal states $z = (z_1, z_2, \ldots, z_k)$, including states where the machine classifies the objects in the manner required.

The task of training such a machine is as follows: A relatively limited number $l \ll N$ of examples of classifying objects is introduced into machine A, thus putting it into one of the states z, in which it would perform the required classification of objects included in the training sequence l.

It has been found that, in the case of establishing the state in which the machine correctly classifies an adequately large number l of objects selected from among cluster D, the machine could (with a certain degree of reliability) classify fairly satisfactory all objects of the given cluster as well. Consequently, the process of training machine A in pattern recognition mainly involves the following:

With its receptor field activated, machine A (FIGURE 2) obtains l objects that have been, for training purposes, selected at random from cluster D. In the process, the model machine, A_m (whose part may be played by man), shows machine A to which category should be assigned each of l objects of the training sequence: d_{j1}, $d_{j2}, \ldots, d_{jl}$.

The training device (TD) compares, in the course of the training process, reactions r_m of the model machine A_m with the reactions r of machine A under training. Thus, the z states of machine A are modified to enable its reactions to correspond as often as possible with the reactions r_m. Theoretically, a training sequence l_t has been established capable of guaranteeing the necessary reliability of classification.[2] However, there is usually no need for the machine to obtain all the l_t situations; instead, the training sequence is limited to a considerably smaller number, l_p. In the process, to establish the degree of training acquired by the machine, it is given, after the completion of the training stage, an equally arbitrarily chosen control series of inconsistencies. When the machine does not commit more errors than is acceptable, training is considered completed. If the opposite is true, the training of machine A is repeated until the necessary reliability of pattern recognition has been attained, or until it has been established that the machine cannot be trained to identify objects on the basis of particular characteristics.

Training efficiency depends to a great extent on three factors: 1) the way in which objects are represented in the receptor field of the machine, that is, which characteristics of real objects (and in which manner) are coded for introduction into machine A; 2) the variety of possible transformations of the input configuration into the output reaction of the machine, that is, the number of its different states; and 3) the algorithm for operation of the training device.

Representation must be selected in such a manner that the entry unit to the classification device contains sufficient information for classification. The variety of transformations (the number K of states Z) must be large enough to make it possible to train the machine to resolve a sufficiently wide range of tasks, although not so wide that it results in the unnecessary extension of the training sequence for the establishment of the requisite transformation. The algorithm for operation of the training unit must ensure the optimal relation between recognition reliability and the length of the inconsistency series.

How About a Theory?

In the past few years, formal methods of classification have been widely used in different fields of science, medicine, and technology. These methods are mostly based on teaching computers to recognize patterns. They were successfully applied to the solution of many important problems, such as medical diagnostics, epidemiology, phenomena research, interpretation of geophysical data, weather forecasting, and character recognition.

Although the ability of machines to recognize patterns has not yet been developed satisfactorily, many unexpected, not quite understandable, and even sensational applications have materialized. It seems, however, that these achievements are due to the heuristics of the algorithm used rather than to the theory.

The classification rules used in the learning of pattern recognition, even with sparse empirical data, are sometimes found to be much stronger than could be expected on the basis of the estimates that are valid for the problem in question (in the sense of their extrapolation power). Possibly, the programmer intuitively chooses for the classification rules an approximation of the simplest class of functions. And if we presume that our world is simple, the simple classification rules prove to be so strong because they are in accordance with nature.

Thus, to organize in a computer a successful learning process for working out a powerful classification rule (as well as in many other problems of computer learning), we do need to inculcate in the program the aspiration for not choosing any but the simplest approximation of the true rule.† Let us note that finding a simple

† A discussion of the meaning of the term "simplest approximation of the true rule" is found in Reference 3.

rule that would satisfy the empirical facts is much more difficult than to find any rule (we are short of simple rules). In our human perception, simplicity is intimately related to beauty. It is may be for this reason that properties like "simplicity," "unexpectedness," or "beauty" are used by scientists to select a law representing different aspects of the same function.

Einstein's statement that "God is ingenious, but not evil" remains the profoundest of all. God is ingenious because it is hard to understand the complexity of the world, and it is difficult to cognize the world. He is not evil because the world can be comprehended with the tools available to us, although it is not so difficult to imagine a world in the cognizance of which we would be eternally short of experience!

This would probably constitute the subject of the science of learning.

References

1. Vapnik, V. & A. Lerner. 1963. Autom. Telemech. (USSR) **24** (6).
2. Vapnik, V., A. Lerner & A. Chervonenkis. 1965. Eng. Cybern. (USSR) **1**.
3. Lerner, A. 1972. Frontiers of Pattern Recognition. Academic Press. New York.

LIGHT FOCUSING BY GRAVITATIONAL FIELDS

Vladimir Dashevsky

Moscow, U.S.S.R.

Light focusing by gravitational fields, or the gravitational lens effect (GLE), has been under investigation in relation to General Relativity and theoretical astrophysics for over half a century. During the 1960s, the study of geometric optics in curved space-time has received a strong new impetus from the works of Ehlers, Sachs, Penrose, and others.[1] These recent developments have resulted in formulation of the optical scalars theory and of the focusing theorem. This theorem, which is a differential equation governing the change of the cross-sectional area of an arbitrary beam in an arbitrary gravitational field, is of singular importance.

In regard to the astrophysical side of the question, one requires more information to interpret the observational data than the mere possession of the equation itself. One needs its solutions, obtained for some realistic situations and expressed in terms of observable quantities.

The only analytic solution of the GLE obtained so far has been the approximate solution, which corresponds to what may be called a strong lens effect in a weak gravitational field. In fact, to my knowledge, all the analytic considerations of the GLE are based on the admission of two simplifying assumptions: that the gravitational field is assumed to be weak everywhere along the ray so that the linearized Einsteinian theory can be used, and that only paraxial bundles of rays are considered, or, in other words, the light source is supposed to lie close to the straight line between the observer and deflector. Kantowski[2] studied light focusing in the Swiss-cheese cosmological models without these restrictions, but the differential equation was integrated numerically.

In this paper, the exact solution of the geometric-optic problem of light focusing is obtained by use of a spherically symmetric gravitational field in a vacuum.

In the next section, light rays in a Schwarzschild-like field are described by means of the parameter θ, which is the angle between the ray and the radial direction.

In the last section, the cross-sectional area of a bundle of rays is expressed in terms of the variation, $\delta\theta$, of that parameter across the beam. For $\delta\theta$, an auxiliary differential equation is constructed and solved, thus yielding the cross-sectional area as a function of a point along the ray. Some properties of this solution are discussed.

LIGHT RAYS IN THE SCHWARZSCHILD-LIKE METRIC

In this section, we shall consider null geodesics in a spherically symmetric gravitational field in a vacuum. In Schwarzschild-like coordinates, the line element is

$$ds^2 = (1 - x) \cdot dt^2 - (1 - x)^{-1} \cdot dr^2 - r^2(d\vartheta^2 + \sin^2 \vartheta \cdot d\varphi^2), \qquad (1)$$

where $x = r_g/r$, $r_g = 2GM/c^2$ is the gravitational radius of the central body with mass M, and the speed of light $c = 1$.

We shall choose the polar axis directed from the center toward the observer (see FIG. 1). Then, any ray reaching the observer lies in a φ = const. plane. To avoid the

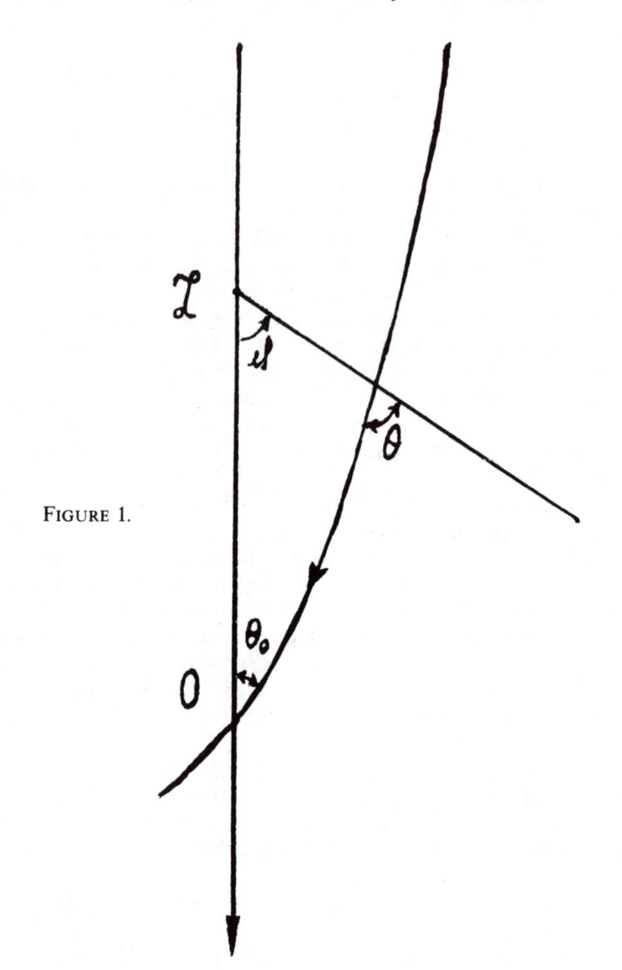

FIGURE 1.

discontinuity of φ as it crosses the polar axis, we shall use slightly modified (spatial) spherical coordinates; the only modification is that here the "latitude" is $\vartheta \in [0, 2\pi)$, and the "longitude" $\varphi \in [0, \pi)$, while they are usually

$$\vartheta_{\text{usual}} \in [0, \pi], \qquad \varphi_{\text{usual}} \in [0, 2\pi). \tag{2}$$

A null geodesic is described by the set of equations,[1]

$$dt/d\lambda = (1 - x)^{-1} \tag{3a}$$

$$(dr/d\lambda)^2 = 1 - (\rho^2/r^2)(1 - x) \tag{3b}$$

$$d\vartheta/d\lambda = -\rho/r^2, \tag{3c}$$

$$\varphi = \text{const.}, \tag{3d}$$

where ρ is the impact parameter of a ray, λ is the affine parameter along the ray, normalized so that the light frequency is unit on the infinity. (Note that our λ differs from that of Reference 1 by the factor ρ, which is insignificant for a single geodesic, but our normalization is more suitable for a consideration of a congruence of geodesics.) To remove arbitrariness of the sign in Equation 3a, we have chosen it to be negative, which means that the ray is propagating in the direction of decreasing ϑ.

Now, we introduce the parameter θ, which turns out to be useful in investigations of light rays. We define θ by the relations

$$(\rho/r)(1-x)^{1/2} = \sin\theta, \tag{4a}$$

$$\mathrm{d}r/\mathrm{d}\lambda = \cos\theta. \tag{4b}$$

In terms of θ, ray Equations 3b and 3c acquire the form

$$\mathrm{d}r/r = -[(1-x)/(1-3x/2)]\cdot\cot\theta\cdot\mathrm{d}\theta \tag{5a}$$

$$\mathrm{d}\vartheta = [(1-x)^{1/2}/(1-3x/2)]\cdot\mathrm{d}\theta. \tag{5b}$$

The parameter θ has a simple meaning. Let us transform at some point of the ray to the local Lorentz observer, instantaneously at rest relative to the Schwarzschild-like reference system; for this observer, the two-dimensional spatial line element in a $\varphi = \text{const.}$ plane is

$$\widehat{\mathrm{d}l}^2 = \widehat{\mathrm{d}r}^2 + r^2\,\mathrm{d}\vartheta^2, \tag{6}$$

where

$$\widehat{\mathrm{d}r} = (1-x)^{-1/2}\cdot\mathrm{d}r.$$

Combining Equations 5 and 6, one obtains:

$$\widehat{\mathrm{d}r}/(r\,\mathrm{d}\vartheta) = -\cot\theta. \tag{7}$$

Thus, θ is just the angle between the ray and the radial (outward) direction, as measured by the local Lorentz, instantaneously immovable observer.

From Equations 5, it is easy to derive all the well-known features of the Schwarzschild-like photon orbits.[1,3] The only circular orbit has the radius $r_c = (3/2)r_g$, which corresponds to the critical value of the impact parameter:

$$\rho_c = (3\cdot\sqrt{3}/2)\cdot r_g. \tag{8}$$

The photons with $\rho < \rho_c$ are gravitationally captured. Along the ray with $\rho < \rho_c$, the angle θ decreases from π at $r = \infty$ to $\theta_{\min} = \pi - \sin^{-1}(\rho/\rho_c)$ at $r = r_c$. After crossing the critical circle $r = r_c$, θ increases again and approaches $\theta = \pi$ as $r \to r_g$. At $r < r_g$, the angle becomes imaginary, in accord with general properties of the Schwarzschild-like coordinate system.

We are interested here in photons that reach a distant observer ($r_o > r_c$), so we shall consider only rays with $\rho > \rho_c$. For any "normal" source of a gravitational field, this inequality is trivial because the body's opaque radius is large in comparison with its r_g. In this case, θ decreases monotonously from π at $r = \infty$ to $\pi/2$ at $r = r_{\min}$; moreover, $\theta \to 0$ as $r \to \infty$. Thus, in this case, the value of θ unambiguously determines a point on the ray.

The dependence $x(\theta)$ is determined by Equation 4a, from which one has

$$x(1-x)^{1/2}/\sin\theta = x_0(1-X_0)^{1/2}/\sin\theta_o \tag{9}$$

(all the values with the subscript "o" refer to the observer), and

$$dx/d\theta = [x(1 - x)/(1 - 3x/2)] \cdot \cot\theta. \tag{10}$$

It may be shown that parameters λ and θ are related by

$$d(\rho \cot\theta)/d\lambda = (1 - 3x/2)/[(1 - x)^{3/2}]. \tag{11}$$

In conclusion, I should point out that parameter θ is not only useful in the rigorous theory but also in the linear approximation. If a gravitational field is weak everywhere (that is, $x \ll 1$), it follows from Equation 11 that the "Euclidean form" distance, $\rho \cdot (\cot\theta_o - \cot\theta)$, differs from the affine parameter only in the second order over x.

Integration of Equation 5b gives in the linear approximation:

$$\vartheta \approx (\theta - \theta_o) + (r_g/\rho)(\cos\theta_o - \cos\theta). \tag{12}$$

In particular, when the parameter angle changes from $\theta = \pi$ to $\theta_o = 0$, the polar angle ϑ is changed by the amount $\pi + 2(r_g/\rho)$, which is the classical Einsteinian result.

Focusing of a Bundle of Rays

Let us consider an infinitesimally narrow bundle of rays, each ray passing through the observer located at point r_o; $\vartheta_o = 0$. A ray in the bundle is specified by the azimuthal angle φ of its plane and by angle θ_o between the ray and the polar axis at the observer.

The cross-sectional area of the beam can be written in the form

$$A = r^2(\sin\vartheta/\cos\theta) \cdot \delta\vartheta \cdot \delta\varphi, \tag{13}$$

where δ is the variation operator across the beam. (This expression defines A as a projection of the area vector $\vec{\delta l}_\vartheta \times \vec{\delta l}_\varphi$ on the ray direction; thus, its absolute value is the usual area, and its sign coincides with that of $\sin\vartheta$.)

Our aim is to obtain this area as a function of a point along the ray: $A(\theta)$. (We shall not discuss the question of astigmatic change of the cross-sectional shape. With $\delta\vartheta(\theta)$ known, this question is answered immediately; it will be discussed elsewhere.)

The dependences $r(\theta)$ and $\vartheta(\theta)$ are known from integrated Equations 5a and 5b; $\delta\varphi = \text{const.}$ along the ray. The problem, therefore, is to determine the function $\delta\vartheta(\theta)$.

The condition that $\vec{\delta l}_\vartheta$ is normal to the ray direction,

$$(1 - x)^{-1}\, dr \cdot \delta r + r^2\, d\vartheta \cdot \delta\vartheta = 0, \tag{14}$$

together with Equations 5a and 5b, yields

$$\delta r/r = (1 - x)^{1/2} \tan\theta \cdot \delta\vartheta. \tag{15}$$

On the other hand, a variation of Equation 4a gives

$$\delta r/r = [1 - (x/2)/(1 - x)]^{-1} \cdot \{\delta\rho/\rho - \cot\theta \cdot \delta\theta\}. \tag{16}$$

Moreover, since the point of observation is common to all the rays, we have the additional condition, $\delta r_o = 0$. Hence,

$$\delta\rho/\rho = \cot\theta_o \cdot \delta\theta_o. \tag{17}$$

Combining Equations 15–17, we have, finally,

$$\delta\vartheta = [(1 - x)^{1/2}/(1 - 3x/2)] \cdot \cot\theta \cdot [\cot\theta_o - u(\theta) \cdot \cot\theta] \cdot \delta\theta_o, \tag{18}$$

where $u(\theta) = \delta\theta/\delta\theta_o$.

To determine $u(\theta)$, let us take the variation of $d\vartheta/d\theta$,

$$\delta(d\vartheta/d\theta) = d(\delta\vartheta)/d\theta - (d\vartheta/d\theta) \cdot [d(\delta\theta)/d\theta]. \tag{19}$$

(Here we have used the commutativity of the d- and δ-operators when applied to the scalar fields θ and ϑ.) Inserting Equations 5b and 18 into Equation 19, we arrive, after some transformations, at the differential equation for $u(\theta)$:

$$du/d\theta - u[2 - F(\theta)] \cdot \cot\theta = -[1 - F(\theta)]\cot\theta_o, \tag{20}$$

where

$$F(\theta) = x(1 - 3x/4/[(1 - (3x/2)^2]. \tag{21}$$

Equation 20 is a linear first-order equation; its solution is

$$u = (\sin\theta/\sin\theta_o)\cos(\theta - \theta_o) + \sin^2\theta \cdot \psi(\theta), \tag{22}$$

where

$$\psi = [\varphi(\theta) - 1] + \varphi(\theta) \cdot I(\theta)\cot\theta_o, \tag{23}$$

$$\varphi = [(1 - x_o)^{1/2}/(1 - x)^{1/2}] \cdot [(1 - 3x/2)/(1 - 3x_o/2)], \tag{24}$$

and

$$1 = \int_{\theta_o}^{\theta} [F(\theta)/\varphi(\theta)]\, d\theta. \tag{25}$$

Equation 13, with Equation 18 substituted for $\delta\vartheta$ and Equation 22 for u, now represents the required dependence $A(\theta)$.

It is convenient to express the result in terms of the "apparent-size distance," which can be defined as

$$R = \{|A|/\Omega_o\}^{1/2}, \tag{26}$$

where Ω_o is the solid angle of the cone of rays reaching the observer. Unlike area A, which is the same for all the Lorentz observers, angle Ω_o depends on the observer's movement. To avoid the appearance of the factor $(1 + z)$, where z is the red shift, we shall use the Lorentz observer moving so that $z = 0$. All the quantities related to this system will be supplied with a superscript "(o)".

To evaluate the angle $\Omega_o^{(o)}$, we have to take into account that all the distances to the observer are the same when this distance is infinitesimal. In particular, the apparent-size distance coincides in the first order with the affine parameter. To the first order, we have, at $\vartheta \to 0$,

$$\lambda = -[(1 - x_o)/(1 - 3x_o/2)](r_o/\sin\theta_o)\, d\theta,$$

$$A = [\lambda^2/(1 - x_o)\sin\theta_o \cdot \delta\theta_o \cdot \delta\varphi;$$

together with Equation 26, this gives:

$$\Omega_o^{(o)} = [\sin\theta_o/(1 - x_o)] \cdot \delta\theta_o \cdot \delta\varphi. \tag{27}$$

Substitution of Equations 27 and 13, 18, and 22 into Equation 26 leads to the final result in the form:

$$R^{(o)} = [r_o \sin(\theta - \theta_o)/\sin \theta] \cdot [(1 - x)^{3/2}/(1 - 3x/2)] \cdot [|\sin \vartheta|/\sin(\theta - \theta_o)] \cdot [1 - C(\theta)]^{1/2}, \quad \mathbf{(28)}$$

where

$$C(\theta) = [\sin \theta_o \cos \theta/\sin(\theta - \theta_o)] \cdot \psi(\theta).$$

Equation 28 is the exact solution of the spherical gravitational lens problem. Its detailed analysis, as well as astrophysical implications, will be discussed elsewhere. I shall now make some brief remarks.

When the gravitational field is weak everywhere ($x \ll 1$) along the ray, and when the ray runs close to the lens–observer axis (that is, $\pi - \theta + \theta_o \ll 1$), Equation 28 gives, in the lowest (second) order of magnitude, the well-known result[4]:

$$[R^{(o)}(\rho, \theta_o, \theta)/R_{\mathrm{Eucl}}(\rho, \theta_o, \theta)]^2 = 1 - [2r_g r_o r_s/(\rho^2 R)]^2, \quad \mathbf{(29)}$$

where r_s is the radial distance from lens to source, and R is the source–observer distance; regardless of how it is defined (because we need it only in zero order), $R \approx r_o + r_s$.

According to Equation 29, the gravitational lens effect depends on the impact parameter very sharply—it decreases inversely to ρ^4. This is the reason why the lens effect of only the nearest deflector is practically always taken into account.

However, Equation 29 is only applicable within the confines of the two restrictions described above. If any of these conditions is violated, the rigorous consideration, based on Equation 28, is demanded.

If the deflector is a very compact object, the expansion into the series does not work for the rays with $\rho \approx r_g$, and it is necessary to evaluate the two elliptical integrals entering the exact solution in $\vartheta(\theta)$ and $I(\theta)$.

On the other hand, if the ray does not run close to the lens–observer axis, the effect of an individual lens is very small; nevertheless, this case must be thoroughly investigated for the sake of the problem of cosmological observations. During the past 15 years, an extensive literature has appeared on observations in a locally nonhomogeneous universe.[2,4–9] All these authors used either an approximate or a numerical consideration or both. The solution presented in this paper now enables us to investigate the observations, for example, in the "Swiss-cheese" cosmological model, explicitly.

References

1. Misner, C., K. S. Thorne & J. A. Wheeler. 1973. Gravitation.
2. Kantowski, R. 1969. Astrophys. J. **155:** 89.
3. Novikov, I. D. & Ya. B. Zel'dovich. 1965. Relativistic Astrophysics.
4. Refsdal, S. 1970. Astrophys. J. **159:** 357.
5. Zel'dovich, Ya. B. 1964. Sov. Astron.-AJ **9:** 671.
6. Dashevskii, V. M. & Ya. B. Zel'dovich. 1965. Sov. Astron.-AJ. **8:** 854.
7. Dashevskii, V. M. & V. I. Slysh. 1966. Sov. Astron.-AJ **9:** 671.
8. Bertotti, B. 1966. Proc. R. Soc. London Ser. A **269:** 185.
9. Gunn, J. E. 1967. Astrophys. J. **147:** 61.

ON THE POSSIBILITIES OF SOLVING COMBINATORIAL PROBLEMS BY ANALYTIC METHODS

P. L. Buzytsky and Gregory A. Freiman

Moscow, U.S.S.R.

Let us consider the equation

$$a_1 x_1 + \cdots + a_n x_n = b, \tag{1}$$

where a_j, b are positive integers, and $a_j \leq b$, $x_j = 0, 1; j = 1, \ldots, n$.

Assume that $a_1 \leq \cdots \leq a_n$.

In general, to determine whether there is a solution to Equation 1 seems to require the complete enumeration, or consideration, of 2^n variants. If b is relatively small, the number of variants to be enumerated is substantially lower. For instance, dynamic programming is known to have $0(nb)$ as an efficiency estimation.

Let us consider the case when b does not increase too fast in comparison with n; namely, we assume that $n^2 < b < n^\nu$; ν being a positive integer greater than 2.

For any value of b, Equation 1 may or may not have a solution, depending on the values of a_j. It is not difficult to construct appropriate examples. However, the further considerations are based on the fact that, "generally speaking," Equation 1 must have a solution when b is small. Actually, 2^n values of the linear form in the left-hand side of Equation 1 are placed in the interval $[0, \sum_{j=1}^n a_j]$. Since $\sum_{j=1}^n a_j < nb$, each integer from this interval, including b (except for integers too close to the ends of the interval), are among the values of the left-hand side linear form approximately $2^n/nb$ times, of course, only when the values of this form are uniformly distributed.

Such a uniform distribution is realized in the case of, in a sense, regular behavior of the numbers a_j.

Therefore, it is interesting to find the most broad cases of such a regular behavior of a_j, which would imply the solvability of Equation 1. The considerations above let us hope that weak sufficient conditions for solvability of Equation 1 could be deduced, whereas the sets of a_j that do not satisfy these conditions would be "strongly" degenerated and could be treated by some special technique of analyzing Equation 1. Altogether, these conditions could lead to a completion of the solvability problem of Equation 1 with relatively small b, the efficiency being better than for the known methods.

This paper is only the first step of applying the ideas above to combinatorial problems. Further generalizations are quite possible for an inequality and systems of both equations and inequalities as well as for optimization combinatorial problems.

The theorem established by Freiman (this monograph) implies the following result.

Let the real number σ be determined by the equation

$$\sum_{j=1}^{n} a_j/[\exp(\sigma a_j) + 1] = b. \tag{2}$$

Further, let

$$D = \sum_{j=1}^{n} [\exp(\sigma a_j) a_j^2]/[\exp(\sigma a_j) + 1]^2, \tag{3}$$

$$\rho_3 = \sum_{j=1}^{n} a_j^3/[\exp(\sigma a_j) + 1], \tag{4}$$

$$t_1 = \frac{M}{\sqrt{D}}, \; M = \min\left(h^{-1/4}, \frac{D^{1/4}}{\rho_3^{1/6}}\right), \; h = \left(\max_j \frac{a_j^2}{\exp(\sigma a_j) + 1}\right) \Big/ D. \tag{5}$$

There then exist absolute constants n_0, c_0, and h_0, such that if the conditions

$$n > n_0, \tag{6}$$

$$h < h_0, \tag{7}$$

$$\rho_3/D^{3/2} < c_0, \tag{8}$$

$$\sum_{j=1}^{n} \|\alpha a_j\|^2/[\exp(\sigma a_j) + 1] > \tfrac{1}{4} \ln(nb), \qquad \alpha \in [t_1, \tfrac{1}{2}] \tag{9}$$

(where $\|\alpha a_j\|$ is the distance from a_j to the nearest integer) hold, Equation 1 has a solution.

Notice that to treat a specific problem, it is necessary to know the values of the constants. This question requires some special considerations.

Conditions 7–9 are not obvious and have to be properly investigated.

First of all, a class of Equations 1 satisfying these conditions should be studied. We have to show that this class is rather wide, which may be done by reformulating Conditions 7–9 in more natural terms.

Secondly, the computations involved in checking up these conditions should be considered.

This paper deals with these two problems.

Let us show that Conditions 7 and 8, "generally speaking," hold. Namely, we shall describe a class of Equations 1 that satisfy Conditions 7 and 8, and the share of the equations that is out of the class tends to zero when n approaches infinity.

Further, we shall denote by $C_1, C_2, \ldots$ sufficiently large positive constants and by $c_1, c_2, \ldots$ sufficiently small positive constants.

Let C be an arbitrarily large positive constant, $k_0 = k_0(n) = [\ln(n^{1/2} \ln^2 n)/(\ln C)]$, N_k be the number of a_j such that $a_j \in m_k$, where $m_k = [b/C^{k+1}, b/C^k]$, $k = 0, 1, \ldots, k_0 - 1$; $N = \min N_k$.

Denote by $\mathscr{A}$ a class of Equations 1 for which

$$N > \ln^8 n. \tag{10}$$

Then:

Theorem 1. Equations of Class $\mathscr{A}$ satisfy Conditions 7 and 8.

Proof. Consider first the case when σ is small; namely, let

$$\sigma < (n^{1/2} \ln^2 n)/(Cb). \tag{11}$$

If $\sigma > 1/b$, by virtue of Equation 10,

$$h = \frac{\max_j \{a_j^2/[\exp(\sigma a_j) + 1]\}}{\sum_{j=1}^{n} \exp(\sigma a_j) a_j^2/[\exp(\sigma a_j) + 1]^2} \le 2 \frac{\max_j \{\sigma^2 a_j^2/[\exp(\sigma a_j) + 1]\}}{\sum_{j=1}^{n} \{\sigma^2 a_j^2/[\exp(\sigma a_j) + 1]\}}$$

$$\le 4 \Big/ \left\{ \sum_{1/(C\sigma) \le a_j \le C/\sigma} a_j^2/[\exp(\sigma a_j) + 1] \right\} \le 4/Nc_1 \le 4/(c_1 \ln^8 n), \tag{12}$$

since

$$\max_{x\geq 0} x^2/[\exp(x)+1] < 2, \qquad \min_{1/C\leq x\leq C} x^2/[\exp(x)+1] > c_1.$$

Further,

$$\begin{aligned}\rho_3/D^{3/2} &\leq \left\{\sum_{a_j<((3v+1)/\sigma)\ln n} a_j^3/[\exp(\sigma a_j)+1]\right. \\ &\quad + \sum_{a_j\geq[(3v+1)/\sigma]\ln n} a_j^3/[\exp(\sigma a_j)+1)]/D^{3/2} \\ &\leq 2^{3/2}\left\{\sum_{a_j<((3v+1)/\sigma)\ln n} \sigma^3 a_j^3/[\exp(\sigma a_j)+1]\right\}\Bigg/\left\{\sum_{j=1}^{n}\sigma^2 a_j^2/[\exp(\sigma a_j)+1]\right\}^{3/2} \\ &\quad + (nb^3/n^{3v+1})/D^{3/2} \leq [2^{3/2}(3v+1)\ln n]\Bigg/\left\{\sum_{j=1}^{n}\sigma^2 a_j^2/[\exp(\sigma a_j)+1]\right\}^{1/2} \\ &\quad + 1/D^{3/2} < [2^{3/2}(3v+1)\ln n]\Bigg/\left\{\sum_{1/C\sigma\leq a_j\leq C/\sigma}\sigma^2 a_j^2/[\exp(\sigma a_j)+1]\right\}^{1/2} \\ &\quad + 1/D^{3/2}[2^{3/2}(3+1)\ln n]/\sqrt{c_1 N} + 1/D^{3/2} \leq C_1/\ln^3 n.\end{aligned} \tag{13}$$

If $\sigma < 1/b$,

$$h \leq [2(b^2/2)]\Bigg/\left[\sum_{b/2\leq a_j\leq b} a_j^2/(e+1)\right] \leq 4(e+1)/N \leq 4(e+1)/\ln^8 n, \tag{14}$$

and

$$\begin{aligned}\rho_3/D^{3/2} &\leq 2^{3/2}\frac{\sum_{j=1}^{n} a_j^3/[\exp(\sigma a_j)+1]}{\{\sum_{j=1}^{n} a_j^2/[\exp(\sigma a_j)+1]\}^{3/2}} \leq \frac{2^{3/2}b}{\{\sum_{j=1}^{n} a_j^2/[\exp(\sigma a_j)+1]\}^{3/2}} \\ &\leq \frac{2^{3/2}(e+1)^{1/2}b}{(\sum_{a_j\geq b/2} a_j^2)^{1/2}} \leq 2^{3/2}(e+1)^{1/2}/N^{1/2} \leq 2^{3/2}(e+1)^{1/2}/\ln^4 n.\end{aligned} \tag{15}$$

So Case 11 is fully considered, and we can assume that

$$\sigma \geq n^{1/2}\ln^2 n/Cb. \tag{16}$$

Let j_0 be the minimal number for which $a_{j_0}\sigma \geq 1$. First of all, consider the case of

$$a_1 + \cdots + a_{j_0} \geq b/2. \tag{17}$$

The following estimations take place (see Equations 12 and 13):

$$h \leq 4\Bigg/\left(\sum_{a_j\leq 1/\sigma}\sigma^2 a_j^2/[\exp(\sigma a_j)+1]\right) \leq 4(e+1)\Bigg/\left(\sigma^2\sum_{j=1}^{j_0-1} a_j^2\right), \tag{18}$$

and

$$\begin{aligned}\rho_3/D^{3/2} &\leq \frac{2^{3/2}(3v+1)\ln n}{\{\sum_{j=1}^{n}\sigma^2 a_j^2/[\exp(\sigma a_j)+1]\}^{1/2}} + 1/D^{3/2} \\ &\leq \frac{2^{3/2}(3v+1)\ln n}{\{\sum_{j=1}^{j_0-1}\sigma^2 a_j^2/[\exp(\sigma a_j)+1]\}^{1/2}} + 1/D^{3/2} \leq \frac{2^{3/2}(3v+1)(e+1)^{1/2}\ln n}{\sigma(\sum_{j=1}^{j_0-1} a_j^2)^{1/2}}.\end{aligned} \tag{19}$$

It is obvious that to verify Estimations 7 and 8, we must show by virtue of Equations 18 and 19 that $G = \sigma(\sum_{j=1}^{j_0-1} a_j^2)^{1/2}$ is sufficiently large.

Let us prove that the inequality

$$G \geq \ln^4 n \tag{20}$$

is valid if Condition 16 holds.

Consider the intervals $\Delta_i = [\frac{1}{2}^i, \frac{1}{2}^{i-1}]$, where $i = 1, \ldots, i_0$; $i_0 = [\log_2(n^{1/2}/\ln^4 n)]$.

Let p_i be the number of a_j from Δ_i. We shall show that at least for one $i = i_1$ the inequality

$$p_{i_1} \geq 2^{2i_1} \ln^8 n$$

is valid, which implies

$$G = \sigma\left(\sum_{j=1}^{j_0-1} a_j^2\right)^{1/2} \geq \sigma\left(\sum_{a_j \in \Delta_i} a_j^2\right)^{1/2} \geq \sigma(2^{2i_1} \ln^8 n/2^{2i_1}\sigma^2)^{1/2} = \ln^4 n,$$

or Equation 20.

Actually, in the opposite case,

$$p_i < 2^{2i} \ln^8 n, \qquad 1 \leq i \leq i_0. \tag{21}$$

By virtue of Equations 17 and 21, we have

$$\begin{aligned} b/2 &\leq p_1/(2\sigma) + p_2/(2^2\sigma) + \cdots + p_{i_0}/(2^{i_0}\sigma) \\ &\leq (\ln^8 n/\sigma) \sum_{i=0}^{i_0} 2^i < (\ln^8 n/\sigma) 2^{i+1} \leq (2 \ln^4 n \cdot n^{1/2})/\sigma. \end{aligned}$$

Hence, $\sigma < 4n^{1/2} \ln^4 n/b$, which contradicts Equation 16.

Only one case is left to be considered, when

$$a_1 + \cdots + a_{j_0-1} \leq b/2, \tag{22}$$

and Condition 16 holds.

Then,

$$\begin{aligned} b = \sum_{j=1}^{j_0-1} a_j/[\exp(\sigma a_j) + 1] + \sum_{j=j_0}^{n} a_j/[\exp(\sigma a_j) + 1] \\ \frac{1}{2} \sum_{j=1}^{j_0-1} a_j + \frac{\sum_{j=j_0}^{n} a_j}{[\exp(\sigma a_j) + 1]}. \end{aligned}$$

Hence, by virtue of Equation 22,

$$\sum_{j=j_0}^{n} a_j/[\exp(\sigma a_j) + 1] \geq 3b/4. \tag{23}$$

Estimating h and $\rho_3/D^{3/2}$, as in Equations 12 and 13, we have

$$\begin{aligned} h &\leq \frac{4}{\sum_{j=1}^{n} \sigma^2 a_j^2/[\exp(\sigma a_j) + 1]} \leq \frac{4}{\sum_{j=j_0}^{n} a_j/[\exp(\sigma a_j) + 1]} \\ &\leq 16/3b \leq 16C/(3n^{1/2} \ln^2 n), \end{aligned} \tag{24}$$

and

$$\rho_3/D^{3/2} \leq \frac{2^{3/2}(3v+1)\ln n}{\{\sum_{j=1}^{n} \sigma^2 a_j^2/[\exp(\sigma a_j)+1]\}^{1/2}} + \frac{1}{D^{3/2}}$$
$$\leq \frac{2^{3/2}(3v+1)\ln n}{\{\sum_{j=j_0}^{n} \sigma a_j/[\exp(\sigma a_j)+1]\}^{1/2}} + \frac{1}{D^{3/2}}$$
$$\leq \frac{2^{5/2}(3v+1)\ln n}{3^{1/2}(\sigma b)^{1/2}} + \frac{1}{D^{3/2}} < \frac{C_2}{n^{1/4}}. \qquad \textbf{(25)}$$

By Equations 12–15, 18–20, 24, and 25, the theorem is true.

Theorem 2. Let $N(\bar{\mathscr{A}})$ be the number of Equations 1 that do not belong to $\mathscr{A}$, and $N(\mathscr{U})$ be the number of all Equations 1 of n variables with right-hand side b. Then,

$$N(\bar{\mathscr{A}})/N(\mathscr{U}) < \exp(-C_3 n^{1/2}/\ln^2 n)$$

for sufficiently large values of n.

Proof. The number of all Equations 1 is $\binom{n+b-1}{n}$. Let an interval of length l contain no more than q numbers a_j. The number of Equations 1 satisfying this condition is no greater than

$$\binom{1+q-1}{q}\binom{b-1+n-q-1}{n-q}.$$

By definition of the class, the number $N(\bar{\mathscr{A}})$ satisfies the inequality

$$N(\bar{\mathscr{A}}) \leq \sum_{k=0}^{k_0} \sum_{q=0}^{N} \binom{l_k+q-1}{q}\binom{b-l_k+n-q-1}{n-q},$$

where l_k is the number of integers from m_k.

It is easy to show that the expression in the sum takes its greatest value when q is maximal and l_k is minimal.

So,

$$N(\bar{\mathscr{A}}) \leq (k_0+1)(N+1)\binom{l_{k_0}+N-1}{N}\binom{b-l_{k_0}+n-N-1}{n-N}$$
$$< \binom{l_{k_0}+N}{N}\binom{b-l_{k_0}+n}{n} < (2l_{k_0})^N \binom{b-l_{k_0}+n}{n}.$$

Therefore,

$$r = \frac{N(\bar{\mathscr{A}})}{N(\mathscr{U})} < b^N \frac{\binom{b-l_{k_0}+n}{n}}{\binom{b+n-1}{n}} = b^N \frac{(b-l_{k_0}+n)\cdots(b-l_{k_0}+1)}{(b+n-1)\cdots b}$$
$$< b^N \left(\frac{b-l_{k_0}+n}{b+n-1}\right)^n,$$

(26)

which implies

$$\ln(r) < N\ln(b) + n\ln\left(1 - \frac{l_{ko} - 1}{b + n - 1}\right) < N\ln b - n\frac{l_{ko} - 1}{b + n - 1}$$

$$< -(nl_{ko})/2b < -(C_3 n^{1/2})/\ln^2 n;$$

that is, $r < \exp[(-C_3 n^{1/2})/\ln^2 n]$.

Theorem 2 is proved.

Theorem 3. Equation 1 from $\mathscr{A}$ has a solution if

$$\sum_{a_j \leq a_s/2 \ln n} \|\alpha a_j\|^2 > \ln(nb), \qquad \alpha \in [t_2, \tfrac{1}{2}],$$

where $t_2 = n^{1/2} \ln n/C_4 b$; s is determined by the inequalities

$$\sum_{j=1}^{s} a_j < b/2, \qquad \sum_{j=1}^{s+1} a_j \geq b/2. \tag{27}$$

Proof. First of all, let us prove that Equation 9 holds for all $\alpha \in [t_1, t_2]$ in Class $\mathscr{A}$.

Consider the case when $t_1 < \alpha < \sigma/(2C_5 \ln n)$. Then, Equation 9 may be written as

$$\sum_{j=1}^{n} \|\alpha a_j\|^2/[\exp(\sigma a_j) + 1] = \sum_{a_j < (C_5 \ln n)/\sigma} \|\alpha a_j\|^2/[\exp(\sigma a_j) + 1]$$

$$+ \sum_{a_j \geq (C_5 \ln n)/\sigma} \|\alpha a_j\|^2/[\exp(\sigma a_j) + 1]$$

$$> \alpha^2 \sum_{a_j < (C_5 \ln n)/\sigma} a_j^2/[\exp(\sigma a_j) + 1]$$

$$> C_6 \alpha^2 D > \tfrac{1}{4}\ln(nb).$$

Therefore, Equation 9 holds true if

$$C_6 t_1^2 D > \tfrac{1}{4}\ln(nb). \tag{28}$$

However, Equation 28 is valid, since for $\mathscr{A}$ we have

$$h < 1/(C_7 \ln^2 n), \qquad \rho_3/D^{3/2} < 1/(C_7 \ln^3 n)$$

(see the proof of Theorem 1).

Consider now the case when $\sigma/(2C_5 \ln n) \leq \alpha < t_2$.

If $\sigma > (n^{1/2} \ln^2 n)/(2C^2 b)$,

$$\alpha \geq \sigma/(2C_5 \ln n) > t_2 \qquad \text{(when } C_4 > C_5 C^2\text{)}.$$

If $\sigma \leq (n^{1/2} \ln^2 n)/(2C^2 b)$ and if $\alpha < \sigma$,

$$\sum_{j=1}^{n} \|\alpha a_j\|^2/[\exp(\sigma a_j) + 1] > C_8 \sum_{a_j \leq 1/(2\sigma)} \|\alpha a_j\|^2 = C_8 \alpha^2 \sum_{a_j \leq 1/(2\sigma)} a_j^2$$

$$> C_8 \frac{\sigma^2 N}{4C_5^2 \ln^2 n C_9 4\sigma^2} > C_{10} \ln^6 n > \frac{1}{4}\ln(nb),$$

and if $\alpha > \sigma$

$$\sum_{j=1}^{n} \|\alpha a_j\|^2/[\exp(\sigma a_j) + 1] > C_{11} \sum_{a_j \leq 1/(2\alpha)} \|\alpha a_j\|^2 > C_{11} \sum_{a_j \leq 1/(2\alpha)} \alpha^2 a_j^2$$

$$> C_{12} N > C_{12} \ln^8 n > \tfrac{1}{4}\ln(nb).$$

Thus, Equation 9 holds for all $\alpha \in [t_1, t_2]$.
Further,

$$\sum_{j=1}^{n} \|\alpha a_j\|^2/[\exp(\sigma a_j) + 1] > \sum_{a_j \le 1/\sigma} \|\alpha a_j\|^2/[\exp(\sigma a_j) + 1]$$
$$\ge [1/(e+1)] \sum_{a_j \le 1/\sigma} \|\alpha a_j\|^2.$$

Therefore, Equation 9 is implied by the inequality

$$\sum_{a_j \le 1/\sigma} \|\alpha a_j\|^2 > \ln(nb), \qquad \alpha \in [t_2, \tfrac{1}{2}]. \tag{29}$$

For σ, we have the following estimation:

$$\sigma < (2 \ln n)/a_s,$$

where a_s is taken from Equation 27.
Actually, in the opposite case

$$b/2 < \sum_{j=s+1}^{n} a_j/[\exp(2a_j \ln n/a_s) + 1] < nb/n^2 = b/n.$$

Thus, Equation 29 can be replaced by

$$\sum_{a_j < a_s/(2 \ln n)} \|\alpha a_j\|^2 > \ln(nb), \qquad \alpha \in [t_2, \tfrac{1}{2}]. \tag{30}$$

Theorem 3 is proved.

Condition 30 looks relatively clear. At the same time, it is rather general. Actually, let us show that the share of Equations 1 satisfying Condition 30 tends to unity when n approaches infinity.

Theorem 4. Let N^* be the number of Equations 1 satisfying Condition 30. Then,

$$N^*/N(\mathscr{U}) \ge 1 - \exp(-c_2 n^{1/2}/\ln n),$$

where $N(\mathscr{U})$ is the number of all Equations 1.

Proof. Let $a_k = m$ for a given k. The number of Equations 1 under this condition equals

$$\binom{m+k-1}{k}\binom{b-m+n-k}{n-k}.$$

When m increases, this value increases if $m < kb/n$ and decreases if $m > kb/n$.

Let

$$m_0 = [(1-\theta) \cdot bk/n], \qquad m_1 = [(1+\theta) \cdot bk/n].$$

Then, the ratio p of the number of Equations 1 that satisfy the condition

$$|bk/n - a_k| > \theta bk/n$$

(for a given θ, $0 < \theta < 1$) to $N(\mathscr{U})$ is no greater than

$$p \le b[p_1(m_0) + p_1(m_1)],$$

where

$$p_1(m) = \frac{\binom{m+k-1}{k}\binom{b-m+n-k}{n-k}}{\binom{b+n-1}{n}}.$$

Estimating $p_1(m)$ from above, we have

$$p_1(m) < \frac{(m+k)^k n^k}{k!\,(b+n-k)^k}\left[1 - \frac{m}{(b+n-k)}\right]^{n-k}$$

$$< \left[\frac{e(m+k)n}{k(b+n-k)}\right]^k \cdot \exp\left[-(1-\varepsilon)\frac{m(n-k)}{b+n-k}\right]. \tag{31}$$

If

$$|bk/n - m| > \theta bk/n, \qquad k = o(n), \qquad k = o(m), \tag{32}$$

then

$$p_1(m) < \left\{(1+\varepsilon)\frac{emn}{kb}\exp\left[-(1-2\varepsilon)\frac{mn}{kb}\right]\right\}^k$$

$$= \left[(1+\varepsilon)\exp\left(2\varepsilon\frac{mn}{kb}\right)\right]^k \left[\frac{mn}{kb}\Big/\exp\left(\frac{mn}{kb}-1\right)\right]^k. \tag{33}$$

Since the function $x/\exp(x-1)$ attains its maximal value, unity, when $x = 1$, by virtue of Equation 33, we have from Equation 32

$$p_1(m) < \exp(-c_3 k). \tag{34}$$

So, Equations 31 and 34 imply

$$p < \exp(-c_4 k). \tag{35}$$

By virtue of Equation 27,

$$(s+1)a_{s+1} \geq b/2.$$

Hence,

$$a_{s+1} \geq b/2(s+1).$$

Consider the case when

$$s \leq \bar{s} = [\tfrac{1}{4}n^{1/2}]. \tag{36}$$

Then,

$$a_{\bar{s}+1} \geq a_{s+1} \geq b/2(s+1) \geq b/2(\bar{s}+1),$$

and so

$$a_{\bar{s}+1} \geq b/n^{1/2}. \tag{37}$$

By virtue of Equations 35–37, the share of such cases is no greater than $\exp(-c_4 n^{1/2})$.

Therefore, we can assume that

$$s > n^{1/2}/4.$$

Moreover, we assume that

$$a_s > b/5n^{1/2}$$

since the share of the cases with $a_s \leq b/5n^{1/2}$ is no greater than $\exp(-c_4 n^{1/2})$, by virtue of Equation 35.

Replace Equation 30 by a more restrictive condition

$$\sum_{a_j \le b/(5n^{1/2} \ln n)} \|\alpha a_j\|^2 > \ln(nb).$$

Due to the considerations above, we may assume that the sum is taken only with respect to a_j with $j < n^{1/2}/4$.

If $\|\alpha a\| > \frac{1}{4}$, that is, $\frac{1}{4} < \{\alpha a\} < \frac{3}{4}$, the quantity of the numbers a from the interval $[1, l](l = b/[5n^{1/2} \ln n)]$ equals $\frac{1}{2}l + o(l)$, which is easy to show taking into account the fact that $\alpha \in [n^{1/2} \ln^2 n/b, \frac{1}{2}]$.

Let j_0 be maximal among j such that $a_j \le b/(5n^{1/2} \ln n)$. Then, we can assume that

$$j_0 > n^{1/2}/(6 \ln n) \tag{38}$$

since the share of cases with $j_0 \le n^{1/2}/(6 \ln n)$ is no greater than $\exp(-c_4 n^{1/2}/\ln n)$ by virtue of Equation 35.

If there are $Q > 20 \ln(nb)$ numbers a_j for which $\frac{1}{4} < \{\alpha a_j\} < \frac{3}{4}$, Equation 30 holds true.

Thus, we may assume that the number Q of such a_j satisfies the inequality

$$Q \le 20 \ln(nb).$$

Then, the number of sets $a_1, \ldots, a_{j_0}$ corresponding to Equation 32 is no greater than

$$\sum_{Q=0}^{20 \ln(nb)} \binom{\frac{1}{2} + j_0 - Q}{j_0 - Q}\binom{\frac{1}{2}}{Q} < C_{13} \ln(n) \cdot \exp(\max Q) \frac{(\frac{1}{2}l + j_0) \cdots (\frac{1}{2}l + 1)}{j_0!}.$$

The number of all the sets is

$$\binom{\frac{1}{2}l + j_0 - 1}{j_0}.$$

Therefore, the share of Equations 1 $\bar{p}$ corresponding to Equation 32 is estimated as

$$\bar{p} < C_{13} \ln(n) \cdot \exp(C_{13} \ln^2(n) \frac{(\frac{1}{2}l + j_0) \cdots (\frac{1}{2}l + 1)}{(\frac{1}{2}l + j_0 - 1) \cdots \frac{1}{2}l} < \left(\frac{1}{2} - \varepsilon\right)^{j_0}$$
$$< \exp(-c_2 n^{1/2}/\ln n).$$

Theorem 4 is proved.

We have described the class $\mathscr{A}$ of Equations 1 where Conditions 7 and 8 are valid and Condition 9 is replaced by Condition 30, which is true almost always, as is shown in Theorem 4.

Let us consider the question of efficiency of checking Condition 30. We shall show that if the inequality

$$\sum_{a_j \le T} \|\alpha_0 a_j\|^2 > 2 \ln(nb) \tag{39}$$

is true for a fixed T and a certain α_0, then

$$\sum_{a_j \le T} \|\alpha a_j\|^2 > \ln(nb) \tag{40}$$

is true for all α satisfying the condition

$$|\alpha - \alpha_0| < \ln(nb)/TN_T, \tag{41}$$

where N_T is the cardinality of the set $\{j \mid a_j \leq T\}$.

Actually, designating by $S(\alpha) = \sum_{a_j \leq T} \|\alpha a_j\|^2$, we have

$$|S(\alpha) - S(\alpha_0)| \leq \sum_{a_j \leq T} |\|\alpha a_j\|^2 - \|\alpha_0 a_j\|^2| \leq \sum_{a_j \leq T} |\|\alpha a_j\| - \|\alpha_0 a_j\||$$
$$\leq \sum_{a_j \leq T} |\alpha - \alpha_0| a_j \leq TN_T |\alpha - \alpha_0| \leq \ln(nb).$$

Hence, and from Equation 39, we have Equation 40:

$$S(\alpha) \geq S(\alpha_0) - \ln(nb) > \ln(nb).$$

So the efficiency to check Condition 30 is $O[a_s n/(2 \ln n)]$.

Notice that beyond the considered class $\mathscr{A}$ there are Equations 1 for which Conditions 7–9 also could be true. Therefore, it would make sense to consider the question of directly testing these conditions.

The check of Conditions 7 and 8 is not difficult, requiring no more than $O(n)$ operations.

To check Condition 9 in its general form, we could use the technique above (Equations 39–41).

When α is small ($\alpha < \frac{1}{2}T$), the condition

$$\sum_{j=1}^{n} \|\alpha a_j\|^2/[\exp(\sigma a_j) + 1] > \tfrac{1}{4} \ln(nb)$$

could be too restrictive. Then, we may successively enlarge T, for instance, taking $T_i = 2^i T$, $i = 1, 2, \ldots, [\log_2(bn^\beta/T)]$; $\beta > 0$. The efficiency of such a check is estimated as $O(Tn)$ under natural assumption $T > n^{\beta+\varepsilon}$, $\varepsilon > 0$.

AN ANALYTICAL METHOD OF ANALYSIS OF LINEAR BOOLEAN EQUATIONS

Gregory A. Freiman

Moscow, U.S.S.R.

PART I

Let us consider the equation

$$a_1 x_1 + \cdots + a_n x_n = b, \tag{1}$$

where a_j, b are positive integers, $x_j = 0, 1; j = 1, \ldots, n$.

Without loss of generality, assume that

$$a_j \leq b, a_1 \leq \cdots \leq a_n, \qquad b \leq \frac{1}{2} \sum_{j=1}^{n} a_j .$$

Let I_n be the number of solutions of Equation 1. Then I_n may be expressed as follows:

$$I_n = \exp(\sigma b) \prod_{j=1}^{n} [1 + \exp(-\sigma a_j)] \int_0^1 \prod_{j=1}^{n} [p_{1j} + p_{2j} \exp(2\pi i \alpha a_j)] \times \exp(-2\pi i \alpha b) \, d\alpha, \tag{2}$$

where $i = \sqrt{-1}$ and

$$p_{1j} = 1/[1 + \exp(-\sigma a_j)], \qquad p_{2j} = [\exp(-\sigma a_j)]/[1 + \exp(-\sigma a_j)]. \tag{3}$$

Equation 2 is true for any real σ.

Let σ be determined by the equation

$$\sum_{j=1}^{n} a_j/(\exp(\sigma a_j) + 1) = b. \tag{4}$$

It is easy to see that σ is determined by Equation 4 uniquely.

Partition the integral in Equation 2 into two parts, as follows:

$$\int_0^1 = \int_{-1/2}^{1/2} = \int_{-t_1}^{t_1} + \left(\int_{t_1}^{1/2} + \int_{-1/2}^{-t_1} \right) = S_1 + S_2, \tag{5}$$

where

$$t_1 = M/D^{1/2}, \qquad M = \min(1/h^{1/4}, D^{1/4}/\rho_3^{1/6}), \qquad h = (\max p_{2j} a_j^2)/D, \tag{6}$$

$$D = \sum_{j=1}^{n} p_{2j}(1 - p_{2j}) a_j^2, \tag{7}$$

$$\rho_3 = \sum_{j=1}^{n} p_{2j} a_j^3. \tag{8}$$

Let us assume that

$$\rho_3/D^{3/2} = o(1), \tag{9}$$

$$h = o(1). \tag{10}$$

Transforming S_1, we have

$$\begin{aligned} p_{1j} + p_{2j}\exp(2\pi i\alpha a_j) &= p_{1j} + p_{2j}(1 + 2\pi i\alpha a_j - 2\pi^2\alpha^2 a_j^2) + 0[(\alpha a_j)^3] \\ &= 1 + 2\pi i p_{2j}\alpha a_j - 2\pi^2\alpha^2 a_j^2 p_{2j} + 0(\alpha^3 p_{2j} a_j^3) \\ &= \{\exp[2\pi i\alpha p_{2j} a_j - 2\pi^2\alpha^2 p_{2j}(1 - p_{2j})a_j^2]\} \\ &\quad \times [1 + 0(\alpha^3 p_{2j} a_j^3)]. \end{aligned} \tag{11}$$

Let us now transform the integrand in Equation 2 by use of Equation 4 and taking into account that Equations 6 and 9 imply $\alpha^3\rho_3 = o(1)$.

$$\prod_{j=1}^{n} [p_{1j} + p_{2j}\exp(2\pi i\alpha a_j)]\exp(-2\pi i\alpha b) = [\exp(-2\pi^2\alpha^2 D)][1 + 0(\alpha^3\rho_3)]. \tag{12}$$

Transform S_1 by use of Equation 12:

$$\begin{aligned} S_1 &= \int_{-t_1}^{t_1} \exp(-2\pi^2\alpha^2 D)[1 + 0(\alpha^3\rho_3)]\,d\alpha \\ &= \int_{-\infty}^{\infty} \exp(-2\pi^2\alpha^2 D)\,d\alpha + 0\left[\int_{t_1}^{\infty} \exp(-2\pi^2\alpha^2 D)\,d\alpha + I t_1^3 \rho_3\right], \end{aligned}$$

where, as is known, $I = \int_{-\infty}^{\infty} \exp(-2\pi^2\alpha^2 D)\,d\alpha = 1/\sqrt{2\pi D}$.
Thus,

$$S_1 = 1/\sqrt{2\pi D} + 0[(1/\sqrt{D})\exp(-\pi^2 M^2) + (1/\sqrt{D})t_1^3\rho_3].$$

By virtue of Equation 6, we have

$$S_1 = (1/\sqrt{2\pi D})\{1 + 0[\exp(-\pi^2 M^2) + (\rho_3/D^{3/2})^{1/2}]\}. \tag{13}$$

Estimate now the integrand in S_2. We have

$$\begin{aligned} A = |1 + p_{2j}[\exp(2\pi i\alpha a_j) - 1]| &= [1 - 4p_{2j}(1 - p_{2j})\sin^2\pi\alpha a_j]^{1/2} \\ &< 1 - 2p_{2j}(1 - p_{2j})\sin^2\pi\alpha a_j < \exp[-2p_{2j}(1 - p_{2j})\sin^2\pi\alpha a_j]. \end{aligned} \tag{14}$$

By use of the inequality

$$\sin\theta \geq \theta(2/\pi), \qquad 0 \leq \theta \leq \pi/2,$$

we have

$$|\sin\pi\alpha a_j| \geq 2\|\alpha a_j\|,$$

where $\|x\|$ is the distance from x to the nearest integer.
Hence, and by virtue of Equation 14, the following inequality is true:

$$A < \exp[-8p_{2j}(1 - p_{2j})\|\alpha a_j\|^2]. \tag{15}$$

Assume that the following condition holds:

$$\sum_{j=1}^{n} \|\alpha a_j\|^2/\exp[(\sigma a_j) + 1] > \frac{1}{4} \ln(bn). \tag{16}$$

By virtue of Equations 3 and 15, we have

$$B = \left| \prod_{j=1}^{n} [p_{1j} + p_{2j} \exp(2\pi i \alpha a_j)] \right|$$

$$\times \exp[-8 \sum_{j=1}^{n} p_{2j}(1 - p_{2j})\|\alpha a_j\|^2]$$

$$= \exp\{-8 \sum_{j=1}^{n} [\exp(\sigma a_j)\|\alpha a_j\|^2]\}/[\exp(\sigma a_j) + 1]^2.$$

Since $b \geq 0$ implies $[\exp(\sigma a_j)]/[\exp(\sigma a_j) + 1] \geq \frac{1}{2}$, by virtue of Equation 16

$$B < \exp(-4 \sum_{j=1}^{n} \|\alpha a_j\|^2)/[\exp(\sigma a_j) + 1] < \frac{1}{nb}. \tag{17}$$

By use of Equations 2, 5, 13, and 17, we have the following asymptotic formula:

$$I_n = \exp(\sigma b) \prod_{j=1}^{n} [1 + \exp(-\sigma a_j)](1/\sqrt{2\pi D})[1 + 0(\rho_3/D^{3/2})^{1/2}]$$

$$+ 0[\exp(-\pi^2 M^2) + 0(1/n)]. \tag{18}$$

Part II*

Consider the following system of equations:

$$a_1 x_1 + \cdots + a_n x_n = a,$$

$$c_1 x_1 + \cdots + c_n x_n = c, \tag{19}$$

where all the quotients are positive integers, and the variables take the values 0, 1.

Berstein[2] describes a connection between the solvability of Equations 19 and the well-known knapsack problem and presents an analytic approach to the solution of Equation 19.

The number of solutions of Equation 19 is given by the formula

$$I_n = \exp(\sigma a + \rho c) \prod_{j=1}^{n} [1 + \exp(-\sigma a_j - \rho c_j)] \int_0^1 \int_0^1 \prod_{j=1}^{n} (p_{1j} + p_{2j})$$

$$\times \exp[2\pi i(\alpha a_j + \beta c_j)] \exp[-2\pi i(\alpha a + \beta c)] \, d\alpha \, d\beta, \tag{20}$$

* The results in this part were obtained in cooperation with A. A. Berstein.

where σ and ρ are determined by the system

$$\sum_{j=1}^{n} a_j/[\exp(\sigma a_j + \rho c_j) + 1] = a, \tag{21}$$

$$\sum_{j=1}^{n} c_j/[\exp(\sigma a_j + \rho c_j) + 1] = c. \tag{22}$$

Such σ and ρ are shown to exist under some reasonable conditions on a and c.[2] Let

$$p_{1j} = 1/[1 + \exp(-\sigma a_j - \rho c_j)], \tag{23}$$

$$p_{2j} = [\exp(-\sigma a_j - \rho c_j)]/[1 + \exp(-\sigma a_j - \rho c_j)]. \tag{24}$$

Partition the integral in Equation 20 into two parts

$$S = \int_0^1 \int_0^1 = \int_{-1/2}^{1/2} \int_{-1/2}^{1/2} = \int_{-t_1}^{t_1} \int_{-t_2}^{t_2} + \iint = S_1 + S_2, \tag{25}$$

where $S_2 = S \backslash S_1$.

Also, we introduce the following notations:

$$t_1 = M_1/D_1^{1/2} \tag{26}$$

$$t_2 = M_2/D_2^{1/2} \tag{27}$$

$$M_1 = \min(h_1^{-1/4}, D_1^{1/4}/\rho_{13}^{1/6}) \tag{28}$$

$$M_2 = \min(h_2^{-1/4}, D_2^{1/4}/\rho_{23}^{1/6}) \tag{29}$$

$$D_1 = \sum_{j=1}^{n} p_{2j}(1 - p_{2j})a_j^2 \tag{30}$$

$$D_2 = \sum_{j=1}^{n} p_{2j}(1 - p_{2j})c_j^2 \tag{31}$$

$$D_{12} = \sum_{j=1}^{n} p_{2j}(1 - p_{2j})a_j c_j \tag{32}$$

$$\rho_{13} = \sum_{j=1}^{n} a_j^3/[\exp(\sigma a_j + \rho c_j) + 1] \tag{33}$$

$$\rho_{23} = \sum_{j=1}^{n} c_j^3/[\exp(\sigma a_j + \rho c_j) + 1] \tag{34}$$

$$h_1 = \max_j a_j^2 p_{2j}/D_1 \tag{35}$$

$$h_2 = \max_j c_j^2 p_{2j}/D_2. \tag{36}$$

The following conditions are assumed to hold:

$$\rho_{13}/D_1^{3/2} = o(1) \tag{37}$$

$$\rho_{23}/D_2^{3/2} = o(1) \tag{38}$$

$$h_1 = o(1) \tag{39}$$

$$h_2 = o(1) \tag{40}$$

Transform S_1 as follows:

$$\begin{aligned}
&p_{1j} + p_{2j}\exp[2\pi i(\alpha a_j + \beta c_j)] \\
&\quad = p_{1j} + p_{2j}[1 + 2\pi i(\alpha a_j + \beta c_j)] - 2\pi^2(\alpha a_j + \beta c_j)^2 + 0[(\alpha a_j)^3 + (\beta c_j)^3] \\
&\quad = (1 + 2\pi i p_{2j}\alpha a_j - 2\pi^2 p_{2j}\alpha^2 a_j^2)(1 + 2\pi i p_{2j}\beta c_j - 2\pi^2 p_{2j}\beta^2 c_j^2) \\
&\qquad \times (1 + 4\pi^2 p_{2j}^2 \alpha\beta a_j c_j) \times [1 + 0(\alpha^3 p_{2j} a_j^3 + \beta^3 p_{2j} c_j^3)] \\
&\qquad \times \exp(2\pi i p_{2j}\alpha a_j + 2\pi i p_{2j}\beta c_j) \\
&\quad = \exp[-2\pi^2 p_{2j}(1 - p_{2j}) \\
&\qquad \times (a_j^2\alpha^2 + 2a_j c_j\alpha\beta + c_j^2\beta^2)][1 + 0(\alpha^3 p_{2j} a_j^3 + \beta^3 p_{2j} c_j^3)].
\end{aligned}$$

Hence, by transforming the integrand in Equation 20 and using Equations 21 and 22, we have

$$\begin{aligned}
\Pi &= \prod_{j=1}^{n} \{p_{1j} + p_{2j}\exp[2\pi i(\alpha a_j + \beta c_j)]\}\exp[-2\pi i(\alpha a + \beta c)] \\
&= \exp[-2\pi^2(D_1\alpha^2 + 2D_{12}\alpha\beta + D_2\beta^2)][1 + 0(\alpha^3\rho_{13} + \beta^3\rho_{23})].
\end{aligned}$$

Since

$$\exp[-2\pi^2(D_1\alpha^2 + 2D_{12}\alpha\beta + D_2\beta^2)]\, d\alpha\, d\beta = 1/\sqrt{2\pi\,\Delta}, \tag{41}$$

where

$$\Delta = D_1 D_2 - D_{12}^2,$$

it follows that

$$\begin{aligned}
S_1 = \int_{-t_1}^{t_1}\int_{-t_2}^{t_2} &\exp[-2\pi^2(D_1\alpha^2 + 2D_{12}\alpha\beta + D_2\beta^2)] \\
&+ 0\{[1/\sqrt{\Delta}\, t_1^3\rho_{13} + (1/\sqrt{\Delta})t_2^3\rho_{23}]\}.
\end{aligned} \tag{42}$$

So we have

$$\begin{aligned}
&\int_{t_1}^{\infty}\int_{-\infty}^{\infty} \exp[-2\pi^2(D_1\alpha^2 + 2D_{12}\alpha\beta + D_2\beta^2)]\, d\alpha\, d\beta \\
&\qquad = \int_{t_1}^{\infty} \exp(-2\pi^2 D_1\alpha^2 + 2\pi^2\alpha^2 D_{12}^2/D_2) \\
&\qquad\quad \times \int_{-\infty}^{\infty} \exp[-2\pi^2(\beta\sqrt{D_2} + \alpha D_1/\sqrt{D_2})^2]\, d\alpha\, d\beta \\
&\qquad = (1/\sqrt{D_2})\int_{t_1}^{\infty} \exp(-2\pi^2\,\Delta\alpha^2/D_2)\, d\alpha \\
&\qquad = 0[(1/\sqrt{\Delta})\exp(-\pi^2 t_1^2\,\Delta/D_2)].
\end{aligned} \tag{43}$$

The latter is true if

$$\Delta/D_2 = D_1 - D_{12}^2/D_2 = \varepsilon D_1,$$

and this is valid if

$$D_{12}^2/D_1 D_2 < 1 - \varepsilon. \tag{44}$$

Similarly,

$$\int_{-\infty}^{\infty}\int_{t_2}^{\infty} \exp[-2\pi^2(D_1\alpha^2 + 2D_{12}\alpha\beta + D_2\beta^2)]\, d\alpha\, d\beta = 0[(1/\sqrt{\Delta})\exp(-\pi^2 t_2^2 D_2)]. \tag{45}$$

By virtue of Equations 25–27, 42, 43, and 45, we have

$$\begin{aligned} S_1 &= 1/\sqrt{2\pi\,\Delta} + 0\{(1/\sqrt{\Delta})[\exp(-\pi^2\varepsilon t_1^2 D_1) + \exp(-\pi^2\varepsilon t_2^2 D_2) + t_1^3\rho_{13} + t_2^3\rho_{23}]\} \\ &= 1/\sqrt{2\pi\,\Delta}\{1 + 0[\exp(-\pi^2\varepsilon M_1^2) + \exp(-\pi^2\varepsilon M_2^2) \\ &\quad + (\rho_{13}/D_1^{3/2})^{1/2} + (\rho_{23}/D_2^{3/2})^{1/2}]\}. \end{aligned}$$

By virtue of Inequality 15, we have the following estimation:

$$\begin{aligned} |\Pi| &\leq \exp\left[-8\sum_{j=1}^{n} p_{2j}(1-p_{2j})\right]\|\alpha a_j + \beta c_j\|^2 \\ &\leq \exp\left(-4\sum_{j=1}^{n}\|\alpha a_j + \beta c_j\|^2\right)\Big/[\exp(\sigma a_j + \rho c_j) + 1]. \end{aligned}$$

Let us assume that the condition

$$\sum_{j=1}^{n}\|\alpha a_j + \beta c_j\|^2/[\exp(\sigma a_j + \rho c_j) + 1] > \frac{1}{2}\ln(\Delta n) \tag{46}$$

holds.

Then,

$$|S_2| \leq 1/(\Delta n)^{1/2}. \tag{47}$$

By virtue of Equations 20, 26, 31, and 33, we have the following asymptotic formula:

$$\begin{aligned} I_n &= \exp(\sigma a + \rho c)\prod_{j=1}^{n}[1 + \exp(-\sigma a_j - \rho c_j)](1/\sqrt{2\pi\,\Delta}) \\ &\quad \times \{1 + 0[\exp(-\pi^2\varepsilon M_1^2) + \exp(-\pi^2\varepsilon M_2^2) \\ &\quad + (\rho_{13}/D_1^{3/2})^{1/2} + (\rho_{23}/D_2^{3/2})^{1/2} + n^{-1/2}]\}. \end{aligned} \tag{48}$$

The question of determining the area of (a, c) for which Formula 48 is true requires some additional considerations.

References

1. Frieman, G. A. 1958. Waring problem with growing number of summands. *In* Proceedings of Elabuga State Pedagogical Institute. Vol. 3: 105 (in Russian).
2. Berstein, A. A. 1978. A numerical method of solving allocation problem. *In* The problems of program-goal planning and control, Moscow, CEMI (in Russian).

ON FOUNDATIONS OF CONVEX ANALYSIS

Alexander D. Ioffe

Moscow, U.S.S.R.

The novelty of this paper is mainly conceptual, although it does contain some new theorems and proofs. I intend to demonstrate that by using a new class of objects called fans (which are homogeneous subadditive convex-valued mappings), one can unite various methods and approaches used today in convex analysis and cope with certain difficulties facing such analysis.

"Standard" convex analysis[5,9,15,19] deals with (extended) real-valued functions. However, there is increasing interest in convex mappings into more than one- (usually, infinitely) dimensional ordered spaces (see References 1, 16, and 19 for the situations where such mappings appear). A theory comparable to what applies in the scalar case was developed, however, quite recently and only for mappings into conditionally complete vector lattices.[10] Moreover, it is clear that no such theory can exist in the general case unless a new approach is found.

We shall mention two problems that suggest such a conclusion:

(1) The most crucial fact of standard convex analysis is that any real-valued convex function f is the pointwise upper bound of the family of affine functions majorized by f. A natural counterpart of this fact for mappings would be that any convex mapping F into an ordered space is the upper bound of the family of affine mappings majorized by F.

Unfortunately, this is not usually the case unless the range space is a conditionally complete vector lattice. There are even examples of "very nice" convex mappings that majorize no affine mappings.[3,12]

As a result, we cannot use linear mappings to build such objects as subdifferentials of convex mappings, the most important objects in standard convex analysis (again, unless the range space is a conditionally complete vector lattice).

(2) Because of the unfortunate circumstance mentioned above, standard convex analysis does not permit one to study composite mappings $G \cdot F$, where F is a convex mapping from X into Z and $G: Z \to Y$ is a convex nondecreasing mapping, unless both Z and Y are conditionally complete vector lattices. For the latter case, a complete description of subdifferentials of and conjugates to $G \cdot F$ was given by Kutateladze,[10] as was a relatively undetailed description for the case when only Y is a conditionally complete vector lattice.

To emphasize the problem, we observe that all local operations (i.e., such that the value of the resultant mapping is defined by the values of the initial mappings at a given point) used in convex analysis are particular cases of the composition operation.

An important fact is that similar difficulties appear in nonsmooth analysis (local theory of nonsmooth mappings) when one tries to construct surrogates of derivatives from linear operators. The only way to overcome or avoid these difficulties is to stop considering linear operators as the only elementary cells of which all objects of convex analysis are composed and replace them by something more flexible.

An attempt to develop this program for nonsmooth mappings has been undertaken[8,21] with fans as the main approximation tool. This paper is, to a large extent, a translation of convex analysis into the language of fans.

[A curious coincidence is that Halkin[7] also used the term fan to describe objects he introduced for similar purposes. But Halkin's fans have little in common with mine because they are sets (even compact) of linear operators.]

It is, of course, too early to claim that fans are the only right objects for the purpose at hand. However, I believe that they are at least close to the right ones. They have many good properties (for purposes of convex analysis). First of all, the problems that have been mentioned appear to be overcome almost automatically. Secondly, all objects (e.g., linear operators, sets of linear operators, subdifferentials, and convex and sublinear mappings) can be easily represented as fans, and duality correspondences turn into fan-to-fan duality, very similar to the correspondence between a linear operator and its adjoint (this fact is thoroughly discussed in Reference 21). Thirdly, the fundamental principles of convex analysis, such as the Hahn–Banach and Mazur–Orlicz theorems as well as their proofs, have natural fan analogies and extensions. Finally, except for these principles and a few related results, proofs of most facts of convex analysis (which we call theorems rather to follow the tradition than to indicate difficulties in proving them) are almost trivial, and formulas expressed in terms of fans sometimes look even more symmetric.

In this paper, I shall first provide the main definitions, consider extension and selection theorems for fans, and prove the recent results of Kutateladze[10] for mappings into conditionally complete vector lattices to show, in particular, how the theorems established in the preceding section work. (Notice that although the proofs of most results in this section are close to those given by Kutateladze, they are rather geometric as compared with the completely algebraic nature of Kutateladze's theory.) I shall then give certain information about adjoint fans and, last, will introduce subdifferentials of convex mappings into general ordered spaces.

Preliminaries

Let X and Y be linear spaces. A set-valued mapping $\mathscr{A}$ from X into Y is called a fan if

$$\mathscr{A}(x) \text{ is a nonempty convex set for any } x \in X; \tag{1.1}$$

$$\mathscr{A}(0) = \{0\}; \tag{1.2}$$

$$\mathscr{A}(\lambda x) = \lambda \mathscr{A}(x) \quad \text{for all} \quad \lambda > 0, \qquad x \in X; \tag{1.3}$$

$$\mathscr{A}(x_1 + x_2) \subset \mathscr{A}(x_1) + \mathscr{A}(x_2) \quad \text{for all} \quad x_1, x_2 \in X. \tag{1.4}$$

The fan $\mathscr{A}$ is odd if

$$\mathscr{A}(-x) = -\mathscr{A}(x) \quad \text{for all} \quad x \in X. \tag{1.5}$$

In particular, for an odd fan, $\mathscr{A}(\lambda x) = \lambda \mathscr{A}(x)$ for all $\lambda \neq 0$, $x \in X$.

Let $K \subset Y$ be a pointed convex cone. We denote by $\geq_K$, or simply $\geq$ if no confusion arises, the partial ordering in Y defined by K; that is, $y \geq v$ if $y - v \in K$. Usually, we shall augment Y by adding two improper infinite elements $-\infty$ and ∞ such that $-\infty \leq y \leq \infty$ for any y (by definition).

The sets Y, $y + K$, $y - K$, and $(y + K) \cap (v - K)$ are called order intervals. We shall denote them by $[-\infty, \infty]$, $[y, \infty]$, $[-\infty, y]$, and $[y, v]$, respectively. The order interval $[y, v]$ is nonempty if, and only if, $y \leq v$.

A mapping F from X into $Y \cup \{\infty\}$ is called K-convex or, as is usual, simply convex, if

$$F[tx + (1 - t)u] \leq tF(x) + (1 - t)F(u)$$

for all $x, u \in X$ and $0 \le t \le 1$. If, in addition, F is homogeneous, that is, $F(\lambda x) = \lambda F(x)$ for $\lambda > 0$, F is called sublinear. The domain of F is

$$\text{dom } F = \{x \mid F(x) \neq \infty\}.$$

Clearly, the domain of a sublinear mapping is a cone. The mapping F is concave (resp. superlinear) if $-F$ is convex (sublinear). For a concave (superlinear) mapping, we say that dom $F = \text{dom}(-F)$.

Let $\mathscr{A}$ be a set-valued mapping from X into Y whose values are order intervals: $\mathscr{A}(x) = [Q(x), P(x)]$. Then $\mathscr{A}$ is a fan if and only if P is a sublinear mapping, Q is a superlinear mapping, and $P(x) \ge Q(x)$ for all x. Thus, there is a one-to-one correspondence between order-interval-valued fans and pairs (P, Q) of mappings, the first of which is sublinear, the second superlinear, and the first majorizes the second. For such fans, we write $\text{dom}^+ \mathscr{A} = \text{dom } P$, $\text{dom}^- \mathscr{A} = \text{dom } Q$.

A linear operator $A: X \to Y$ is a linear selection of the fan $\mathscr{A}$ if $Ax \in \mathscr{A}(x)$ for all x. If $\mathfrak{A}$ is a collection of linear operators from X into Y, the fan $\mathscr{A}$ is generated by $\mathfrak{A}$ if $\mathscr{A}(x) = \{y \in Y \mid y = Ax \text{ for some } A \in \mathfrak{A}\}$.

Let $C \subset X$ be a convex set. A point $x \in C$ is called a relative interior point of C if for any u belonging to the affine hull of C there is a $t > 0$ such that $x + t(u - x) \in C$. The collection of all such points is denoted ri C (this set may be empty).

A vector lattice Y is called conditionally complete if any bounded set has a supremum (the least upper bound). In other words, any collection of order intervals has a nonempty intersection, provided that any two intervals meet each other and there are two intervals in the collection, one bounded from below and one bounded from above.

It is said that two cones K_1 and K_2 are in general position if $K_1 - K_2 = K_2 - K_1$. This statement can be extended to any finite collection of cones; namely, cones $K_1, \ldots, K_n$ are in general position if for some permutation $\sigma(i)$ of indexes, $K_{\sigma(i)}$ and $\bigcap_{s=i+1}^{n} K_{\sigma(s)}$ are in general position for any $i = 1, \ldots, n - 1$.

A necessary and sufficient condition for the cones K_1 and K_2 to be in general position is that

$$K_1 \subset K_2 - K_1 \quad \text{and} \quad K_2 \subset K_1 - K_2 .$$

The most important cases that satisfy this condition are when $(\text{ri } K_1) \cap (\text{ri } K_2) \neq \varnothing$, particularly when both cones are subspaces, and when one cone (e.g., K_1) meets the algebraic interior of the other (the set of points x such that $K_2 - x$ is an absorbing set).

If $C \subset X$ is a convex set and $x \in C$, the cone of admissible directions of C at x denoted by Ad(C, x) is the conic hull of $C - x$.

Extension and Selection Theorems

In this section, Y is a conditionally complete vector lattice if nothing else is said about it.

Lemma. Let $\mathscr{A}$ be an order-interval-valued odd fan from X into Y, let $L \subset X$ be a subspace of codimension one, and let $B: L \to X$ be a linear operator such that $Bx \in \mathscr{A}(x)$ for all $x \in L$.

For a linear selection of $\mathscr{A}$ to exist coinciding with B on L, it is necessary and sufficient that for any $e \notin L$

$$\bigcap_{x \in L} [\mathscr{A}(x + e) - Bx] \neq \varnothing. \tag{1}$$

Proof. If Equation 1 holds and y belongs to the intersection, we set

$$A(x + \lambda e) = Bx + \lambda y,$$

so that

$$\mathscr{A}(x + \lambda e) - Bx = \lambda[\mathscr{A}(x/\lambda + e) - B(x/\lambda)] \ni \lambda y,$$

and hence

$$A(x + \lambda e) \in \mathscr{A}(x + \lambda e).$$

Conversely, if an extension A exists, setting $y = Ae$, we have $A(x + e) = Bx + y \in \mathscr{A}(x + e)$, so that $y \in \mathscr{A}(x + e) - Bx$.

Theorem 2.1. Let $\mathscr{A}$ be an order-interval-valued odd fan from X into Y. Assume that there are a subspace $L \subset X$ and a linear operator $B: L \to Y$ such that $L \cap \text{ri}(\text{dom}^+ \mathscr{A}) \neq \varnothing$ and $Bx \in \mathscr{A}(x)$ for all $x \in L$.

Then, there is a linear selection of $\mathscr{A}$ coinciding with B on L.

Proof. To prove the theorem, it suffices to show that whenever $e \notin L$, there is a linear mapping B' from L' into Y, with L' the linear hull of $L \cup \{e\}$, such that $B'x \in \mathscr{A}(x)$ for $x \in L'$ and $B'x = Bx$ for $x \in L$. Once we have established that this linear mapping exists, the usual transfinite induction arguments yield the desired result.

According to the lemma, B' will exist if Equation 1 holds. First of all, we observe that

$$[\mathscr{A}(x + e) - Bx] \cap [\mathscr{A}(u + e) - Bu] \neq \varnothing$$

for any $x, u \in L$. Indeed (since $\mathscr{A}$ is odd),

$$\begin{aligned} 0 \in \mathscr{A}(x - u) - B(x - u) &\subset \mathscr{A}(x + e) - \mathscr{A}(u + e) - B(x - u) \\ &= [\mathscr{A}(x + e) - Bx] - [\mathscr{A}(u + e) - Bu] \end{aligned}$$

Furthermore, again since $\mathscr{A}$ is odd, $\text{dom}^+ \mathscr{A} = -\text{dom}^- \mathscr{A}$, and the subspace $S = \text{dom}^+ \mathscr{A} - \text{dom}^- \mathscr{A}$ is the affine hull of each of them. We consider two cases.

If $S \cap (L + e) = \varnothing$, $\mathscr{A}(x + e) = Y$ for all $x \in L$, and hence the intersection in Equation 1 also equals Y.

If $S \cap (Le) \neq \varnothing$, we can choose $x \in L \cap \text{ri}(\text{dom}^+ \mathscr{A})$, $w \in L$, $t^+ > 0$, $t^- > 0$ in such a way that $u + t^+e \in \text{dom}^+ \mathscr{A}$, $v + t^-e \in \text{dom}^- \mathscr{A}$, where $u = x + t^+w$, $v = -x + t^-w$. Then, $u/t^+ + e \in \text{dom}^+ \mathscr{A}$, $v/t^- + e \in \text{dom}^- \mathscr{A}$. In other words, among the sets $\mathscr{A}(x + e) - Bx (x \in L)$, there is at least one bounded from above and at least one bounded from below. Since these sets are order intervals and any two meet each other, it follows that Equation 1 holds.

Corollary 2.1. Let $\mathscr{A}$ be an order-interval-valued odd fan from X into Y such that $\text{dom}^+ \mathscr{A}$ is a subspace. Then, $\mathscr{A}$ is generated by the set of its linear selections.

Proof. In this case, $\text{ri}(\text{dom}^+ \mathscr{A}) = \text{dom}^+ \mathscr{A}$. Let $y \in \mathscr{A}(x)$. If $x \in \text{dom}^+ \mathscr{A}$, we take $L = \{tx \mid t \in R\}$ and $B(tx) = ty$. If $x \notin \text{dom}^+ \mathscr{A}$, taking any linear selection A of $\mathscr{A}$, we set $L =$ linear hull of $(\text{dom}^+ \mathscr{A}) \cup \{x\}$ and define B by $B(u + tx) = Au + ty$ for $u \in \text{dom}^+ \mathscr{A}$, $t \in R$.

Remark 1. Obviously, the theorem will remain valid if, instead of order intervals, we use any system of convex sets that is closed under translation, summation, and scalar multiplication and has the property that any collection of elements of the system has a nonempty intersection, provided that any two meet each other.

Theorem 2.2. Let Y be a linear space, and let $\mathscr{C}$ be a collection of convex subsets of Y such that

$$\text{(i) } C + y \in \mathscr{C} \quad \text{for any} \quad C \in \mathscr{C}, y \in Y;$$

$$\text{(ii) } C + D \in \mathscr{C} \quad \text{for any} \quad C, D \in \mathscr{C};$$

$$\text{(iii) } tC \in \mathscr{C} \quad \text{for any} \quad C \in \mathscr{C}, t \in R.$$

Assume that for any linear space X and any $\mathscr{C}$-valued odd fan $\mathscr{A}$ from X into Y the following extension property holds: if $L \subset X$ is a subspace and B: $L \to Y$ is linear such that $Bx \in \mathscr{A}(x)$ for all $x \in L$, there is a linear selection of $\mathscr{A}$ that coincides with B on L.

Then, any collection of elements of $\mathscr{C}$ with pairwise nonempty intersections has a common point.

Proof. Let $\{C_\nu\}_{\nu \in N}$ be a family of elements of $\mathscr{C}$ such that $C_\nu \cap C_\mu \neq \varnothing$ for any two elements. Let X be the linear space of all real-valued functions $x(\nu)$ on N that assume nonzero values only at finitely many points (depending on the function of course). Consider the set-valued mapping

$$\mathscr{A}[x(\cdot)] = \sum_{\nu \in N} x(\nu) C_\nu .$$

Then, $\mathscr{A}$ is an odd fan. Indeed, properties 1.1–1.3 are obvious, the latter for all t, and not necessarily positive. Property 1.4 is also easy to verify if one takes into consideration that $(t + s)C \subset tC + sC$.

Let

$$L = \left\{ x(\cdot) \in X \,\middle|\, \sum_\nu x(\nu) = 0 \right\}.$$

Then, L is a subspace of X of codimension one. Let $x(\cdot) \in L$. Then,

$$x(\nu) = \sum_i t_i \mathscr{E}_{\nu_i}(\nu) - \sum_j s_j \mathscr{E}_{\mu_j}(\nu),$$

where $t_i > 0$, $s_j > 0$, $\sum t_i = \sum s_j = r$, and

$$\mathscr{E}_\mu(\nu) = \begin{cases} 1, & \text{if } \nu = \mu \\ 0, & \text{for all other } \nu. \end{cases}$$

Let $a_{ij} = (t_i s_j)/r$. Then, $a_{ij} > 0$, and

$$x(\nu) = \sum_{i,j} a_{ij} [\mathscr{E}_{\nu_i}(\nu) - \mathscr{E}_{\mu_j}(\nu)].$$

Since all ν_i and μ_j are distinct and all $a_{ij} > 0$, we have

$$[x(\cdot)] = \sum_{i,j} a_{ij} (C_{\nu_i} - C_{\mu_j}),$$

and hence $0 \in \mathscr{A}[x(\cdot)]$ for any $x(\cdot) \in L$.

By the assumptions, there is a linear selection of $\mathscr{A}$ that vanishes on L. According to the lemma, this implies that, for any $e \in L$,

$$\bigcap_{x(\cdot) \in L} \mathscr{A}[x(\cdot) + e] \neq \varnothing.$$

Arbitrarily take a $\mu \in N$ and set $e = \mathscr{E}_\mu$. Then (since $\mathscr{E}_\nu - \mathscr{E}_\mu \in L$ for any $\nu \in N$),

$$\bigcap_{\nu \in N} C_\nu = \bigcap_{\nu \in N} \mathscr{A}(\mathscr{E}_\nu) = \bigcap_{\nu \in N} \mathscr{A}[(\mathscr{E}_\nu - \mathscr{E}_\mu) + \mathscr{E}_\mu] \supset \bigcap_{x(\cdot) \in L} \mathscr{A}[x(\cdot) + \mathscr{E}_\mu],$$

hence $\cap C_\nu \neq \varnothing$.

Now we shall consider fans without the oddness assumption.

Theorem 2.3. Let $\mathscr{A}$ be an order-interval-valued fan such that $\operatorname{dom}^+ \mathscr{A}$ and $\operatorname{dom}^- \mathscr{A}$ are in general position. Then, there is an odd order-interval-valued fan $\mathscr{B}$ such that

$$\mathscr{B}(x) \subset \mathscr{A}(x) \quad \text{for all } x.$$

In particular, $\mathscr{A}$ has a linear selection.

Proof. We shall define $\mathscr{B}$ by

$$\mathscr{B}(x) = \cap \left[\sum t_i \mathscr{A}(x_i)\right]\left(\sum t_i x_i = x\right).$$

We have

$$\begin{aligned}\mathscr{B}(-x) &= \cap \left[\sum t_i \mathscr{A}(x_i)\right]\left(\sum t_i x_i = -x\right)\\ &= - \cap \left[\sum (-t_i)\mathscr{A}(x_i)\right]\left[\sum (-t_i)x_i = x\right]\\ &= - \cap \left[\sum t_i \mathscr{A}(x_i)\right]\left(\sum t_i x_i = x\right),\end{aligned}$$

hence $\mathscr{B}$ is odd if it is a fan. The equality $\mathscr{B}(tx) = t(x)$ is obvious. We have also

$$\begin{aligned}\mathscr{B}(u+v) &= \cap \left[\sum t_i \mathscr{A}(x_i)\right]\left(\sum t_i x_i = u+v\right)\\ &\subset \cap \left[\sum (s_i + r_i)\mathscr{A}(x_i)\right]\left(\sum s_i x_i = u, \sum r_i x_i = v\right)\\ &\subset \mathscr{B}(u) + \mathscr{B}(v).\end{aligned}$$

Note also that any set $\mathscr{B}(x)$, if it is nonempty, is an order interval.

Thus, it remains to verify that $\mathscr{B}(x) \neq \varnothing$ for all x. This will imply in particular that $\mathscr{B}(0) = 0$.

Since $\operatorname{dom}^+ \mathscr{A}$ and $\operatorname{dom}^- \mathscr{A}$ are in general position, their difference (denoted by L) is a linear subspace. It follows that whenever $x \notin L$ and $x = \sum t_i x_i$ (with all $t_i \neq 0$), there is at least one x_i not belonging to L. For this x_i, $\mathscr{A}(x_i) = Y$, and we conclude that $\mathscr{B}(x) = \mathscr{A}(x) = Y$ for every $x \in L$.

Now let $x \in L$, and let

$$x = \sum t_i u_i = \sum s_j v_j.$$

Then, setting

$$w^+ = \sum t_i^+ u_i, \quad w^- = \sum t_i^- u_i, \quad z^+ = \sum s_j^+ v_j, \quad z^- = \sum s_j^- v_j$$

(where $t^+ = \max(t, 0)$, $t^- = \max(-t, 0)$), we have

$$w^+ - w^- = z^+ - z^-,$$

that is,

$$w^+ + z^- = w^- + z^+,$$

and

$$\begin{aligned}0 &\in \mathscr{A}(w^+ + z^-) - \mathscr{A}(w^- + z^+)\\ &\subset \mathscr{A}(w^+) + \mathscr{A}(z^-) - \mathscr{A}(w^-) - \mathscr{A}(z^+)\\ &\subset \sum t_i^+ \mathscr{A}(u_i) + \sum s_j^- \mathscr{A}(v_j) - \sum t_i^- \mathscr{A}(u_i) - \sum s_j^+ \mathscr{A}(v_j)\\ &= \sum t_i \mathscr{A}(u_i) - \sum s_j \mathscr{A}(v_j)\end{aligned}$$

(since both t and t^- cannot be positive).

Thus, any two sets $\sum t_i \mathscr{A}(u_i)$ and $\sum s_j \mathscr{A}(v_j)$ meet each other if $\sum t_i u_i = \sum s_j v_j$. Finally, if $x \in L$,

$$x = u' - v' = v'' - u''$$

for some $u', u'' \in \text{dom}^+ \mathscr{A}$, $v', v'' \in \text{dom}^- \mathscr{A}$. It follows that the set $\mathscr{A}(u') - \mathscr{A}(v')$ is bounded from above and $\mathscr{A}(v'') - \mathscr{A}(u'')$ is bounded from below. Therefore, $\mathscr{B}(x)$ is a nonempty bounded set for any $x \in L$.

This proves the first part of the theorem. To prove the second, it suffices to observe that $\text{ri}(\text{dom}^+ \mathscr{A}) = \text{dom}^+ \mathscr{A} = L$ and apply Corollary 2.1.

Remarks 2. Theorem 2.1 is, of course, a reformulation of the Hahn–Banach–Kantorovic–Nachbin extension theorem and, if one compares the proofs,[4] one will easily discover that they are essentially the same.

3. Theorem 2.2 extends the result of Bonnice–Silverman[2] and To.[20] The advantage of using fans is easily seen here since we do not even touch upon such questions as linear closedness, which has made the proof much simpler.

4. It is difficult to relate certain well-known facts to Theorem 2.3. However, it may be considered an extension of the theorem of Mazur–Orlicz.[6,10,14]

Linear Subdifferentials

Slightly changing the existing terminology, we shall call the set of linear operators $A: X \to Y$, satisfying

$$Ah \le F(x + h) - F(x), \qquad \forall h \in X$$

(F being a convex mapping from X into $Y \cup \{\infty\}$), the linear subdifferential of F at x, and denote the set, as usual, by $\partial F(x)$.

For a sublinear mapping P, we shall use the symbol ∂P to denote $\partial P(0)$.

In this section, Y is, again, a conditionally complete vector lattice.

Theorem 3.1 (the sandwich theorem). Let F and G be a convex and a concave mapping from X into $Y \cup \{\infty\}$ and $Y \cup \{-\infty\}$, respectively, such that $F(x) \ge G(x)$ for all x, and that there is $u \in X$ such that the cones of admissible directions $\text{Ad}(\text{dom } F, u)$ and $\text{Ad}(\text{dom } G, u)$ are in general position. Then, there are $y \in Y$ and a linear operator $A: X \to Y$ such that

$$G(x) \le Ax + y \le F(x), \qquad \forall x \in X.$$

Proof. Denote by H_F the Hörmander transform of F, which is a sublinear operator from $R \times X$ into Y, defined by

$$H_F(t, x) = \begin{cases} tF(x/t), & \text{if } \ t > 0, \\ 0, & \text{if } \ t = 0, x = 0, \\ \infty, & \text{if otherwise.} \end{cases}$$

Then, $\text{dom } H_F = \{(t, x) \,|\, t > 0,\ x/t \in \text{dom } F\} \cup \{(0, 0)\}$. The Hörmander transform of G (which is superlinear since G is concave) and $\text{dom } H_G$ are defined similarly.

We shall show that $\text{dom } H_F$ and $\text{dom } H_G$ are in general position, which, by the assumptions, will follow from the formulas

$$\text{dom } H_F - \text{dom } H_G = R[\text{Ad}(\text{dom } F, 0) - \text{Ad}(\text{dom } G, 0)], \tag{2}$$

$$\text{dom } H_G - \text{dom } H_F = R[\text{Ad}(\text{dom } G, 0) - \text{Ad}(\text{dom } F, 0)], \tag{3}$$

which we shall prove assuming $u = 0$ (which would change nothing since we always can consider $F(xu)$ instead of $F(x)$, and so on).

Since $0 \in (\text{dom } F) \cap (\text{dom } G)$,

$$K = \text{dom } H_F - \text{dom } H_G \supset \text{dom } H_F - (1, 0).$$

It follows (K is a convex cone!) that $(t, 0) \in K$ for any $t \in R$, $(0, x) \in K$ for any $x \in \text{Ad}(\text{dom } F, 0)$, and hence

$$R \times \text{Ad}(\text{dom } F, 0) \subset K.$$

Likewise, from $(1, 0) - \text{dom } G \subset K$, we derive that

$$-[R \times \text{Ad}(\text{dom } G, 0)] \subset K,$$

so that the right-hand side of Equation 2 belongs to the left-hand side. On the other hand, if $(t, x) \in \text{dom } H_F$ and $x \neq 0$, $x/t \in \text{dom } F \subset \text{Ad}(\text{dom } F, 0)$. Therefore, $\text{dom } H_F \subset R \times \text{Ad}(\text{dom } F, 0)$, and so on. This proves Equations 2 and 3.

It remains to consider the fan

$$\mathscr{A}(t, x) = [G(x), F(x)].$$

Take an arbitrary linear selection B of $\mathscr{A}$ (which exists thanks to Theorem 2.3) and set $A = B(0, \cdot)$, $y = B(1, 0)$.

All other results of this section follow from Theorem 3.1.

Theorem 3.2. Let F be a convex mapping from X into $Y \cup \{\infty\}$. If $x \in \text{ri}(\text{dom } F)$, $\partial F(x) \neq \varnothing$.

Proof. Let

$$G(u) = \begin{cases} F(x), & \text{if } u = x, \\ -\infty, & \text{if } u \neq x. \end{cases}$$

Then, $\text{Ad}(\text{dom } F, x)$ is a subspace, and $\text{Ad}(\text{dom } G, x) = \{0\}$ is also a subspace. Thus, the two cones are in general position. Apply Theorem 3.1.

Theorem 3.3. Let $F_1, \ldots, F_n$ be convex mappings from X into $Y \cup \{\infty\}$ such that for certain $u \in X$, the cones $\text{Ad}(\text{dom } F_i, u)$, $i = 1, \ldots, n$, are in general position. Then,

$$\partial(F_1 + \cdots + F_n)(x) = \partial F_1(x) + \cdots + \partial F_n(x)$$

for any $x \in X$.

Proof. Since the domain of the sum of convex mappings is the intersection of their domains, it suffices to consider only the case where $n = 2$. The inclusion $\partial(F_1 + F_2)(x) \supset \partial F_1(x) + \partial F_2(x)$ can be verified directly. Let $A \in \partial(F_1 + F_2)(x)$. Then,

$$F_1(x + h) - F_1(x) \geq Ah - F_2(x + h) - F_2(x), \forall h \in X.$$

By the sandwich theorem, there is a linear operator $A_1 \colon X \to Y$ such that

$$F_1(x + h) - F_1(x) \geq A_1 h \geq Ah - F_2(x + h) - F_2(x), \forall h \in X.$$

This means that $A_1 \in \partial F_1(x)$ (left inequality) and that $A_2 = A - A_1 \in \partial F_2(x)$ (right inequality).

Theorem 3.4. Let F be a convex mapping from a linear space Z into $Y \cup \{\infty\}$, and let $A \colon X \to Z$ be a linear operator such that for certain $u \in (\text{Range } A) \cap (\text{dom } F)$ the sets $L = \text{Range } A$, and $\text{Ad}(\text{dom } F, u)$ are in general position. Then, for any $x \in X$,

$$\partial(F \circ A)(x) = [\partial F(Ax)] \circ A = \{C \circ A \mid C \in \partial F(Ax)\}.$$

Proof. Again, only the inclusion $(FA)(x)[F(Ax)]A$ needs to be proved. Let $D \in \partial(FA)(x)$; that is,

$$Dh \leq F(Ah + Ax) - F(Ax), \forall h \in X.$$

Let

$$G(z) = \sup\{Dh \mid Ah = z\}$$

(where we assume, as usual, $\sup \phi = -\infty$). Then, G is a superlinear mapping with dom $G = L$. (Indeed, $G(z) = -\infty$ for $z \in L$, by definition. On the other hand, $Dh \leq 0$ for $h \in \operatorname{Ker} A$ so that $Dh_1 = Dh_2$ if $Ah_1 = Ah_2$. It follows that G is a linear mapping on L.) We also have

$$G(z) \leq F(z + Ax) - F(Ax), \forall z \in Z,$$

so that, by the sandwich theorem, there is a linear operator $C: Z \to Y$ satisfying

$$G(z) \leq Cz \leq F(z + Ax) - F(Ax), \forall z \in Z.$$

Thus, $C \in \partial F(Ax)$, and $Dh \leq G(Ah) \leq C(Ah)$ for all $h \in X$, which implies that $D = C \circ A$.

Theorem 3.5. Let Z be an ordered space, let F_1 be a convex mapping from X into $Z \cup \{\infty\}$, and let F_2 be a nondecreasing convex mapping from Z into Y (i.e., $F_2(z) \geq F_2(w)$ if $z \geq w$). Then, for any $x \in X$,

$$(F_2 F_1)(x) = \bigcup_{A \in \partial F_2[F_1(x)]} \partial(A \circ F_1)(x).$$

Proof. The proof is essentially the same as in the preceding theorem. The only difference is that we define G by

$$G(z) = \sup\{Dx \mid F_1(x + h) - F_1(x) \leq z.$$

The sandwich theorem can be applied because dom $F_2 = Z$.

Denote by M the collection of multipliers in Y, that is, such linear operators $m: Y \to Y$ that satisfy $0 \leq my \leq y$ for every $y \geq 0$.

Theorem 3.6. Let F_1 and F_2 be convex mappings from X into $Y \cup \{\infty\}$. Assume that there is a $u \in X$ such that the cones Ad(dom F_1, u) and Ad(dom F_2, u) are in general position. Then, for any $x \in X$,

$$\partial(F_1 \vee F_2)(x) = \bigcup_{m \in M} [m \circ \partial F_1(x) + (I - m) \circ \partial F_2(x)],$$

where I is the identity operator.

Proof. This follows from Theorem 3.5.[10]

To conclude this section, we shall consider topological versions of the theorems. For this purpose, we shall assume that X is a locally convex topological space and that Y is a locally convex conditionally complete vector lattice (i.e., the topology in Y is defined by a family of monotone seminorms). Then, any linear mapping $A: X \to Y$ bounded (from below or above) on an open set is continuous. In particular, if a convex mapping F is continuous at a certain point x, any element of $\partial F(x)$ is a continuous linear operator. We shall denote by $L(X, Y)$ the space of all continuous linear operators from X into Y and shall write $\partial_c F(x) = \partial F(x) \cap L(X, Y)$.

We shall also say that a convex mapping F is lower semicontinuous (l.s.c.) if every level set $\{x \mid F(x) \leq y\}$ is closed. Observe that F is l.s.c. if it is weakly l.s.c. in the sense that for any positive functional $y^* \in Y^*$ (the asterisk denotes topological duality), the extended real-valued function $x \to \langle y^*, F(x) \rangle$ is l.s.c.

Theorem 3.7. Let F be a convex l.s.c. mapping from X into $Y \cup \{\infty\}$. Then F coincides with the pointwise supremum of the family of continuous affine mappings from X into Y majorized by F.

Proof. First, we note that the statement is based on the implicit assumption that any bounded set has a unique supremum. What we shall actually prove does not require this assumption. Namely, we shall show that whenever $y \leq F(x)$ and $y \ngeq F(x)$, there is an affine mapping $u \to Au + a$ from X into Y such that

$$y \leq Ax + a \leq F(x) \tag{4}$$

and

$$Au + a \leq F(u) \quad \text{for all} \quad u \subset X. \tag{5}$$

Indeed, if y is such as above, there is an open $U \subset X$ containing x such that $y \leq F(u)$ for all $u \subset U$ (due to the fact that F is l.s.c.).

If U dom $F \neq \phi$, the conditions of the sandwich theorem will be fulfilled for F and G defined by

$$G(u) = \begin{cases} y, & \text{if} \quad u \in U, \\ -\infty, & \text{if} \quad u \notin U, \end{cases}$$

so that Equations 4 and 5 will be satisfied for certain $A \in L(X, Y)$ and $a \in Y$.

If U dom $F = \phi$, there is $x^* \in X^*$ separating U and dom F or, if we assume without loss of generality that $0 \in U$, such an $x^* \in X^*$ that $\langle x^*, w\rangle \leq 0 < \langle x^*, u\rangle$ for all $w \in \text{dom}\, F$ and $u \in U$. Take an arbitrary continuous affine mapping $S: X \to Y$ majorized by F (which exists according to what has been just proved) and for any positive $t \in R$, $v \in Y$ consider the affine continuous mapping

$$S_{t,v}(u) = Su + t\langle x^*, u\rangle v.$$

Then, $S_{t,v}(u) \leq Su \leq F(u)$ if $u \in \text{dom}\, F$, so that F majorizes $S_{t,v}$ for any $t \geq 0$, $v \geq 0$. On the other hand, we can take $\bar{t} > 0$, $\bar{v} \geq 0$ to ensure that $Sx + \bar{t}\langle x^*, x\rangle\bar{v} \geq y$ (since $x \in U$), so that Equations 4 and 5 will again be satisfied.

Theorem 3.8. Let F_1 and F_2 be convex mappings from X into $Y \cup \{\infty\}$ such that F_1 is continuous at all points of dom F_2. Then, for any $x \in X$,

$$\partial_c(F_1 + F_2)(x) = \partial_c F_1(x) + \partial_c F_2(x).$$

Proof. We have $[\text{int}(\text{dom}\, F_1)] \cap (\text{dom}\, F_2) \neq \phi$, so that the assumptions of Theorem 3.3 are satisfied, and hence

$$\partial(F_1 + F_2)(x) = \partial F_1(x) + \partial F_2(x).$$

However, $\partial F_1(x) = \partial_c F_1(x)$, and if $A \in \partial_c(F_1 + F_2)(x)$, $A = A_1 + A_2$, $A_i \in \partial F_i(x)$, $A_2 = A - A_1 \in L(X, Y)$.

Theorem 3.9. Let F be a convex mapping from Z into $Y \cup \{\infty\}$, and let $A \in L(X, Z)$ be such that F is continuous at all points of the range of A. Then,

$$\partial_c(FA)(x) = [\partial_c F(x)] \circ A.$$

Proof. This follows immediately from Theorem 3.4 since $\partial F(x) = \partial_c F(x)$, and so on.

Topological versions of the other theorems can be derived in the same way.

Remark 5. Of course, all these theorems imply corresponding and many other facts of standard convex analysis. In what follows, we shall use those facts freely without special explanations.

Adjoint Fans

In this section, we shall consider certain simple general results concerning fans, not necessarily order-interval-valued. Formally, the results will not be used later; we include them mainly to outline the framework of the notions to be introduced.

Here, X and Y are locally convex Hausdorff topological spaces. Note that in the topological setting, it is natural to consider only fans with closed values and to use a weaker form of condition 1.4, namely,

$$\mathscr{A}(x+u) \subset \overline{\mathscr{A}(x) + \mathscr{A}(u)},$$

where the bar denotes the closure.

Let $\mathscr{A}$ be a fan from X into Y (Y is not necessarily ordered). The function

$$s_{\mathscr{A}}(y^*, x) = \sup\{\langle y^*, y\rangle \mid y \in \mathscr{A}(x)\}$$

is called the support function of $\mathscr{A}$. (As a rule, we shall omit the subscript $\mathscr{A}$.) We shall assume that there are convex cones $K \subset X$ and $L^* \subset Y^*$ such that

$$\{(y^*, x) \mid s(y^*, x) < \infty\} = (L^* \times K) \cup (\{0\} \times X) \cup (Y^* \times \{0\})$$

(note that $s(0, x) = s(y^*, 0) = 0$ for all y^*, x by definition).

The cone K will be called the domain of $\mathscr{A}$ and L will be called the orientation of $\mathscr{A}$: $K = \operatorname{dom} \mathscr{A}$, $L = \operatorname{or} \mathscr{A}$.

The fan will be called closed if the extended real-valued function $s(y, \cdot)$ is l.s.c. on X for any $y^* \in \operatorname{or} \mathscr{A}$ (and hence for any $y^* \in Y^*$). If $\mathscr{A}$ is a closed fan, the set

$$\mathscr{A}^*(y^*) = \{x^* \in X^* \mid s(y^*, x) \geq \langle x^*, x\rangle, \forall x\}$$

is nonempty and closed and

$$s(y^*, x) = \sup\{\langle x^*, x\rangle \mid x^* \in \mathscr{A}^*(y^*)\}.$$

(This result follows, for instance, from Theorem 3.7.) In other words, $\mathscr{A}^*(y^*)$ is the linear subdifferential of the sublinear function $s(y^*, \cdot)$.

The set-valued mapping $\mathscr{A}^*$ is a fan from Y^* into X^* with $\operatorname{dom} \mathscr{A}^* = L^*$, $\operatorname{or} \mathscr{A}^* = K$. This fan is called the adjoint to $\mathscr{A}$. The reader is referred to Ioffe[21] for a more thorough discussion concerning adjoint fans. Here, we shall only list some of their properties relevant to the notion of the subdifferential defined below.

Consider the following operations with fans:

(1) addition: $\mathscr{A}_1 + \mathscr{A}_2$ is defined by

$$(\mathscr{A}_1 + \mathscr{A}_2)(x) = \overline{\mathscr{A}_1(x) + \mathscr{A}_2(x)};$$

(2) supremum: $\mathscr{A}_1 \vee \mathscr{A}_2$ is defined by

$$(\mathscr{A}_1 \vee \mathscr{A}_2)(x) = \overline{\operatorname{conv}}\,[\mathscr{A}_1(x) \cup \mathscr{A}_2(x)];$$

(3) right composition with a linear operator:

$$(\mathscr{A} \circ A)(x) = \mathscr{A}[A(x)];$$

(4) left composition with a linear operator:

$$(A \circ \mathscr{A})(x) = \overline{A[\mathscr{A}(x)]};$$

(5) composition: if $\mathscr{A}$ has property 1.4, then $\mathscr{B} \circ \mathscr{A}$, defined by

$$(\mathscr{B} \circ \mathscr{A})(x) = \operatorname{conv}\left[\bigcup_{z \in \mathscr{A}(x)} \mathscr{B}(z)\right],$$

is a fan.

Proposition 4.1. We have

$$s_{\mathscr{A}_1+\mathscr{A}_2}(y^*, x) = s_{\mathscr{A}_1}(y^*, x) + s_{\mathscr{A}_2}(y^*, x).$$

In particular, $\mathrm{dom}(\mathscr{A}_1 + \mathscr{A}_2) = \mathrm{dom}\ \mathscr{A}_1 \cap \mathrm{dom}\ \mathscr{A}_2$, or

$$(\mathscr{A}_1 + \mathscr{A}_2) = \mathrm{or}\ \mathscr{A}_1 \cap \mathrm{or}\ \mathscr{A}_2$$

and

$$(\mathscr{A}_1 + \mathscr{A}_2)^* = \mathscr{A}_1^* + \mathscr{A}_2^*.$$

If, in addition, there is a $u \in \mathrm{dom}\ \mathscr{A}_2$ such that $s_{\mathscr{A}_1}(y^*, \cdot)$ is continuous at u for any $y^* \in \mathrm{or}\ \mathscr{A}_1$ and the orientations of both fans coincide, then

$$\mathscr{A}_1^*(y^*) + \mathscr{A}_2^*(y^*) = (\mathscr{A}_1 + \mathscr{A}_2)^*(y^*),\ \forall y^*.$$

Proposition 4.2. We have

$$s_{\mathscr{A}_1 \vee \mathscr{A}_2}(y^*, x) = \max[s_{\mathscr{A}_1}(y^*, x), s_{\mathscr{A}_2}(y^*, x)].$$

In particular, $\mathrm{dom}(\mathscr{A}_1 \vee \mathscr{A}_2) = \mathrm{dom}\ \mathscr{A}_1 \cap \mathrm{dom}\ \mathscr{A}_2$, or

$$(\mathscr{A}_1 \vee \mathscr{A}_2) = \mathrm{or}\ \mathscr{A}_1 \cap \mathrm{or}\ \mathscr{A}_2$$

and

$$(\mathscr{A}_1 \vee \mathscr{A}_2)^* = \mathscr{A}_1^* \vee \mathscr{A}_2^*.$$

If, in addition, the orientations of both fans coincide and there is $u \in \mathrm{dom}\ \mathscr{A}_2$ such that $s_{\mathscr{A}_1}(y^*, \cdot)$ is continuous at u for any $y^* \in \mathrm{or}\ \mathscr{A}_1$, then

$$(\mathscr{A}_1 \vee \mathscr{A}_2)^*(y^*) = \mathrm{conv}[\mathscr{A}_1(y^*) \cup \mathscr{A}_2(y^*)],\ \forall y^*.$$

Proposition 4.3. We have

$$s_{\mathscr{A} \circ A}(y^*, x) = s_{\mathscr{A}}(y^*, Ax),$$

so that $\mathrm{dom}\ \mathscr{A} \circ A = A^{-1}(\mathrm{dom}\ \mathscr{A})$, or $(\mathscr{A} \circ A) = \mathrm{or}\ \mathscr{A}$ and

$$(\mathscr{A} \circ A)^* = A^* \circ \mathscr{A}^*.$$

If, in addition, $s(y^*, \cdot)$ is continuous at certain points of the range of A,

$$(\mathscr{A} \circ A)^*(y^*) = A^*[\mathscr{A}^*(y^*)],\ \forall y^*.$$

Proposition 4.4. We have

$$s_{A \circ \mathscr{A}}(y^*, x) = s_{\mathscr{A}}(A^* y^*, x),$$

so that $\mathrm{dom}(A \circ \mathscr{A}) = \mathrm{dom}\ \mathscr{A}$, or $(A \circ \mathscr{A}) = A^{*-1}(\mathrm{or}\ \mathscr{A})$ and

$$(A \circ \mathscr{A})^* = \mathscr{A}^* \circ A^*.$$

Proofs of all four propositions follow immediately from well-known properties of support functions of convex sets. It is possible to prove also that under certain assumptions,

$$(\mathscr{B} \circ \mathscr{A})^* = \mathscr{A}^* \circ \mathscr{B}^*$$

(see Reference 21).

SUBDIFFERENTIALS: GENERAL APPROACH

We shall now assume that Y is a locally convex ordered space, in particular, that the positive cone in Y is closed.

For $y^* \geq 0$, we set

$$F_{y^*}(x) = \langle y^*, F(x) \rangle.$$

Let F be a convex mapping from X into $Y \cup \{\infty\}$, and let $\partial F_{y^*}(x)$ be the linear subdifferential of F_{y^*} at x, that is, the set of all continuous linear functionals x^* on X^* that satisfy

$$\langle x^*, h \rangle \leq \langle y^*, F(x+h) - F(x) \rangle \ \forall h \in X.$$

The set-valued mapping

$$y^* \to D_x F(y^*) = \begin{cases} \partial F_{y^*}(x), & \text{if } y^* \geq 0 \\ X, & \text{for all other } y^* \end{cases}$$

will be called the subdifferential of F at x. The mapping F is said to be subdifferentiable at x if $D_x F(y^*) \neq \phi$ for any y^*. This is definitely true if F_{y^*} is continuous at x for any $y^* \geq 0$; in particular, if F itself is continuous at x.

Proposition 5.1. Assume that F is subdifferentiable at x. Then, $D_x F$ is a fan from Y^* into X^* with dom $D_x F$ equal to the positive cone in Y^* or $D_x F$ lying between Ad(dom F, x) and its closure. Also,

$$D_x F(y_1^* + y_2^*) = D_x F(y_1^*) + D_x F(y_2^*).$$

Moreover, if the support function of $D_x F$ is weakly l.s.c. in y^* for any x or $D_x F$, $D_x F$ is the adjoint to the fan $\mathscr{A}$, defined as follows:

$$\mathscr{A}(h) = \begin{cases} B(h), & \text{if } h \neq 0 \\ \{0\}, & \text{if } h = 0 \end{cases},$$

where $B(h) = \bigcap\limits_{t>0} \{y \mid y \leq t^{-1}(F(x+th) - F(x))\}$.

Proof. Clearly, $D_x F(0) = \{0\}$ and $D_x F(\lambda y^*) = D_x F(y^*)$ for $\lambda > 0$. We have also

$$F_{y^*+v^*}(x) = F_{y^*}(x) + F_{v^*}(x),$$

which implies Equation 6.

Furthermore,

$$s(y^*, h) = \inf_{t>0} t^{-1} \langle y^*, F(x+th) - F(x) \rangle$$

$$= \inf_{t>0} \sup\{\langle y^*, y \rangle \mid y \leq t^{-1}(F(x+th) - F(x)\} \tag{6}$$

Since $s(.,.)$ is l.s.c. in y^*, the sets

$$\mathscr{A}(h) = \{y^* \mid \langle y^*, y \rangle \leq s(y^*, h), \forall\ y^* \geq 0\}$$

are nonempty and

$$s(y^*, h) = \sup\{\langle y^*, y \rangle \mid y \in \mathscr{A}(h)\}. \tag{7}$$

If $y^* \in \mathscr{A}(h)$, then

$$\langle y^*, y \rangle \le s(y^*, h) \le t^{-1}(F(x + th) - F(x)), \forall\, y^* \ge 0$$

and hence $y \le t^{-1}(F(x + th) - (x))$.

On the other hand, since both $\mathscr{A}(h)$ and $B(h)$ are convex and closed and, as follows from Equation 6,

$$\sup\{\langle y^*, y \rangle \mid y \in B(h)\} \le s(y^*, h) = \sup\{\langle y^*, y \rangle \mid y \le \mathscr{A}(h)\},$$

we have $B(h) \subset \mathscr{A}(h)$ and hence $B(h) = \mathscr{A}(h)$.

The fact that $\mathscr{A}$ is a fan follows from Equation 6 because $s(.,.)$ is sublinear in each argument.

Note that, if

$$\inf_{t>0} t^{-1}[F(x + th) - F(x)] = y(h)$$

exists, then, obviously, $s(y^*, h) = \langle y^*, y(h) \rangle$, and $\mathscr{A}(h) = \{y \mid y \le y(h)\}$. However, despite the fact that $t^{-1}[F(x + th) - F(x)]$ decreases as t approaches zero, the lower bound may not exist.

If the subdifferential is the adjoint to the fan defined in the proposition, all the results of the preceding section can be reformulated for subdifferentials. However, corresponding results hold also without the assumption.

Theorem 5.2. Let F_1 and F_2 be convex mappings from X into $Y \cup \{\infty\}$. Then, for any $x \in X$,

$$D_x(F_1 + F_2) = D_x F_1 + D_x F_2,$$

$$D_x(F_1 \vee F_2) = (D_x F_1) \vee (D_x F_2).$$

If, in addition, there is a $u \in \operatorname{dom} F_2$ such that $F_{1y^*}(\cdot)$ is continuous at u for any $y^* \ge 0$, then

$$D_x(F_1 + F_2)(y^*) = D_x F_1(y^*) + D_x F_2(y^*),$$

$$D_x(F_1 \vee F_2)(y^*) = \operatorname{conv}[D_x F_1(y^*) \cup D_x F_2(y^*)]$$

for any $x \in X$, $y^* \ge 0$.

Theorem 5.3. Let Z be another locally convex space, let F be a convex mapping from Z into $Y \cup \{\infty\}$, and let $A \in L(Z, Y)$. Then, for any x,

$$D_x(F \circ A) = A^* \circ D_{Ax} F.$$

If, in addition, there is a $u \in X$ such that $F_{y^*}(\cdot)$ is continuous at Au for any $y^* \ge 0$,

$$D_x(F \circ A)(y^*) = A^* \circ [D_{Ax} F(y^*)]$$

for any $x \in X$, $y^* \ge 0$.

In the first parts of both theorems, operations with fans are understood, of course, in the sense defined in the preceding section. Both theorems are immediate corollaries of the corresponding results of the third section or, equivalently, from the corresponding results of standard convex analysis. The following formula is, however, new.

Theorem 5.4. Let Z be an ordered locally convex space, let F be a convex mapping from X into $Z \cup \{\infty\}$, and let G be a convex nondecreasing mapping from Z into Y such that the functions $G_{y^*}(\cdot)$ are continuous for all $y^* \ge 0$. Then,

$$D_x(G \circ F) = D_{F(x)} G \circ D_x F,$$

and, moreover, for any $x \in X$, $y^* \ge 0$

$$D_x(G \circ F)(y^*) = \bigcup_{z^* \in D_{F(x)} G(y^*)} D_x F(z^*).$$

Proof. We have

$$(G \circ F)_{y^*}(u) = (G_{y^*} \circ F)(u),$$

where G_{y^*} is a mapping into a conditionally complete vector lattice R, so that we can apply the topological version of Theorem 3.5 and get

$$D_x(G \circ F)(y^*) = \bigcup_{z^* \in \partial G_{y^*}[F(x)]} \partial F_{z^*}(x)$$
$$= \bigcup_{z^* \in D_{F(x)}G(y^*)} D_x F(z^*).$$

To conclude, we note that the set-valued mappings defined here as subdifferentials were explicitly considered by Linke[13] to establish the existence of linear subdifferentials of convex mappings into the space of continuous functions.

References

1. Amann, H. 1976. Fixed points equations and non-linear eigenvalue problems in ordered spaces. SIAM (Soc. Ind. Appl. Math.) Rev. **18:** 620–709.
2. Bonnice, W. & R. Silverman, 1967. The Hahn-Banach extension and the least upper bound properties are equivalent. Proc. Am. Math. Soc. **18:** 843–850.
3. Corson, H. H. & J. Lindenstrauss. 1966. Continuous selections with non-metrizable range. Trans. Am. Math. Soc. **121:** 492–504.
4. Day, M. M. Normed Linear Spaces.
5. Ekeland, I. & R. Temam. 1972. Analyse Convexe et Problèmes Variationelles. Dunod. Paris.
6. Feldman, M. M. 1975. On sublinear operators defined on a cone. Sib. Math. J. **16:** 1308–1321 (in Russian).
7. Halkin, H. 1978. Lecture Notes in Mathematics: 680. Springer. New York.
8. Ioffe, A. D. 1979. Gradients géneralisée d'une application localement lipschitzienne d'un espace de Banach dans un autre. C.R. Acad. Sci.
9. Ioffe, A. D. & V. M. Tihomirov. 1974. Theory of Extremal Problems. Nauka. Moscow; English translation published by North-Holland, 1979.
10. Kutateladze, S. S. 1979. Convex operators. Usp. Mat. Nauk **34**(1): 167–196 (in Russian).
11. Levin, V. L. 1972. Subdifferentials of convex mappings and composed functions. Sib. Math. J. **13:** 1295–1303 (in Russian).
12. Linke, Ju. 1972. On support sets of sublinear operators. Dokl. Akad. Nauk SSSR **207:** 531–533.
13. Linke, Ju. 1976. Sublinear operators with values in spaces of continuous functions. Dokl. Akad. Nauk SSSR, **228:** 540–542.
14. Mazur, S. & W. Orlicz. 1953. Sur les espaces metriques. Stud. Math. **13:** 137–173.
15. Moreau, J.-J. 1966. Fonctionelles Convexes. Collége de France. Paris.
16. Zeleny, M. (Ed.). 1976. Multiple Criteria Decision Making. Lecture Notes in Economics and Mathematical Systems: 123. Springer. New York.
17. Nachbin, L. 1950. A theorem of Hahn-Banach type for linear transformations. Trans. Am. Math. Soc. **68:** 28–46.
18. Rockafellar, R. T. 1970. Convex Analysis. Princeton University Press. Princeton, N.J.
19. Rubinov, A. M. 1977. Sublinear operators and their applications. Usp. Mat. Nauk **32**(4): 113–174; English translation published in Russ. Math. Surveys.
20. To, T.-O. 1971. The equivalence of the least upper bound property and the Hahn-Banach extension property in ordered vector spaces. Proc. Am. Math. Soc. **30:** 287–296.
21. Ioffe, A. D. Nonsmooth analysis: differential calculus of nondifferentiable mappings. To be published.

FINITE-DIFFERENCE SCHEMES FOR THE NAVIER–STOKES SYSTEM

Irina Brailovsky

Moscow, U.S.S.R.

It is well known that the Navier–Stokes system consists of a set of parabolic quasilinear second-order partial equations and a hyperbolic (mass-conservation) first-order equation.

In this paper, however, I shall describe the properties of finite-difference schemes to solve the system for simple linearized models of heat transfer and wave propagation equations. This approach will be taken because it is easy in each case to write an analogous finite-difference scheme for the real Navier–Stokes system and because the stability and quality of approximation of the schemes in question are almost the same for such simple models and for the real Navier–Stokes system, as demonstrated by many numerical experiments with computers at Moscow State University. In investigating the simple linear models, we must consider the schemes for different values of the coefficients of our linear equations, so that the range of values of the coefficients is the same as the range of variation of corresponding functions of the Navier–Stokes system for problems that interest us.

This author's interest lies in flow around bodies of supersonic compressible viscous gases. The Reynolds numbers are rather large for such a flow, and the corresponding coefficients in the simple models (i.e., heat-transfer coefficients in parabolic equations) must approach relatively small values. Since it is possible that computers could be used to study the problems that interest us, we should consider the situation where the magnitude of the heat-transfer coefficient is comparable to the magnitude of the unit cell in our set, or even less. We therefore must search for schemes that will provide sufficient approximation and that will be stable under the conditions described above. Note that this author's main interest lies in consideration of stationary flow; use of nonstationary equations is a method of approaching the steady state, i.e., the solution of the stationary problem.

First of all, let us consider a one-dimensional equation with constant coefficients:

$$\frac{\partial u}{\partial t} = a\frac{\partial u}{\partial x} + \nu\frac{\partial^2 u}{\partial x^2}, \tag{1}$$

where ν may be sufficiently small to be of the same order as space step Δx or less, and a is of the order of unity.

Consider the following finite-difference scheme for Equation 1:

$$\begin{aligned}\frac{\bar{U}_m^{n+1} - U_m^n}{\tau} &= a\frac{U_{m+1}^n - U_{m-1}^n}{2h} + \nu\frac{U_{m+1}^n - 2\bar{U}_m^{n+1} + U_{m-1}^n}{h^2}\\ \frac{U_m^{n+1} - U_m^n}{\tau} &= a\frac{\bar{U}_{m+1}^{n+1} - \bar{U}_{m-1}^{n+1}}{2h} + \nu\frac{U_{m+1}^n - 2U_m^{n+1} + U_{m-1}^n}{h^2},\end{aligned} \tag{2}$$

where $U_m^n = U(mh, nh) = U(m\,\Delta x, n\,\Delta t) = U(x, t)$. By investigating stability in the usual way using a Fourier method, one obtains (if $U_m^n = \lambda^h e^{ikmh}$) that the scheme is stable if $\tau \le h/|a|$ at any value of ν.

Concerning the error (ε) in the approximation of Scheme 2, one can show that $\varepsilon = 0(h^2) + 0[(U^{n+1} - U^n)/h^2]$, so that, for a steady-state solution, the second term vanishes, and one obtains the usual second-order approximation $\varepsilon = 0(h^2)$.

Scheme 2 can be written for two- and three-dimensional cases of a heat equation, and the same kind of stability and quality of approximation can be proved for these cases. It is easy to write the difference approximation according to Scheme 2 for the full Navier–Stokes system. This approximation has been written[1] and, as numerical calculations for a two-dimensional viscous gas flow have shown, the properties of the scheme remain the same. There appeared to be analogous conditions of stability for the time process; namely, there was convergence if

$$\tau \leq \frac{\min(\Delta x, \Delta y)}{(|u| + |v| + c\sqrt{2})},$$

where u and v are velocity components, and c is the sonic velocity. The approximation was of the second order in spatial steps and did not depend on the time step for steady-state flow, obtained as a limit of the solution of the nonstationary equations.

I shall now consider other similar schemes, that is, schemes with good stability conditions. The procedure for calculating these schemes is explicit, as one can see from Scheme 2, for example, yet the schemes are not explicit in the usual sense.

We can obtain such schemes by taking the values of unknown functions from different time levels in an unusual way, so that additional errors of approximation are admitted, namely, of the type $A(U^{n+1} - U^n)/h^2$, where n is the number of a time step and A is a spatial-difference operator. Such terms of an error vanish as a solution approaches its steady-state point and U^{n+1} becomes equal to U^n; at the same time, the process of convergence to the steady-state point is more rapid than with the usual approximation.

Consider the following scheme for Equation 1:

$$\frac{U_m^{n+1} - U_m^n}{\tau} = a\frac{U_{m+1}^n - U_{m-1}^{n+1}}{2h} + v\frac{U_{m+1}^n - 2U_m^{n+1} + U_{m-1}^{n+1}}{h^2}. \quad (3)$$

This scheme is stable (according to the Fourier method of investigation) if $\tau \leq h/|a|$.

Modification of Scheme 3 for the two-dimensional case, that is, for the equation

$$\frac{\partial u}{\partial t} = a\frac{\partial u}{\partial x} + b\frac{\partial u}{\partial y} + v\left(\frac{\partial^2 u}{\partial x^2} + \frac{\partial^2 u}{\partial y^2}\right), \quad (4)$$

takes the following form:

$$\frac{U_{m,l}^{n+1} - U_{m,l}^n}{\tau} = a\frac{U_{m+1,l}^n - U_{m-1,l}^{n+1}}{2h} + b\frac{U_{m,l+1}^n - U_{m,l-1}^{n+1}}{2h_1}$$
$$+ v\left(\frac{U_{m+1,l}^n - 2U_{m,l}^{n+1} + U_{m-1,l}^{n+1}}{h^2} + \frac{U_{m,l+1}^n - 2U_{m,l}^{n+1} + U_{m,l-1}^{n+1}}{h_1^2}\right), \quad (5)$$

where $U(x, y, t) = U(mh, lh_1, n\tau)$.

Again, we can see, using the Fourier method of investigation, that the scheme is stable if

$$\tau \leq \min\left(\frac{h}{|a|}, \frac{h_1}{|b|}\right). \quad (6)$$

Analogous conditions can also be obtained for the three-dimensional case. Approximation is of second order for the steady-state solution; that is,

$$\varepsilon = ([\max(h, h_1)]^2). \tag{7}$$

Scheme 5 has been used[2] in the Navier–Stokes system to calculate the two-dimensional cases (plane and axisymmetric) of a compressible gas. The process was stable with a condition like Inequality 6, and the approximation for a steady-state solution was of second-order in space steps, as in Equation 7.

The process led to a steady-state solution 1.5 times faster than with Scheme 2; since in the case of Scheme 5 there was no need for recalculation at each time step.

In some problems, the functions change more rapidly in one direction (e.g., in the y direction) than in another because of peculiarities of the problem. In such cases, it may be helpful to use the following kind of scheme for Equation 4:

$$\frac{U^n_{m,l} - U^n_{m,l}}{\tau} = a\frac{U^n_{m+1,l} - U^{n+1}_{m-1,l}}{2h} + b\frac{U^n_{m,l+1} - U^{n+1}_{m,l-1}}{2h_1}$$
$$+ \nu\frac{U^n_{m+1,l} - 2U^{n+1}_{m,l} + U^n_{m-1,l}}{h^2} + \nu\frac{U^{n+1}_{m,l+1} - 2U^{n+1}_{m,l} + U^{n+1}_{m,l-1}}{h_1^2}. \tag{8}$$

The scheme is implicit, but only in the y direction, and it is relatively easy to solve one-dimensional implicit equations using well-known methods.

On investigating the scheme using the Fourier method, we can find (if $U^n_{m,l} = U(n\,\Delta t, m\,\Delta x, l\,\Delta y) = \lambda^n \exp[i(kmh + k_1\,lh_1)]$:

$$\lambda = \frac{A + iB}{A + 4\dfrac{\nu\tau}{h^2}\sin^2\dfrac{kh}{2} + 4\dfrac{\nu\tau}{h_1^2}\sin^2\dfrac{k_1 h_1}{2} - iB},$$

where

$$A = 1 + \left(2\frac{\nu\tau}{h^2} + \frac{1}{2}\frac{a\tau}{h}\right)\cos kh + \frac{1}{2}\frac{b\tau}{h_1}\cos k_1 h_1,$$

$$B = \frac{1}{2}\frac{a\tau}{h}\sin kh + \frac{1}{2}\frac{b\tau}{h_1}\sin k_1 h_1.$$

Therefore, $|\lambda| < 1$, and the process leads to the steady-state solution if

$$2A + 4\frac{\nu\tau}{h^2}\sin^2\frac{kh}{2} + 4\frac{\nu\tau}{h_1^2}\sin^2\frac{k_1 h_1}{2} > 0,$$

that is, if

$$\left(1 + \frac{a\tau}{h}\cos kh\right) + \left(1 + \frac{b\tau}{h_1}\cos k_1 h_1\right) + 4\frac{\nu\tau}{h^2}\cos^2\frac{kh}{2} + 4\frac{\nu\tau}{h_1^2}\sin^2\frac{k_1 h_1}{2} > 0,$$

and a sufficient condition for it is:

$$\tau \le \min\left(\frac{h}{|a|}, \frac{h_1}{|b|}\right).$$

The magnitudes of the modulus for different wave numbers k and k_1 show that one may expect rather rapid convergence to a steady-state point, especially with our supposition that the function changes more rapidly in the y direction, that is, that high frequencies are important only for y-direction dependence of the function.

It is relatively easy to generalize the scheme to the three-dimensional case and also to the Navier–Stokes system. The investigation for the three-dimensional case gives the same kind of results as for the two-dimensional one given above. As for the Navier–Stokes system, there is need for numerical calculations.

The quality of the solution (and, for the real Navier–Stokes system, even the possibility of obtaining the solution) depends not only on the convergence of the time process but also on the properties and accuracy of the spatial approximation. The quality of approximation of the steady-state solution for the schemes mentioned above is of second order in spatial steps, does not depend on time steps, and coincides with the quality of approximation of the usual second-order centered spatial differences. We shall therefore now consider centered differences.

As mentioned above, we must consider situations where v (the heat-transfer coefficient in our models or viscosity in the Navier–Stokes system) is rather small, of the order of the space-step size or even less. In this case, a steady-state solution, or a solution of stationary finite-difference equations, is formally of second order, yet really (for all practically possible mesh sizes) has little in common with the exact solution of the initial differential equations. This is due to the fact that the coefficient of the approximation error depends on the value of v, which is, as mentioned above, the same as or less than the size of a cell. Consequently, one obtains parasitic pulsations, sawlike additions of sufficiently large amplitude that they can completely distort the solution. This effect is not due to an instability of time-step process but, rather, to peculiarities of the spatial finite-difference approximation, namely, peculiarities of the exact solution of the stationary finite-difference equations with centered second-order differences. As shown by Ilgin,[3] the nature of the effect can be described by a simple example: the solution of the differential equation $vU_{xx} + aU_x = 0$ is $U(x) = C_1 + C_2 \exp(-ax/v)$, whereas the solution of its centered second-order approximation is $U_h(kh) = C_1 + C_2[(2v - ah)/(2v + ah)]^k$ and gives very poor approximation if $v = 0(h)$.

There is a finite-difference spatial approximation of the equation[4] that has maximum principle at any value of v and approaches in a good way the rapidly changing part of the solution of the differential equation. The approximation in the simple case of the equation $vU_{xx} + a(x)U_x = 0$ is the following:

$$\frac{ah}{2} \operatorname{cth} \frac{ah}{2v} \frac{U_{m+1} - 2U_m + U_{m-1}}{h^2} + a\frac{U_{m+1} - U_{m-1}}{2h} = 0, \tag{9}$$

and this approximation is easily generalized to multi-dimensional cases.

However, the approximation error[3] involved is of first order if v is sufficiently small, so that it approaches the smooth, slow changing part of the solution in a worse way than the usual centered scheme.

The solution to this dilemma is to use an approximation like Equation 9 in small regions where functions are changing rapidly and to use the usual centered approximation in other regions. For the model Equation 1 (if $a(x) > 0$), it looks like this:

$$\frac{U_m^{n+1} - U_m^n}{\tau} = a\frac{U_{m+1}^n - U_{m-1}^{n+1}}{2h} + \gamma\frac{U_{m+1}^n - 2U_m^{n+1} + U_{m-1}^{n+1}}{h^2}, \quad m_o \geq m > 0, \tag{10}$$

where $\gamma = (ah/2) \operatorname{cth} (ah/2v)$ and Equation 3 for $M > m > m_o$ (if $m = 0, 1, 2 \cdots M$).

The steady-state solution of Equation 1 is the solution of the stationary equation

$$vU_{xx} + aU_x = 0. \tag{11}$$

Let us take, for example, the first boundary problem for Equation 11. Then, it can be proved that if $\exp(-ahm_o/v) < h^2$, the solution of the boundary problem for the difference system (Equations 10′ and 3′) approximates the genuine solution with an accuracy $0(h^2)$. More exactly: if $U^n = U(m, h) = U_m$ and

$$a\frac{U_{m+1} - U_{m-1}}{2h} + \gamma\frac{U_{m+1} - 2U_m + U_{m-1}}{h^2} = 0, \qquad 0 < m \le m_o \qquad \mathbf{(10')}$$

$$a\frac{U_{m+1} - U_{m-1}}{2h} + v\frac{U_{m+1} - 2U_m + U_{m-1}}{h^2} = 0, \qquad m_o < m < M \qquad \mathbf{(3')}$$

$$U(0) = U_0,\ U(Mh) = U(1) = U_1,$$

then $|(U^h - U)| < Ch^2$, where C does not depend on $v (v > 0, a > 0)$.

The Navier–Stokes system for a compressible gas has, in addition to parabolic equations, an equation of first order, namely, the mass-conservation equation. Schemes 5 and 8 for this equation look like the following:

$$\frac{\rho_{m,l}^{n+1} - \rho_{m,l}^{n}}{\tau} = \frac{(\rho U)_{m+1,l}^{n} - (\rho U)_{m-1,l}^{n+1}}{2h} + \frac{(\rho v)_{m,l+1}^{n} - (\rho v)_{m,l-1}^{n+1}}{2h_1}. \qquad \mathbf{(12)}$$

As a linearized model for the equation, we take the equation:

$$\frac{\partial u}{\partial t} = a\frac{\partial u}{\partial x} + b\frac{\partial u}{\partial y},$$

and its difference approximation:

$$\frac{U_{m,l}^{n+1} - U_{m,l}^{n}}{\tau} = a\frac{U_{m+1,l}^{n} - U_{m-1,l}^{n+1}}{2h} + b\frac{U_{m,l+1}^{n} - U_{m,l-1}^{n+1}}{2h_1}.$$

The time process is stable, as can be shown using the Fourier method on the linearized model and as it appeared to be with the genuine Equation 12 during numerical calculations.

However, the same problem of parasitic pulsations arises. This problem, again, does not relate to an instability, and it does not relate to fulfillment of the conservation laws in the difference equations, as some workers think. This effect relates to peculiarities of centered differences, and the nature of the effect can be described by the simple equation[5]

$$U_x + aU = 0, \qquad \mathbf{(13)}$$

where a is sufficiently large.

The general solution of the centered second-order approximation of Equation 13, namely, the solution of $(U_{m+1} - U_{m-1}) + 2ahU_m = 0$, is $U_h = C_1 q_1^m + C_2 q_2^m$, where $q_{1,2} = -ah \pm \sqrt{1 + a^2h^2}$, so $q_1 \simeq 1 - ah + a^2h^2/2$, $q_2 \simeq -(1 + ah) - a^2h^2/2$, and the extra root q_2 spoils the solution.

The way out of this dilemma is the same—to take another kind of spatial approximation of the mass-conservation equation in small areas where the functions are rapidly changing. The other approximation may be of first-order accuracy since it is used in small (two- or three-step) areas; for example, it may be the well-known left or right first-order approximation (according to the sign of the velocity).

For the first boundary problem of Equation 13 (if $a > 0$ and $U(0) = 1$), the use of a first-order approximation $(U_{m+1} - U_m) + ahU_m = 0$ up to $m = m_o$ and a centered one after it gives

$$C_2 = \frac{1 - \sqrt{1 + a^2h^2}}{q_2 - q_1} \left(\frac{1 - ah}{q_2} \right)^{m_o - 1},$$

so that coefficient C_2 for the parasitic root becomes negligibly small, $C_2 = kh^2(1 - ah)^{m_o - 1}$. (If $ah > 1$, one should take an approximation: $(U_m - U_{m-1}) + ahU_m = 0$.)

References

1. Brailovsky, I. 1971. Near-wake flow. Dokl. Akad. Nauk SSSR **197**(3) (in Russian).
2. Brailovsky, I. 1971. Explicit difference methods for solutions of separation of viscous gas flow. Trans. Computer Center Moscow State Univ. **4** (in Russian).
3. Iljin, A. 1969. Difference scheme for differential equation with small parameter at highest derivative. Math. Zametki **6**(2) (in Russian).
4. Buleev, N. I. & V. S. Petritchev. 1966. Numerical method of solution of hydrodynamics equations for plane flow. Dokl. Akad. Nauk SSSR **169**(6) (in Russian).
5. Godunov, S. K. & V. S. Riabenkiy. 1962. Introduction to the Theory of Difference Schemes. Moscow (in Russian).

ON THE INFLUENCE OF MISSING DATA IN A SAMPLE SET ON THE QUALITY OF A STATISTICAL DECISION RULE

Viktor Brailovsky

Moscow, U.S.S.R.

In many practical cases, problems involved in processing statistical data can be considered problems of multivariate statistical analysis. There are many well-studied models of such problems, but, in general, they are studied under the condition that the sample set used to estimate unknown parameters is regular; that is, for each sample, all components of a multivariate random variable are registered. Often, however, some observations may not have been made or were lost; in other words, one has data with missing values. The effect of such sample-set irregularities on the quality of a decision rule, obtained with the help of the sample set, is what I shall discuss in this paper.

In the second section of the paper, I shall consider the case where an approximation function must be obtained with the help of a sample set lacking some values. Under these conditions, it is impossible to use the least mean-squares (LMS) method literally, so I shall consider a modification that can be called an LMS-like method. Some of the new properties of the latter method will be compared with those of the LMS method. One of the problems considered is the following: to obtain an approximation function, one has two sample sets of different sizes and different qualities. In this problem, we measure the quality of sample set with the help of the frequency ratio of the missing values. When is it better to use both methods and when is it better to use only one? Such problems are typical for applications. In the second section, the criterion that answers this question for the LMS-like approximation method is obtained and discussed.

In the third section, the special case of a model of the two-category classifier is considered, and the effect of missing values in the sample set on the quality of a decision rule is investigated. This effect leads to effects similar to those obtained for the LMS-like approximation. Even the criterion that tells when it is better to use sample sets of different size and quality and when it is better to use only one sample set is similar to that which is obtained in the second section. This criterion demonstrates that the results obtained for the LMS-like approximation function are, in fact, of a more general nature.

There is another aspect of this problem. Usually, an optimal decision rule can be described by a formula that depends on some unknown parameters. To obtain the decision rule, one must perform two procedures: estimate the unknown parameters with the help of a sample set and use these estimations instead of the parameters themselves in the formula for the decision rule. In fact, after such a substitution, the decision rule is not the optimal one, and an additional error arises as a result of the substitution. An important point is that one would expect that improvement of the quality of the parameter estimations automatically leads to improvement of the quality of the decision rule obtained with these estimations. However, the quality of the decision rule does not always improve. In this paper, the situation will be considered in which one adds new samples to a sample set and thus improves the statistical estimations of the unknown parameters; at the same time, however, the quality of the decision rule becomes worse. This effect is due to the influence of the irregularities in the sample set.

Least Mean-squares-like Procedure for Approximation of a Function Based on the Use of Experimental Data

1. In the first part of this section, I shall limit myself to the model of function approximation developed in Reference 1. The outline of the model is as follows.

Let both R ($x \in R$) and R' ($\omega \in R'$) be finite-dimensional spaces of arguments.* Let $f(x, \omega)$ be a function of the product of spaces R and R' ($R \times R'$), and let $\mu(x, \omega)$ be a probability distribution on ($R \times R'$). Let us define the probability distribution on R by

$$\mu(x) = \sum_{\omega \in R'} \mu(x, \omega). \tag{2.1}$$

Here and later we use the symbol $\sum$ to denote the operations of summation and integration. The problem is to approximate the function $f(x, \omega)$ (by an optimal way in a sense) using experimental data. In this section, we will consider the mean squared error criterion of approximation. Let us denote

$$f(x) = \sum_{\omega \in R'} f(x, \omega)\mu(\omega/x), \tag{2.2}$$

$$f(x, \omega) = f(x) + \delta(x, \omega); \tag{2.3}$$

that is, for each $x \in R$, $f(x)$ is the mean value of $f(x, \omega)$ over space R'. $\mu(\omega/x)$ denotes the conditional-probability distribution of ω if x is fixed.

Let $\phi_1(x), \phi_2(x), \ldots, \phi_n(x)$ be a set of n linearly independent functions such that $f(x)$, defined by Equation 2.2, is a linear combination of these functions:

$$f(x) = \sum_{i=1}^{n} a_i \phi_i(x). \tag{2.4}$$

As was shown in Reference 1, the problem of finding the best approximation to the function $f(x, \omega)$ with the help of the system $\{\phi_i(x)\}$ comes down to the problem of finding the best approximation to the function $f(x)$; that is, taking into account Equation 2.4, one has to find the corresponding values of the coefficients $\{a_i\}$. To find these values, let us denote

$$\left.\begin{aligned} P_{ij} &= \sum_{x \in R} \phi_i(x)\phi_j(x)\mu(x) \\ C_j &= \sum_{x \in R} \sum_{\omega \in R'} f(x, \omega)\phi_j(x)\mu(x, \omega) \\ &= \sum_{x \in R} f(x)\phi_j(x)\mu(x) \end{aligned}\right\}. \tag{2.5}$$

From Equations 2.4 and 2.5, it is easy to obtain the following system of equations:

$$\sum_{i=1}^{n} a_i P_{ij} = C_j; \qquad j = 1, 2, \ldots, n. \tag{2.6}$$

The matrix of the Equations 2.6 $\|P_{ij}\|$ is Gram's matrix for the system of linearly independent functions $\{\phi_i(x)\}$; as a result, $d = \det\|P_{ij}\| > 0$. According to Cramer's rule, one obtains

$$a_i = \frac{d_i}{d}; \qquad i = 1, 2, \ldots, n. \tag{2.7}$$

* R denotes the space of arguments (parameters) that are available to the observer, and R' the space of parameters for which information is inaccessible to the observer.

Here, d_i is the determinant obtained from determinant d after replacing the ith column in it by a column of the constant terms of Equations 2.6: $C_1, C_2, \ldots, C_n$. From Equations 2.4 and 2.7, it follows that

$$f(x) = \sum_{i=1}^{n} \frac{d_i}{d} \phi_i(x). \tag{2.8}$$

Now let us determine what kind of information we need to approximate f(x). In Reference 1, I considered the following information.

Let $x^1, x^2, \ldots, x^N$ be N independently drawn samples from space R according to the probability distribution $\mu(x)$. Let us assume that the values of x^i are known exactly. For each x^i, let us draw a sample ω^i from space R' according to the conditional-probability distribution $\mu(\omega/x^i)$. The value of ω^i is unknown, but as a result of the procedure, we know the values of f(x^i, ω^i). We therefore have the set of N sample pairs x^1, f(x^1, ω^1); x^2, f(x^2, ω^2); ..., x^N, f(x^N, ω^N), which we denote by Q_N.

With such information, we can estimate the unknown values of P_{ij} and C_i in Equation 2.5 and use the estimations instead of the parameters themselves in Equations 2.6 and 2.7. We thus use the following estimators:

$$\left.\begin{aligned} \hat{C}_j &= \frac{1}{N} \sum_{i=1}^{N} f(x^i, \omega^i)\phi_j(x^i); \qquad j = 1, 2, \ldots, n \\ \hat{P}_{ij} &= \frac{1}{N} \sum_{r=1}^{N} \phi_i(x^r)\phi_j(x^r); \qquad i, j = 1, 2, \ldots, n \end{aligned}\right\}. \tag{2.9}$$

From these estimations, we obtain the determinants $\hat{d}$ and $\hat{d}_i$ ($i = 1, 2, \ldots, n$) instead of d and d_i.

We therefore obtain the following approximation function:

$$\hat{f}(Q_N, x) = \sum_{i=1}^{n} \hat{a}_i \phi_i(x) = \sum_{i=1}^{n} \frac{\hat{d}_i}{\hat{d}} \phi_i(x). \tag{2.10}$$

As is well known,[2] the same approximation can be obtained with the help of the LMS method using the sample set Q_N. We shall call Formula 2.10 the LMS approximation.

The information needed to calculate f(Q_N, x) in Equation 2.10 may be represented in the form of a matrix: n columns of this matrix consist of sample values of n functions $\phi_1(x), \phi_2(x), \ldots, \phi_n(x)$; we shall call them columns of argument, so that the (n + 1)th column of the matrix consists of sample values of the function f(x, ω), which is a column of function. Each line of the matrix corresponds to a sample vector: $\phi_1(x^i), \phi_2(x^i), \ldots, \phi_n(x^i)$; f($x^i$, ω^i), where $i = 1, 2, \ldots, N$; we shall call this matrix the information matrix.

2. Now let us consider the following model of missing data in the information matrix. Let us assume that the λth part of the elements of each column of argument consists of missing values. The column of function has no missing values: $0 \leq \lambda < 1$. Now let the information matrix have $N/1 - \lambda$ lines (instead of N ones as in the first part of this section); each column of argument therefore consists of N data and $\lambda/1 - \lambda\ N$ missing values. Distributions of the missing values within different columns of argument are independent.

Now let us consider the problem of how to obtain a function that approximates f(x, ω) well enough with respect to the mean squared-error criterion. It is clear that it is impossible to use the LMS principle, since the missing data do not allow us to perform the necessary calculations. We shall discuss two ways to obtain the approximation function. The first one is the following: let us exclude from the information matrix all the lines (observations) that contain one or more missing values. As a result, we obtain only $(1-\lambda)^{n-1} \cdot N$ lines in the information matrix. We can find the LMS approximation on the basis of these data. The second way suggests how we can use the observations with missing values. For the case without missing data (first part of this section), we considered the LMS approximation in the form of Equations 2.10 with the elements of determinants in Equations 2.9. Now we cannot calculate the estimators in Equations 2.9, but, instead, we can calculate other estimators of the same values

$$\hat{C}_j = \frac{1}{N} \sum_{i=1}^{N} f(x^i, \omega^i)\phi_j(x^i); \qquad j = 1, 2, \ldots, n, \tag{2.11}$$

$$\hat{P}_{jj} = \frac{1}{N} \sum_{i=1}^{N} \phi_j^2(x^i); \qquad j = 1, 2, \ldots, n. \tag{2.12}$$

The sums in Equations 2.11 and 2.12 are extended over N lines of the information matrix where column of argument ϕ_j contains no missing values:

$$\hat{P}_{ij} = \frac{1}{N(1-\lambda)} \sum_{r=1}^{N(1-\lambda)} \phi_i(x^r)\phi_j(x^r); \qquad \begin{matrix} i, j = 1, 2, \ldots, n \\ i \neq j \end{matrix} . \tag{2.13}$$

The sum in Equation 2.13 is extended over $(1-\lambda) \cdot N$ lines of the information matrix where columns ϕ_i and ϕ_j contain no missing values.

Let us now consider the determinants $\hat{d}$ and $\hat{d}_i$ $(i = 1, 2, \ldots, n)$ with the structure described above (see Equations 2.9 and 2.10 and the corresponding text) but with the elements in Equations 2.11–2.13 instead of Equation 2.9. Let us determine an approximation function of the form

$$\hat{f}(Q_N, \lambda, x) = \sum_{i=1}^{n} \hat{a}_i \phi_i(x) = \sum_{i=1}^{n} \frac{\hat{d}_i}{\hat{d}} \phi_i(x). \tag{2.14}$$

Here $\hat{d}$ and $\hat{d}_i$ are merely the determinants described above with the elements in Equations 2.11–2.13. Unlike Formula 2.10, the approximation in Equations 2.14 is not a result of the application of the LMS method to experimental data. When $\lambda = 0$, the approximation in Equation 2.14 turns into the LMS approximation in Equation 2.10. If $N \to \infty$ from Equations 2.11–2.13 and 2.5, it follows that $\hat{c}_j \to c_j$; $\hat{p}_{ij} \to p_{ij}$.

Taking into account the similarity of the structures of Equations 2.10 and 2.14, we shall call Equations 2.14 the LMS-like approximation for the case with missing values.

3. Now we shall consider the quality of the LMS-like approximation in Equation 2.14 as compared with that of the LMS-approximation in Equation 2.10. The calculations and full analysis have been given.[3]

We are therefore now interested in estimation of the value:

$$R_\lambda = \mathscr{E} \sum_{x \in R} [\hat{f}(Q_N, \lambda, x) - f(x)]^2 \mu(x). \tag{2.15}$$

Here and later we use $\mathscr{E}$ to denote averaging over all possible sample sets with the size and structure described in the second part of this section. Let us consider R_λ as a function of c_i $(i = 1, 2, \ldots, n)$ and p_{ij} $(i \geq j)$, where $i, j = 1, 2, \ldots, n$,† and we will

† We consider $\hat{p}_{ij}$ when $i \geq j$, because $\hat{p}_{ij} = \hat{p}_{ji}$ (and $p_{ij} = p_{ji}$).

represent R_λ in the form of Taylor's formula with the center point c_i ($i = 1, 2, \ldots, n$) and p_{ij} ($i, j = 1, 2, \ldots, n$; $i \geq j$):

$$\begin{aligned} R_\lambda &= \mathscr{E} \sum_{x \in R} [\hat{f}(Q_N, \lambda, x) - f(x)]^2 \mu(x) \\ &= \mathscr{E} \sum_{x \in R} \left\{ \left[\sum_{i=1}^{n} \frac{\partial \hat{f}}{\partial \hat{c}_i} (\hat{c}_i - c_i) + \sum_{\substack{i, j \\ i \geq j}} \frac{\partial \hat{f}}{\partial \hat{p}_{ij}} (\hat{p}_{ij} - p_{ij}) \right] + R_1 \right\}^2 \mu(x) \\ &= \sum_{x \in R} \mathscr{E}[\ ]^2 \mu(x) + 2 \sum_{x \in R} \mathscr{E}[\ ] \cdot R_1 \mu(x) + \sum_{x \in R} \mathscr{E} R_1^2 \mu(x) \\ &= \sum' + \sum'' + \sum'''. \end{aligned} \qquad \textbf{(2.16)}$$

R_1 denotes the remainder; the partial derivatives written out in Equations 2.16 are to be calculated in the center point of the expansion, and therefore they do not depend on the sample variables. By performing the averaging operations $\mathscr{E}$ for the functions that depend on the samples and taking into account that the estimators in Equations 2.11–2.13 are unbiased and that the value of $f(Q_N, \lambda, x)$ in the center point equals $f(x)$ (see Equation 2.8), one obtains the following expression for the term $\sum'$ from Equations 2.16:

$$\begin{aligned} \sum' &= \sum_{i=1}^{n} \left[\sum_{x \in R} \left(\frac{\partial \hat{f}}{\partial \hat{c}_i} \right)^2 \mu(x) \right] \operatorname{var} \hat{c}_i + \sum_{\substack{i, j \\ i \geq j}} \left[\sum_{x \in R} \left(\frac{\partial \hat{f}}{\partial \hat{p}_{ij}} \right)^2 \mu(x) \right] \operatorname{var} \hat{p}_{ij} \\ &+ 2 \sum_{\substack{i, j \\ i > j}} \left[\sum_{x \in R} \frac{\partial \hat{f}}{\partial \hat{c}_i} \cdot \frac{\partial \hat{f}}{\partial \hat{c}_j} \mu(x) \right] \operatorname{cov}(\hat{c}_i, \hat{c}_j) \\ &+ 2 \sum_{\substack{i, j, k, 2 \\ i \geq j,\, k \geq 2,\, [(i > k) \vee (i = k)(j > r)]}} \left[\sum_{x \in R} \frac{\partial \hat{f}}{\partial \hat{p}_{ij}} \cdot \frac{\partial \hat{f}}{\partial \hat{p}_{kr}} \mu(x) \right] \cdot \operatorname{cov}(\hat{p}_{ij}, \hat{p}_{kr}) \\ &+ 2 \sum_{\substack{i, k, r \\ k \geq r}} \left[\sum_{x \in R} \frac{\partial \hat{f}}{\partial \hat{c}_i} \cdot \frac{\partial \hat{f}}{\partial \hat{p}_{kr}} \mu(x) \right] \operatorname{cov}(\hat{c}_i, \hat{p}_{kr}). \end{aligned} \qquad \textbf{(2.17)}$$

The indexes i, j, k, and r all vary from 1 up to n, and special restrictions are mentioned under each outward symbol $\sum$. For the fourth term on the right-hand side of Equation 2.17, the restrictions are the following: $i \geq j$; $k \geq r$; at the same time, either $i > k$ or $i = k$, $j > r$ simultaneously.

It is easy to prove[3] that the expression $\sum'$ in Equation 2.17 converges to zero with the order $1/N$ (while $N \to \infty$). In Appendix 2 of Reference 1, we proved that terms like $\sum''$ and $\sum'''$ from Equation 2.16 converge to zero with the order no smaller than $1/N^2$ (while $N \to \infty$). Therefore, in estimating the approximation quality, we have only to estimate the term $\sum'$ in Equation 2.17; our calculation being adequately correct for sufficiently large sample sets. In Reference 3, all the calculations are described according to Equation 2.17. Here we shall present the final result.

First of all, let us note that here and later in this section we shall assume that

$$\sum_{x \in R} \delta^2(x, \omega) \mu(\omega/x) = \sigma^2 = \text{const}. \qquad \textbf{(2.18)}$$

This equation is the usual assumption of the regression analysis. In Reference 1, an estimation of the quality of the LMS approximation was obtained using the same approach. With the assumption in Equation 2.18, the quality of the LMS approximation takes the form:

$$R_{\text{LMS}} = \frac{n\sigma^2}{N}; \tag{2.19}$$

as was shown in Reference 3, the quality of the LMS-like approximation takes the form:

$$R_\lambda = R_{\text{LMS}} + \Delta R = \frac{n\sigma^2}{N} + \frac{\lambda V}{N} + \frac{\lambda^2 W}{(1-\lambda)N} + \frac{\lambda\sigma^2 S^2}{N}.\ddagger \tag{2.20}$$

I shall not use here the exact expressions for the parameters V, W, and S^2; only the following properties will be given, which were proved in Reference 3:

(1) $V \geq 0$, $W \geq 0$, and $S^2 \geq 0$, so $\Delta R \geq 0$. Therefore, the LMS-like approximation is not better (in general, it is worse) than the LMS one, which uses a sample set of size N without missing data.

(2) Let us consider the case where a system of the function $\{\phi_i(x)\}$ is orthogonal (we define the scalar product of the functions $\phi_i(x)$ and $\phi_j(x)$ by the formula for p_{ij} [Equation 2.5]). It means that the matrix $\|p_{ij}\|$ is diagonal. For this case, $S^2 = 0$, and the last component in the right-hand side of Equation 2.20 equals zero as well. Moreover, let us assume that the functions $\{\phi_i\}$ are such that $\phi_i(x) \cdot \phi_j(x) \equiv 0$ $(i \neq j)$. Such a situation takes place if, for example, $\phi_i(x)$ are the characteristic functions of some mutual disjoint sets (i.e., the sets that have no common points) in space R. For this case, $V = 0$ and $W = 0$. Therefore, for this case, $R_\lambda = R_{\text{LMS}}$.

(3) In the case of the LMS approximation with the condition that Equation 2.4 holds, the quality of the approximation in Equation 2.19 does not depend on the choice of a function system $\{\phi_i\}$. For the case of the LMS-like approximation, the converse is true. As mentioned above, there are systems of functions in which $V = 0$, $W = 0$, $S^2 = 0$, and $R_\lambda = R_{\text{LMS}}$. However, with the help of a linear transformation, one can find another system of function in which $\Delta R \neq 0$. Taking into account that for all these systems $R_{\text{LMS}} = \text{const.}$, it follows that for different systems of functions, one obtains different qualities of the LMS-like approximations. Hence, for the case with missing data, the choice of the function system is important for obtaining an approximation of good quality. Let us note that this choice must be performed before the sample set is obtained, because it is impossible to perform a linear transformation of data with missing values (after such a transformation, the quantity of missing values increases substantially).

‡ I shall give the formulas for the parameters V, W, and S^2, although they are not necessary for the following text:

$$V = \sum_{x \in R} \sum_{r=1}^{n} a_2^2 \text{var}_x \sum_{\substack{k=1 \\ k \neq r}}^{n} \xi_k(x)\phi_k(y)\phi_2(y) \quad \mu(x),$$

$$W = \sum_{\substack{i,j=1 \\ i>j}}^{n} \sum_{x \in R} \{\text{var}_x[(a_j \xi_i(x) + a_i \xi_j(x))\phi_i(y)\phi_j(y)]\}\mu(x),$$

$$S^2 = \sum_{i=1}^{n} p_{ii}(p_{ii}^{-1}) - n.$$

Here, the variance in the right-hand side of the equations for parameters V and W is calculated; provided that x is fixed $\{\xi_i(x)\}$, $i = 1, 2, \ldots, n$ is such a system of functions that their Gram's matrix is the inverse of $\|p_{ij}\|$. p_{ii}^{-1} denotes the element of the inverse matrix of $\|p_{ij}\|$.

In the second part of this section, we mentioned another way of obtaining an approximation when all the lines of the information matrix that contain one or more missing values are excluded. As a result, one has $(1-\lambda)^{n-1} \cdot N$ samples to obtain the LMS approximation. From Equation 2.19, it follows that the quality of such an approximation is:

$$R = \frac{n\sigma^2}{(1-\lambda)^{n-1} \cdot N};$$

it is easy to present examples when this method of approximation is better than the LMS-like method in Equation 2.20 (it is so if σ^2 is relatively small and, on the other hand, if the values of V, W, and S^2 are relatively large) and vice versa (when, for example, $V, W, S^2 = 0$). We shall not give further consideration to this method of approximation because it does not seem to be interesting.

4. Let us say that we have two sample sets with missing data. Let the first one have a frequency ratio of missing values of λ_1 and a sample size of $N/(1-\lambda_1)$. Let the second sample set have those values, respectively, of λ_2 and

$$\frac{KN}{(1-\lambda_2)}; \qquad (K > 0).$$

If one obtains the LMS-like approximation with only the help of the first sample set, its quality may be estimated with the help of Equation 2.20. Let us use both sample sets and calculate the estimations $\hat{c}_i, \hat{p}_{jj}, \hat{p}_{ij}$ $(i \neq j)$ with the help of both sets of data.

We shall use the following estimators:

$$\hat{C}_j = \frac{1}{N(1+K)} \sum_{i=1}^{N(1+K)} f(x^i, \omega^i)\phi_j(x^i); \qquad j = 1, 2, \ldots, n, \tag{2.21}$$

$$\hat{P}_{jj} = \frac{1}{N(1+K)} \sum_{i=1}^{N(1+K)} \phi_j^2(x^i); \qquad j = 1, 2, \ldots, n. \tag{2.22}$$

In Equations 2.21 and 2.22, the sums are extended over $N(1+k)$ lines in the combined information matrix where argument ϕ_j is registered:

$$\hat{P}_{ij} = \frac{1}{N[(1-\lambda_1) + K(1-\lambda_2)]} \sum_{r=1}^{\substack{N[(1-\lambda_1)\\ +K(1-\lambda_2)]}} \phi_i(x^2)\phi_j(x^2); \qquad i > j. \tag{2.23}$$

Here, the sum is extended over $[(1-\lambda_1)N + (1-\lambda_2)KN]$ lines in the combined information matrix where both arguments ϕ_i and ϕ_j are registered. By using the estimators in Equations 2.21–2.23 instead of those in Equations 2.11–2.12 in the expression 2.14, we obtain the LMS-like approximation function for the combined sample set. I shall not provide the details of the calculations (see Reference 3) but note only that here λ_1 and λ_2 are considered to be parameters having small values, and all the calculations are performed with an accuracy of the order of λ_1 or λ_2 (as well as $1/N$, as was done above). We shall neglect terms of higher order. Hence, if we use only the first sample set to obtain the LMS-like approximation, its quality will have the following form:

$$R_{\lambda_1} = \frac{n\sigma^2}{N} + \frac{\lambda_1 V}{N} + \frac{\lambda_1 \sigma^2 S^2}{N}; \tag{2.24}$$

if we use the combined sample set, the quality of the LMS-like approximation (as was shown in Reference 3) takes the form:

$$R_{\lambda_1\lambda_2} = \frac{n\sigma^2}{N(1+K)} + \frac{(\lambda_1 + K\lambda_2)V}{N(1+K)^2} + \frac{(\lambda_1 + K\lambda_2)\sigma^2 S^2}{N(1+K)^2}. \qquad \textbf{(2.25)}$$

From Equations 2.24 and 2.25, it follows that we obtain a better result when we use the combined sample set than when we use only the first sample set, if $R_{\lambda_1\lambda_2} < R_{\lambda_1}$; that is,

$$K > \frac{(V + \sigma^2 S^2)(\lambda_2 - 2\lambda_1) - n\sigma^2}{n\sigma^2 + \lambda_1(V + \sigma^2 S^2)}. \qquad \textbf{(2.26)}$$

Let us discuss some properties of this result.

(1) If $V = 0$ and $S^2 = 0$ (we discussed this case in the remarks after Equation 2.20), Equation 2.26 takes the form $K > -1$. In other words, in this case the use of the combined sample set leads to a better result for any value of K.

(2) As was mentioned, $V \geq 0$ and $S^2 \geq 0$, thus the denominator of the right-hand side of Equation 2.26 is always positive. Hence, if the numerator of this equation is nonpositive, the use of the combined sample set leads to a better result for any value of K. This situation certainly takes place if $\lambda_2 - 2\lambda_1 \leq 0$, in other words when the second sample set is of better (or the same) quality than the first one (i.e., $\lambda_2 \leq \lambda_1$) or when the second sample set is of worse quality than the first one but not of too bad quality (i.e., $2\lambda_1 \geq \lambda_2 > \lambda_1$). A more precise value of the constant, which would show to what extent the parameters λ_1 and λ_2 may differ from each other, for this situation still takes place, depends on the values of V, S^2, σ^2, and n, as can be seen from Equation 2.26.

(3) If λ_2 is greater than λ_1, in accordance with Equation 2.26, one obtains $K > K^* > 0$. In other words, the use of the second sample set, including parameter λ_2 and a sample size larger than K^*N, as well as the first sample set leads to a better result; however, if one uses the second sample set having a size smaller than K^*N, one obtains a worse result than one obtains using only the first sample set.

A Model of a Linear Two-Category Classifier

1. In this section, we shall see that the effect of missing data on the quality of the decision rule, described for the LMS-like procedure in the previous section, also takes place (under certain conditions) in a model of a two-category classifier. Here, we shall briefly consider a special case rather than the general solution of the problem.

Let both categories be described by the multivariate normal density, with the covariance matrices $\|p_{ij}\|$ identical and nonsingular. Let $\mu_1(\mu_{11}, \mu_{12}, \ldots, \mu_{1n})$ and $\mu_2(\mu_{21}, \mu_{22}, \ldots, \mu_{2n})$ denote the n-component mean vectors for the first and second categories, respectively. The normal densities described by these categories are written as

$$P_i(y_1, y_2, \ldots, y_n) = \frac{1}{(2\pi)^{n/2}|\lambda|^{1/2}} \times \exp\left[-\frac{1}{2|\lambda|}\sum_{m,r=1}^{n} |\lambda_{mr}|(y_m - \mu_{im})(y_r - \mu_{ir})\right]; \qquad i = 1, 2, \qquad \textbf{(3.1)}$$

where $|\lambda| = \det\|P_{ij}\|$; $|\lambda_{mr}|$ is the cofactor of element P_{mr} of the covariance matrix.

Let us suppose that probabilities P_1 and P_2, that is, nature in state 1 or 2, respectively, are equal to $\frac{1}{2}$.

For this case, as is well known,[4] classification of the minimum error rate may be achieved by use of the discriminant function

$$g(y) = \sum_{m,r=1}^{n} |\lambda_{mr}|[(\mu_{2r} - \mu_{1r})y_m + (\mu_{2m} - \mu_{1m})y_r + (\mu_{1m}\mu_{1r} - \mu_{2m}\mu_{2r})]. \quad \textbf{(3.2)}$$

The decision rule has the following form: if $g(y) > 0$, $y \in R_1$; otherwise, $y \in R_2$, with R_1 and R_2 being the decision regions into which the feature space is divided. The equation $g(y) = 0$ describes the decision boundary; in this case, it is a hyperplane. The equation of the hyperplane may be written as

$$y_1 = \frac{-\sum_{m=2}^{n}\left[\sum_{r=1}^{n}|\lambda_{mr}|(\mu_{2r} - \mu_{1r})\right]y_m}{\sum_{r=1}^{n}|\lambda_{1r}|(\mu_{2r} - \mu_{1r})} + \frac{\sum_{m,r=1}^{n}|\lambda_{mr}|(\mu_{2m}\mu_{2r} - \mu_{1m}\mu_{1r})}{2\sum_{r=1}^{n}|\lambda_{1r}|(\mu_{2r} - \mu_{1r})}. \quad \textbf{(3.3)}$$

Here and later we shall assume that

$$\mu_{1i} = 0 (i = 1, 2, \ldots, n); \qquad \mu_{21} > 0; \qquad \mu_{2i} = 0 (i = 2, 3, \ldots, n). \quad \textbf{(3.4)}$$

When we take into account Equation 3.4, Equation 3.3 takes the form:

$$y_1 = -\sum_{m=2}^{n} \frac{|\lambda_{m1}|}{|\lambda_{11}|} y_m + \frac{\mu_{21}}{2} = \sum_{m=2}^{n} a_m y_m + c. \quad \textbf{(3.5)}$$

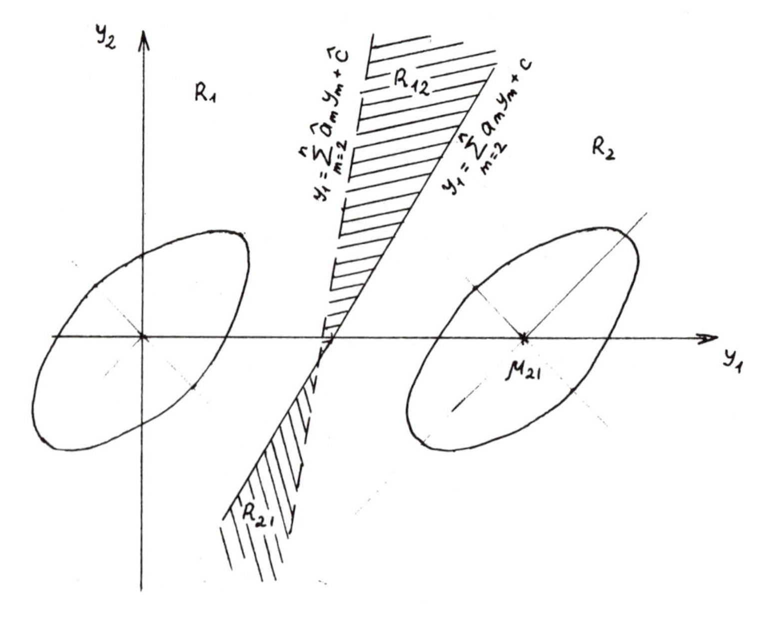

FIGURE 1.

2. Let us note that this paper is concerned with the case when the covariance matrix is known, but the mean vectors μ_1 and μ_2 are unknown and must be estimated with the help of a sample set. Hence, to obtain a decision rule, we use Equation 3.3, statistical estimations of the parameters μ_{ij} $(i = 1, 2; j = 1, 2, \ldots, n)$ instead of the parameters themselves, as the equation of the boundary hyperplane. It takes the form:

$$y_1 = \frac{-\sum_{m=2}^{n}\left[\sum_{r=1}^{n} |\lambda_{mr}|(\hat{\mu}_{2r} - \hat{\mu}_{1r})\right] y_m}{\sum_{r=1}^{n} |\lambda_{1r}|(\hat{\mu}_{2r} - \hat{\mu}_{1r})} + \frac{\sum_{m,r=1}^{n} |\lambda_{mr}|(\hat{\mu}_{2m}\hat{\mu}_{2r} - \hat{\mu}_{1m}\hat{\mu}_{1r})}{2\sum_{r=1}^{n} |\lambda_{1r}|(\hat{\mu}_{2r} - \hat{\mu}_{1r})}$$

$$= \sum_{m=2}^{n} \hat{a}_m y_m + \hat{c}. \qquad \textbf{(3.6)}$$

As a result, we obtain a decision rule that differs from the classification of the minimum error rate (Equation 3.3). As shown by FIGURE 1, there are two regions of misclassification as compared with the case of the minimum error rate, R_{12}, where instead of the optimal classification R_1, we obtain R_2 and R_{21}, and where instead of R_2, we get R_1.

It is clear that the classification performed with the help of the boundary hyperplane in Equation 3.6 leads to an additional error rate of the form:

$$\mathrm{f}(\hat{\mu}_1, \hat{\mu}_2) = \frac{1}{2}\int_{R_{21}} [N(y_1, y_2, \ldots, y_n; \mu_2; \lambda) - N(y_1, y_2, \ldots, y_n; \mu_1; \lambda)]\, dy_1, \ldots, dy_n$$

$$+ \frac{1}{2}\int_{R_{12}} [N(y_1, y_2, \ldots, y_n; \mu_1; \lambda) - N(y_1, y_2, \ldots, y_n; \mu_2; \lambda)]\, dy_1, \ldots, dy_n$$

$$= \frac{1}{2}\int_{-\infty}^{\infty} \cdots \int_{-\infty}^{\infty} \left\{\int_{\sum_{m=2}^{n} a_m y_m + c}^{\sum_{m=2}^{n} \hat{a}_m y_m + c} [N(y_1, y_2, \ldots, y_n; \mu_2; \lambda) - N(y_1, y_2, \ldots, y_n; \mu_1; \lambda)]\, dy_1\right\} dy_2, \ldots, dy_n, \qquad \textbf{(3.7)}$$

where $N(y_1, y_2, \ldots, y_n; \mu_i; \lambda)$ stands for the normal density with the mean vector μ_i and the covariance matrix $\|P_{ij}\|$. The factor $\frac{1}{2}$ represents the probabilities $P_1 = P_2 = \frac{1}{2}$.

Further, we shall consider only the additional error rate in Equation 3.7; its value depends on statistical estimations, irregularities in sample sets, and so on. At the same time, the error rate of the optimal decision rule in Equation 3.3 cannot be changed, and we shall not consider it.

First of all, let us note that Equation 3.7 is a function of the upper limit of the integral over values of variable y_1. Because the upper limit depends on estimations $\hat{\mu}_1$ and $\hat{\mu}_2$, Equation 3.7 is a function $\mathrm{f}(\hat{\mu}_1, \hat{\mu}_2)$. Let us represent the function $\mathrm{f}(\hat{\mu}_1, \hat{\mu}_2)$ in Equation 3.7 in the form of Taylor's formula with the center point μ_1, μ_2:

$$
\begin{aligned}
f(\hat{\mu}_1, \hat{\mu}_2) = & \sum_{i=1,2} \sum_{j=1}^{n} \frac{\partial f}{\partial \hat{\mu}_{ij}} (\hat{\mu}_{ij} - \mu_{ij}) \\
& + \frac{1}{2} \Bigg\{ \sum_{i=1,2} \sum_{j=1}^{n} \frac{\partial^2 f}{\partial \hat{\mu}_{ij}^2} (\hat{\mu}_{ij} - \mu_{ij})^2 \\
& + 2 \Bigg[\sum_{i=1,2} \sum_{\substack{j,k=1 \\ j>k}}^{n} \frac{\partial^2 f}{\partial \hat{\mu}_{ij}\, \partial \hat{\mu}_{ik}} (\hat{\mu}_{ij} - \mu_{ij})(\hat{\mu}_{ik} - \mu_{ik}) \\
& + \sum_{j,k=1}^{n} \frac{\partial^2 f}{\partial \hat{\mu}_{ij}\, \partial \hat{\mu}_{2k}} (\hat{\mu}_{ij} - \mu_{ij})(\hat{\mu}_{2k} - \mu_{2k}) \Bigg] \Bigg\} + R_2,
\end{aligned} \tag{3.8}
$$

where R_2 stands for remainder. All the partial derivatives written in Equation 3.8 are calculated in the center point μ_1, μ_2. We are not interested in calculating the value of the additional error rate in Equations 3.7 and 3.8 averaged over all possible sample sets with a fixed size. By averaging both sides of Equation 3.8, one obtains

$$
\begin{aligned}
\mathscr{E} f(\hat{\mu}_1, \hat{\mu}_2) = & \sum_{i=1,2} \sum_{j=1}^{n} \frac{\partial f}{\partial \hat{\mu}_{ij}} \mathscr{E}(\hat{\mu}_{ij} - \mu_{ij}) \\
& + \frac{1}{2} \Bigg\{ \sum_{i=1,2} \sum_{j=1}^{n} \frac{\partial^2 f}{\partial \hat{\mu}_{ij}^2} \operatorname{var} \hat{\mu}_{ij} \\
& + 2 \Bigg[\sum_{i=1,2} \sum_{\substack{j,k=1 \\ j>k}}^{n} \frac{\partial^2 f}{\partial \hat{\mu}_{ij}\, \partial \hat{\mu}_{ik}} \operatorname{cov}(\hat{\mu}_{ij}, \hat{\mu}_{ik}) \\
& + \sum_{j,k=1}^{n} \frac{\partial^2 f}{\partial \hat{\mu}_{ij}\, \partial \hat{\mu}_{2k}} \operatorname{cov}(\hat{\mu}_{ij}, \hat{\mu}_{2k}) \Bigg] \Bigg\} + \mathscr{E} R_2.
\end{aligned} \tag{3.9}
$$

Here, we take into account the fact that we shall use only the unbiased statistical estimations $\hat{\mu}_{ij}$. It follows that the first sum in the right-hand side of Equation 3.9 (which is outside the brackets) equals zero.

3. To calculate the values of variances and covariances from Equation 3.9, we must describe sample sets used to obtain the estimations $\hat{\mu}_{ij}$ ($i = 1, 2$; $j = 1, 2, \ldots, n$). Let us first consider sample sets of regular structure.

We shall use two sample sets. The first one consists of N independently drawn samples from the first category according to the probability density $P_1(y_1, \ldots, y_n) = N(y_1, \ldots, y_n; \mu_1; \lambda)$: $(y_1^1, y_2^1, \ldots, y_n^1)$, $(y_1^2, y_2^2, \ldots, y_n^2)$, $\ldots$, $(y_1^N, y_2^N, \ldots, y_n^N)$. The second one is drawn in the same way from the second category with the probability density $P_2(y_1, \ldots, y_n) = N(y_1, \ldots, y_n; \mu_2; \lambda)$: $(\tilde{y}_1^1, \tilde{y}_2^1, \ldots, \tilde{y}_n^1)$, $(\tilde{y}_1^2, \tilde{y}_2^2, \ldots, \tilde{y}_n^2)$, $\ldots$, $(\tilde{y}_1^N, \tilde{y}_2^N, \ldots, \tilde{y}_n^N)$.

We shall consider the following unbiased statistical estimators:

$$
\hat{\mu}_{1i} = \frac{1}{N} \sum_{j=1}^{N} y_i^j; \qquad \hat{\mu}_{2i} = \frac{1}{N} \sum_{j=1}^{N} \tilde{y}_i^j; \qquad i = 1, 2, \ldots, n. \tag{3.10}
$$

It is well known[2] that for our case,

$$
\operatorname{var} \hat{\mu}_{1i} = \operatorname{var} \hat{\mu}_{2i} = \frac{P_{ii}}{N}; \qquad i = 1, 2, \ldots, n, \tag{3.11}
$$

$$
\operatorname{cov}(\hat{\mu}_{1i}, \hat{\mu}_{2j}) = 0; \qquad i, j = 1, 2, \ldots, n, \tag{3.12}
$$

$$
\operatorname{cov}(\hat{\mu}_{ik}, \hat{\mu}_{il}) = \frac{P_{kl}}{N}; \qquad \begin{matrix} i = 1, 2 \\ k, l = 1, 2, \ldots, n \end{matrix}. \tag{3.13}
$$

From Equations 3.11–3.13, it follows that the terms inside the braces in Equation 3.9 are of $1/N$ order. It is easy to show that the term R_2 in Equation 3.9 is of $1/N^2$ order. Hence, after calculating the terms inside the braces in Equation 3.9, we obtain the value of the additional error rate $\mathscr{E}\mathrm{f}(\hat{\mu}_1, \hat{\mu}_2)$ with an accuracy of a magnitude of $1/N$ order.

From Equation 3.12, it follows that the second sum inside the brackets in Equation 3.9 equals zero. To obtain this result, we must calculate the first sum inside the braces (which we shall call the sum of variance terms) and the first sum inside the brackets (which we shall call the sum of covariance terms). The final result of these calculations has the form:

$$\mathscr{E}\mathrm{f}(\hat{\mu}_1, \hat{\mu}_2) = \frac{1}{4N}\left[\frac{D}{2} + \frac{2(n-1)}{D}\right]\phi\left(-\frac{D}{2}\right), \tag{3.14}$$

where D is the Mahalanobis distance;[4] in our case, $D^2 = \mu_{21}^2 |\lambda_{11}|/|\lambda|$; and $\phi(-D/2)$ is the univariable normal density $N(0, 1)$ in the point $(-D/2)$.

Equation 3.14 was given by Okamoto.[5] In this paper, we shall study separately the contributions of the variance terms and the covariance terms from Equation 3.9 into the final formula. In this way, we will have an opportunity to estimate the influence of the missing data in the sample sets on the value of the additional error rate $\mathscr{E}\mathrm{f}(\hat{\mu}_1, \hat{\mu}_2)$. For our case,

$$\mathscr{E}\mathrm{f}(\hat{\mu}_1, \hat{\mu}_2) = \frac{V}{N} + \frac{C}{N}, \tag{3.15}$$

where V/N is the contribution of the variance terms and C/N is that of the covariance ones.

Let us present the formulas for V and C, although they are not necessary for further considerations:

$$V = R\left\{\frac{P_{11}}{4} + \sum_{r=2}^{n}\left[\frac{q_{22}}{\mu_{21}^2(P_{11}^{-1})^2} + \frac{1}{4}\frac{|\lambda_{12}|^2}{|\lambda_{11}|^2}\right]P_{rr}\right\};$$

$$C = R\left\{\frac{1}{2|\lambda_{11}|}\sum_{k=2}^{n}|\lambda_{lk}|P_{lk} + 2\sum_{\substack{r,k=2\\ r>k}}^{n}\left[\frac{q_{kr}}{\mu_{21}^2(P_{11}^{-1})^2} + \frac{1}{4}\frac{|\lambda_{1r}|\,|\lambda_{lk}|}{|\lambda_{ll}|^2}\right]P_{kr}\right\}.$$

Here
$$R = \frac{1}{2}\frac{|\lambda_{ll}|^{3/2}\mu_{21}}{(2_\pi)^{1/2}|\lambda|^{3/2}}\exp\left[-\frac{\mu_{21}^2|\lambda_{ll}|}{8|\lambda|}\right];$$

$$P_{ij}^{-1} = \frac{|\lambda_{ij}|}{|\lambda|}; \qquad q_{ij} = P_{ij}^{-1} - \frac{P_{1i}^{-1}\cdot P_{1j}^{-1}}{P_{11}^{-1}}.$$

It is possible to prove that if $\|P_{ij}\|$ is a nonsingular matrix, $V > 0$ and $C \leq 0$ ($C = 0$ if, and only if, $\|P_{ij}\|$ is a diagonal matrix). The fact that $C \leq 0$ was proved for the case when Equations 3.4 apply.

4. Let us now consider the case where the sample sets have missing values. We shall use two sample sets, as described at the beginning of the third part of this section. However, unlike that case, the sample size of each set is now $N/1 - \lambda$; $(0 \leq \lambda \leq 1)$; for each variable, the value is missing in $\lambda[N/(1-\lambda)]$ samples for each sample set. Hence, for each sample set, each variable is registered in only $\{[N/(1-\lambda)] - \lambda[N/(1-\lambda)] = N\}\cdot N$ samples, or in just the same number of samples as was considered above. Let these missing values be distributed randomly and independently for the different variables.

To estimate the unknown mean vectors, we shall again use Equations 3.10; summation in these equations is performed over all registered observations. Equations 3.11 and 3.12 are correct for our case as well, and the contribution of the variance terms in Equation 3.9 equals V/N, as for the previous cases. To obtain a new formula for $\operatorname{cov}(\hat{\mu}_{ik}, \hat{\mu}_{il})$; ($i = 1, 2$; $k \neq 1$), we must take into account the fact that for any pair of variables for each sample set, there are $(l - \lambda) \cdot N$ samples in which both values are registered. It follows that

$$\operatorname{cov}(\hat{\mu}_{ik}, \hat{\mu}_{il}) = \frac{(1 - \lambda)P_{kl}}{N}; \qquad \begin{array}{l} i = 1, 2 \\ k, l = 1, 2, \ldots, n. \\ k \neq l \end{array} \tag{3.16}$$

By comparing Equations 3.13 and 3.16 and taking into account the form of Equation 3.9, we obtain for our case:

$$\mathscr{E}f(\hat{\mu}_1, \hat{\mu}_2) = \frac{V}{N} + \frac{C}{N}(1 - \lambda). \tag{3.17}$$

Hence, even though the unknown mean vectors are estimated with the help of the same estimators in Equations 3.10, as was done in the third part, the additional error rate in our case is, in general, greater than in the case described in the third part. This greater error rate is due to the effect of the missing data. Both Equations 3.17 and 3.15 are identical if the matrix $\|P_{ij}\|$ is diagonal (in this case, $C = 0$; for our case, $C < 0$).

5. Now, we shall use two sample sets from the first and second categories, respectively, of the kind described in the fourth part of this section. Let the frequency ratio of missing values for the sample sets equal λ_1 and the sample size of each set be $N/1 - \lambda_1$.

Let us take two more sample sets from the first and second categories, respectively. Let these sets be of the same kind and have a frequency ratio of missing values λ_2 and a sample size of each set $KN/1 - \lambda_2$; $K > 0$. If we use only the former sample sets to obtain a decision rule, we get the additional error rate value from Equation 3.17 as:

$$\Delta' = \mathscr{E}f(\hat{\mu}_1, \hat{\mu}_2) = \frac{V}{N} + \frac{C}{N}(1 - \lambda_1). \tag{3.18}$$

Now let us combine the former and the latter sample sets and estimate the unknown mean vectors from the combined data. Hence, instead of the estimators in Equations 3.10, we shall use the following ones:

$$\left.\begin{array}{l} \hat{\mu}_{1i} = \dfrac{1}{N(1 + K)}\left[\displaystyle\sum_{j=1}^{N} y_i^j + \sum_{j=N+1}^{N(1+K)} y_i^j\right] \\ \hat{\mu}_{2i} = \dfrac{1}{N(1 + K)}\left[\displaystyle\sum_{j=1}^{N} \tilde{y}_i^j + \sum_{j=N+1}^{N(1+K)} \tilde{y}_i^j\right] \end{array}\right\} i = 1, 2, \ldots, n. \tag{3.19}$$

In these formulas, the first summation is performed over these samples from the former sample sets where the ith variable is registered; the second summation is performed over the samples from the latter sample sets.

It is easy to obtain the following formulas for the estimators in Equations 3.19:

$$\operatorname{var} \hat{\mu}_{1i} = \operatorname{var} \hat{\mu}_{2i} = \frac{P_{ii}}{N(1 + K)}; \qquad i = 1, 2, \ldots, n. \tag{3.20}$$

Hence, the variances of the estimators in Equations 3.19 are less than those of the respective estimators in Equations 3.10:

$$\operatorname{cov}(\hat{\mu}_{1i}, \hat{\mu}_{2j}) = 0; \qquad i, j = 1, 2, \ldots, n. \tag{3.21}$$

$$\operatorname{cov}(\hat{\mu}_{it}, \hat{\mu}_{il}) = \frac{(1 + K) - (\lambda_1 + K\lambda_2)}{N(1 + K)^2} P_{tl}; \qquad \begin{matrix} i = 1, 2 \\ t, l = 1, 2, \ldots, n. \\ t \neq l \end{matrix} \tag{3.22}$$

If we take into account Equations 3.9, 3.11–3.13, and 3.15 as well as 3.20–3.22, we obtain the formula for the additional error rate value for our case with the combined data:

$$\Delta_m = \frac{V}{N(1 + K)} + \frac{(1 + K) - (\lambda_1 + K\lambda_2)}{(1 + K)^2} \frac{C}{N}. \tag{3.23}$$

If we compare Equations 3.23 and 3.17, we come to the conclusion that the use of combined data leads to a better result than does the use of only the former sample sets, when one of the following inequalities holds:

$$\Delta_m - \Delta' < 0.$$

or

$$K > \frac{|C|(\lambda_2 - 2\lambda_1) - (V - |C|)}{(V - |C|) + |C|\lambda_1}. \tag{3.24}$$

The properties of Inequalities 3.24 are similar to those of Inequality 2.26, namely:

(1) If $C = 0$, in other words, if we are dealing with a diagonal covariance matrix, $K > -1$. In this case, combining the former and the latter sample sets therefore always leads to a better result.

(2) If the numerator of the right-hand side of Inequality 3.24 is nonpositive, it also means that combining the former and the latter sample sets always leads to a better result than does that of only the former sample sets. (Because for our case, $V > |C|$, and the denominator of the right-hand side of Inequality 3.24 is always positive). This situation certainly holds if $\lambda_2 - 2\lambda_1 \leq 0$. It has the following meanings:

(a) If $\lambda_1 \geq \lambda_2$, $\lambda_2 - 2\lambda_1 < 0$. It is therefore always useful to add samples of better quality (i.e., with a smaller frequency ratio of missing values $\lambda_2 \leq \lambda_1$).

(b) If $\lambda_2 > \lambda_1 \geq \lambda_2/2$, $\lambda_2 - 2\lambda_1 \leq 0$. It is therefore always useful to add samples of even worse quality but not too bad quality ($\lambda_2 \leq 2\lambda_1$). It is interesting to note that if the quality of the added samples is no more than two times worse than the quality of the former sample set, it is useful to add any quantity of these samples. A more precise value of the constant of the relation λ_1 and λ_2, which shows to what extent these parameters may differ from each other for this situation still takes place, depends on V and C, as is shown by Inequality 3.24.

(3) If λ_2 is greater than λ_1 such that, in accordance with Inequality 3.24, one obtains $K > K^* > 0$, adding sample sets with the parameter λ_2 and a sample size greater than $K^* \cdot N$ therefore leads to a better result. However, adding sample sets with a sample size smaller than $K^* \cdot N$ leads to a worse (or equal) result than does the use of only the former sample sets, even though the variances of the statistical estimators in Equations 3.19 are improved (see Equation 3.20).

References

1. Brailovsky, V. 1979. Comparative analysis of some procedures of function approximation based on use of sample data. To be published.
2. Kendall, M. & A. Stuart. The Advanced Theory of Statistics. Vols. 1–3. Charles K. Griffin Ltd. London.
3. Brailovsky, V. 1979. LMS-like procedure of function approximation based on use of experimental data with missing values. To be published.
4. Duda, R. & P. Hart. 1973. Pattern Classification and Scene Analysis. John Wiley & Sons. New York, N.Y.
5. Okamoto, M. 1963. An asymptotic expansion for distribution of the linear discriminant function. Ann. Math. Stat. **34**(4): 1286–1301.

FLOW IN THE WAKE OF A BODY AND NEAR ITS SIDE SURFACE

Irina Brailovsky

Moscow, USSR

Consider the steady supersonic flow of a viscous gas around a blunt symmetric body. The flows near the side surface away from the front stagnation point and in the wake downstream of the separated zone, i.e., in regions II and IV of FIGURE 1, will be investigated.

Experimental results and also numerical solution[1,2] of the Navier-Stokes equations in the regions I and III show that, except for two more or less small zones near the front stagnation point and the recirculation at the base, longitudinal gradients are much smaller than transverse ones. Therefore we can, in regions II and IV, use a simplification of the Navier-Stokes equations in which all equations are retained but certain second derivatives are eliminated from the viscous terms. The simplification is accurate to $0(\mathrm{Re}^{-1})$, where Re is the Reynolds number based on a longitudinal velocity component.

Note that it is not possible to use the classical boundary-layer equations in regions II and IV for the following reasons:

1. The pressure varies in the transverse direction because of compression and expansion zones there.
2. Conditions at the upper boundary of either region are unknown, whereas classical boundary-layer theory requires the pressure distribution.
3. The thickness of either region is not small enough.
4. The transverse component of velocity is (generally speaking) not very small there, so that the transverse momentum is not negligible, as it is in classical boundary-layer theory.

When simplified in the way mentioned above, the Navier-Stokes equations become

$$u\frac{\partial u}{\partial x} + v\frac{\partial u}{\partial y} + \frac{1}{\rho}\frac{\partial p}{\partial x} = \frac{1}{\mathrm{Re}\,\rho}\left[\frac{\partial}{\partial y}\left(\mu\frac{\partial u}{\partial y}\right) + j\frac{\mu}{y}\left(\frac{\partial u}{\partial y}\right)\right], \tag{1}$$

$$\begin{aligned} u\frac{\partial v}{\partial x} + v\frac{\partial v}{\partial y} + \frac{1}{\rho}\frac{\partial p}{\partial y} = \frac{1}{\mathrm{Re}\,\rho}\left\{\frac{4}{3}\frac{\partial}{\partial y}\left(\mu\frac{\partial v}{\partial y}\right) + \frac{\partial}{\partial x}\left(\mu\frac{\partial u}{\partial y}\right) - \frac{2}{3}\frac{\partial}{\partial y}\left(\mu\frac{\partial u}{\partial x}\right)\right. \\ \left. + j\left[2\mu\frac{\partial}{\partial y}\left(\frac{v}{y}\right) - \frac{2}{3}\frac{\partial}{\partial y}\left(\mu\frac{v}{y}\right)\right]\right\}, \end{aligned} \tag{2}$$

$$\begin{aligned} u\frac{\partial T}{\partial x} + v\frac{\partial T}{\partial y} + (\gamma - 1)T\left(\frac{\partial u}{\partial x} + \frac{\partial v}{\partial y} + j\frac{v}{y}\right) \\ = \frac{1}{\mathrm{Re}\,\rho}\left\{\frac{\gamma}{\mathrm{Pr}}\left[\frac{\partial}{\partial y}\left(\lambda\frac{\partial T}{\partial y}\right) + j\frac{\lambda}{y}\frac{\partial T}{\partial y}\right] + M_0^2(\gamma - 1)\gamma\left(\frac{\partial u}{\partial y}\right)^2\right\}, \end{aligned} \tag{3}$$

$$\frac{\partial(\rho u)}{\partial x} + \frac{\partial(\rho v)}{\partial y} + j\frac{\rho v}{y} = 0, \qquad p = \frac{1}{\gamma M_0^2}\rho T. \tag{4}$$

Here, x and y are longitudinal and transverse coordinates, with u and v the corresponding components of velocity; p, ρ and T are the pressure, density, and

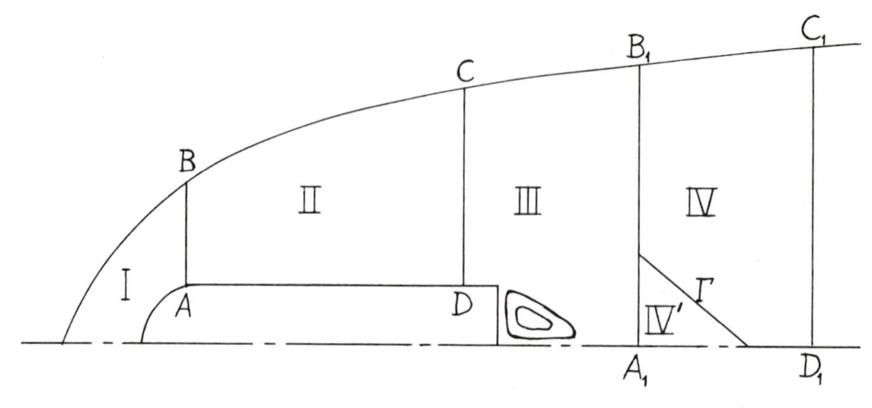

FIGURE 1.

temperature; μ and λ are the viscosity and thermal conductivity coefficients; γ, Pr and M_0 are the heat capacity ratio (c_p/c_v), and the Prandtl and Mach numbers. The values of the variables at infinity have been used to non-dimensionalize the system, so that Re, Pr and M_0 are referred to those values. The parameter j has the value 0 for plane flow and 1 for axisymmetric flow.

The problem is therefore to solve the system of Equations 1–4 under the following mixed boundary conditions:

1. Initial values on the boundary AB or A_1B_1, to be found by numerical solution of the full Navier-Stokes equations in I or III, respectively.
2. Usual boundary conditions on the surface AD of the body or symmetry conditions on A_1D_1.
3. Smoothness of the solution on the upper boundary BC or B_1C_1, i.e., conditions of the type $\partial^2 f/\partial y^2 = 0$.

If the solution of this mixed problem for the system of Equations 1–4 exists for a suitable class of functions, then it will satisfy the full Navier-Stokes system with an accuracy of $0(\mathrm{Re}^{-1})$. That is easily seen by inserting it into the Navier-Stokes equations.

There are difference methods for the numerical solution of the mixed problem for the system in which the unknown functions are found step by step from $x = x_i$ to $x = x_i + \Delta x$. These are much easier than those for solving the full Navier-Stokes system; even when available at large Reynolds numbers for regions as wide and long as II and IV usually are, the latter are very costly.

Unfortunately, it can be shown that the problem for the system of Equations 1–4 is improperly posed for $u^2 < c^2$, where c is the sound speed (in the same sense as in the Cauchy problem for the Laplace equation discussed by Hadamard). We shall show this fact for a model, i.e., that the initial (Cauchy) problem for a linearized system with so-called frozen coefficients is improper for $u^2 < c^2$. The system describes the behavior of small variations from constant values, so it would not be suitable for investigating the nature of the flow at supersonic speeds. However, we are interested here only in the correctness of the Cauchy problem and the conclusions about the model continue to hold when we move to variable coefficients or even to a quasilinear system such as Equations 1–4, or when we move from Cauchy's problem to a mixed one. We shall not try to prove these assertions with proper mathematical rigor here, but merely refer to numerous numerical experiments with mixed problems for partial differential systems, and return to a consideration of the model.

For simplicity in calculation suppose that p is a function of ρ only and that the flow is plane, neither of which simplifications changes conclusions about correctness. One then considers the linearized system

$$u\frac{\partial u'}{\partial x} + v\frac{\partial u'}{\partial y} + \frac{c^2}{\rho}\frac{\partial \rho'}{\partial x} = \mu\frac{\partial^2 u'}{\partial y^2}, \tag{5}$$

$$u\frac{\partial v'}{\partial x} + v\frac{\partial v'}{\partial y} + \frac{c^2}{\rho}\frac{\partial \rho'}{\partial y} = \mu\frac{\partial^2 v'}{\partial y^2}, \tag{6}$$

$$u\frac{\partial \rho'}{\partial x} + v\frac{\partial \rho'}{\partial y} + \rho\left(\frac{\partial u'}{\partial x} + \frac{\partial v'}{\partial x}\right) = 0, \tag{7}$$

where $\mu = 1/\mathrm{Re}\ \rho$ and the primes denote variations from the constant values without primes.

We seek solutions of the system of Equations 5–7 of the form $(u', v', \rho') = (U, V, \mathrm{P})\exp(\lambda x + iky)$ and find that λ must be a root of the characteristic equation of the system, namely

$$\det \mathbf{A} \equiv \begin{vmatrix} u\lambda + iv\kappa + \mu\kappa^2 & 0 & c^2\lambda/\rho \\ 0 & u\lambda + iv\kappa + \mu\kappa^2 & ic^2\kappa/\rho \\ \rho\lambda & i\rho\kappa & u\lambda + iv\kappa \end{vmatrix}$$
$$= (u\lambda + iv\kappa + \mu\kappa^2)^2(u\lambda + iv\kappa) - c^2\lambda^2(u\lambda + iv\kappa + \mu\kappa^2) + c^2\kappa^2(u\lambda + iv\kappa + \mu\kappa^2) = 0. \tag{8}$$

The vector $\mathbf{X} = (U, V, \mathrm{P})$ must then be a solution of the algebraic system $\mathbf{AX} = 0$. Equation 8 has three distinct roots, two of them (λ_1, λ_2) being roots of the quadratic equation $(u\lambda + iv\kappa + \mu\kappa^2)(u\lambda + iv\kappa) - c^2\lambda^2 + c^2\kappa^2 = 0$ and the third (λ_3) being the root of the linear equation $u\lambda + iv\kappa + \mu\kappa^2 = 0$.

In the case of region II or IV, u is positive, so $\mathrm{R}(\lambda_3) = -a\kappa^2$, with $a = \mu/u > 0$ and

$$|\exp(\lambda_3 x + i\kappa y)| = \exp(-a\kappa^2 x) \to 0 \quad \text{as} \quad \kappa \to \infty \tag{9}$$

for any $x > 0$. Next, consider the roots

$$\lambda_{1,2} = \frac{u\kappa(iv + \mu\kappa/2) \mp \sqrt{D}}{c^2 - u^2},$$

where

$$D = c^2\kappa^2(c^2 - u^2 - v^2) + \mu^2u^2\kappa^4/4 + i\mu c^2\kappa^3 v$$
$$= [\mu^2u^2 + 4i\mu c^2 v\kappa^{-1} - 4c^2(u^2 + v^2 - c^2)\kappa^{-2}]\kappa^4/4.$$

Since

$$\mathrm{R}(\sqrt{D}) = \mu u\kappa^2/2 - c^2[u^2 + v^2 - c^2(1 + \mu^2v^2)] + 0(\kappa^{-2}),$$

we find

$$\mathrm{R}(\lambda_1) = c^2[u^2 + v^2 - c^2(1 + \mu^2v^2)]/(c^2 - u^2) + 0(\kappa^{-2}), \tag{10}$$

$$\mathrm{R}(\lambda_2) = \mu u\kappa^2/(c^2 - u^2) + 0(1). \tag{11}$$

In particular, for $u^2 < c^2$, we have

$$R(\lambda_2) > \alpha\kappa^2, \quad \text{with } \alpha = \mu u/2(c^2 - u^2) > 0, \tag{12}$$

when κ is large enough.

Let $\mathbf{X}^*$ be an $0(1)$ solution of $\mathbf{AX} = 0$; then $e^{-\kappa}\mathbf{X}^*$ is also a solution and

$$(u', v', \rho') = e^{-\kappa}\mathbf{X}^* \exp(\lambda_2 x + i\kappa y)$$

is a solution of the system of Equations 5–7 with initial values

$$(u', v', \rho')_0 = e^{-\kappa}\mathbf{X}^* \exp(i\kappa y)$$

on $x = 0$. As $\kappa \to \infty$, these initial values and those of derivatives up to any order tend to zero. But, for any $x > 0$,

$$|e^{-\kappa} \exp(\lambda_2 x + i\kappa y)| = e^{-\kappa} \exp(R(\lambda_2)x) > \exp(-\kappa + \alpha\kappa^2 x) \to \infty \quad \text{as} \quad \kappa \to \infty,$$

since α is positive for $u^2 < c^2$, according to Equation 12. We conclude that the Cauchy problem for the system of Equations 5–7 is improperly posed for $u^2 < c^2$ in the sense of the backward heat equation. It is interesting that the correctness is not associated with the fact that, for $\mu = 0$, the system of Equations 5–7 is elliptic if $u^2 + v^2 < c^2$ and hyperbolic if $u^2 + v^2 > c^2$.

In fact, an example of a system can be given that becomes hyperbolic if the analog of viscosity $\mu = 0$ but has an improperly posed Cauchy problem if $\mu > 0$, in the same sense as for the system of Equations 5–7. Indeed, let us consider the following system:

$$u\frac{\partial u'}{\partial x} + v\frac{\partial u'}{\partial y} + \frac{c^2}{\rho}\frac{\partial \rho'}{\partial x} = \mu\frac{\partial^2 u'}{\partial y^2}, \tag{13}$$

$$u\frac{\partial \rho'}{\partial x} + v\frac{\partial \rho'}{\partial y} + \rho\frac{\partial u'}{\partial x} = 0. \tag{14}$$

Here, also, primed variables are the unknowns. The corresponding characteristic equation is

$$\begin{vmatrix} u\lambda + iv\kappa + \mu\kappa^2 & c^2\lambda/\rho \\ \rho\lambda & u\lambda + iv\kappa \end{vmatrix} = (u\lambda + iv\kappa + \mu\kappa^2)(u\lambda + iv\kappa) - c^2\lambda^2 = 0. \tag{15}$$

It is easy to see that, for $\mu = 0$, the system of Equations 13 and 14 is hyperbolic. The roots of Equation 15 are:

$$\lambda_{1,2} = \frac{u\kappa(iv + \mu\kappa/2) \mp \sqrt{D}}{c^2 - u^2},$$

where

$$D = [\mu^2 u^2 + 4i\mu c^2 v\kappa^{-1} - 4c^2 v^2 \kappa^{-2}]\kappa^4/4.$$

Since

$$R(\sqrt{D}) = \mu u\kappa^2/2 + c^2(c^2\mu^2 - 1)v^2 + 0(\kappa^{-2}),$$

we find

$$R(\lambda_1) = c^2(1 - \mu^2 c^2)v^2/(c^2 - u^2) + 0(\kappa^{-2}),$$

$$R(\lambda_2) = \mu u\kappa^2/(c^2 - u^2) + 0(1),$$

the latter being the same as the result of Equation 11. We conclude that Inequality 12 still holds. Thus, the Cauchy problem for the system of Equations 13 and 14 is improperly posed if $u^2 < c^2$, even though, for $\mu = 0$, the system is hyperbolic for such values of u and c.

The simplest example of a system of this kind is

$$u\frac{\partial u'}{\partial x} + \frac{c^2}{\rho}\frac{\partial \rho'}{\partial x} = \mu\frac{\partial^2 u'}{\partial y^2}, \qquad \textbf{(13*)}$$

$$u\frac{\partial \rho'}{\partial x} + \rho\frac{\partial u'}{\partial x} = 0. \qquad \textbf{(14*)}$$

On using the same variation $\exp(\lambda x + \kappa y)$, we get:

$$\lambda_1 = 0;\ \lambda_2 = \mu u \kappa^2/(c^2 - u^2).$$

Returning to the system of Equations 5–7, we note that, if $u^2 > c^2$, then

$$R(\lambda_2) < -\alpha\kappa^2,$$

according to the result of Equation 11, so that

$$|\exp(\lambda_2 x + i\kappa y)| < \exp(-\alpha\kappa^2 x) \to 0 \text{ as } \kappa \to \infty$$

for any $x > 0$. As for λ_1, the result of Equation 10 shows that

$$|\exp(\lambda_1 x + i\kappa y)| < N \text{ for any } x > 0,$$

where N does not depend on κ. In addition, Equation 9 holds. It is, therefore, possible to prove that the Cauchy problem for the system of Equations 5–7 is properly posed in the case $u^2 > c^2$ by the usual method of Fourier analysis.

Experiments on the numerical solution of the original quasilinear system of Equations 1–4 in the region IV by a finite-difference method have shown that the mixed problem mentioned above can be rather easily solved when the inequality

$$u^2 > c^2 \qquad \textbf{(16)}$$

holds.

Returning to the regions II and IV in their entirety, we note that Inequality 16 does not hold everywhere. There is a subsonic layer in II close to the surface of the body. In IV, it is desirable to put the boundary A_1B_1 as close as possible to the separated zone subject to the requirement that neglected terms in the Navier-Stokes equations are small enough. There is then a part IV′ near the axis of symmetry A_1D_1, where Inequality 16 does not hold. But, in such subregions, classical boundary-layer equations are valid, since they are thin, the pressure is constant across them ($\partial p/\partial y = 0$), and the transverse velocity is small.

Accordingly, the following computational algorithm is offered (though it is described here only for (IV). Separate off the part IV′ so that Inequality 16 holds above the boundary Γ. Under Γ, boundary-layer equations are used, namely the system of Equations 1–4, with Equation 2 replaced by

$$\frac{\partial}{\partial x}p(x, y) = \frac{\partial}{\partial x}p(x, y_0), \qquad \textbf{(2*)}$$

where $p(x, y_0)$ is the pressure just above Γ, and $y < y_0$. The function $p(x, y_0)$ is not known beforehand but is to be found while simultaneously solving the system of Equations 1–4 above Γ and the boundary-layer equations underneath. A numerical method will be described below for this joint solution, which gives the values of the unknowns step by step from x_i to $x_i + \Delta x$.

For both the original system and the boundary-layer equations, we choose the following implicit finite-difference approximation, given here for the simple heat-transfer equation,

$$\frac{\partial u}{\partial x} = \sigma \frac{\partial^2 u}{\partial y^2} + a \frac{\partial u}{\partial y} \, (\sigma > 0), \tag{17}$$

since it can be written in just the same way. The finite-difference approximation of Equation 17 is taken to be

$$\frac{u_m^{n+1} - u_m^n}{\Delta x} = \sigma_m^n \frac{u_{m+1}^{n+1} - 2u_m^{n+1} + u_{m-1}^{n+1}}{(\Delta y)^2} + a_m^n \frac{u_{m+1}^{n+1} - u_{m-1}^{n+1}}{2\,\Delta y}, \tag{18}$$

where $u_m^n = u(n\,\Delta x,\, m\,\Delta y) = u(x_n, y_m)$. For all values of $\Delta x/(\Delta y)^2$ and $\Delta x/\Delta y$, the approximation is stable, the error being $\varepsilon = 0(\Delta x) + 0(\Delta y)^2$ (which is good enough, since the unknowns are changing slowly in the x-direction in our case).

So we use the finite-difference approximation of Equations 1–4 at points $y_m \geq y_{m_0}$, and that of Equations 1, 2*, 3, and 4 under such points, i.e., for $y_m < y_{m_0}$, where m_0 is chosen so that $(u_m^n)^2 > (c_m^n)^2$ if $m \geq m_0$. Pressure Equation 2* has the following finite-difference form (since $p = \rho T/\gamma M_0^2$):

$$\rho_m^n(T_m^{n+1} - T_m^n) + T_m^n(\rho_m^{n+1} - \rho_m^n) = \rho_{m_0}^n(T_{m_0}^{n+1} - T_{m_0}^n) + T_{m_0}^n(\rho_{m_0}^{n+1} - \rho_{m_0}^n). \tag{19}$$

Thus, if values at the station n (i.e., $f_m^n = f(n\,\Delta x,\, m\,\Delta y)$) are known, we have a system of linear algebraic equations from which to calculate $f_m^{n+1} = f[(n+1)\,\Delta x,\, m\,\Delta y]$.

A typical equation can be written as a relation between three adjacent vectors $\mathbf{f}_{m-1}^{n+1}, \mathbf{f}_m^{n+1}, \mathbf{f}_{m+1}^{n+1}$, where $\mathbf{f}_m^n = (u_m^n, v_m^n, \rho_m^n, T_m^n)$. The only exception is Equation 19. However, it appears to be possible to change slightly a well-known vectorial trial-run method for tridiagonal matrices so as to enable our joint system to be solved. We will demonstrate the possibility for a simple example involving scalars, since all the calculations are quite similar.

Consider the difference equations

$$\begin{aligned} &\alpha_m U_{m+1} + \beta_m U_m + \gamma_m U_{m-1} = \delta_m + a_{m_0} U_{m_0} \qquad (1 \leq m < m_0), \\ &\alpha_m U_{m+1} + \beta_m U_m + \gamma_m U_{m-1} = \delta_m \qquad (m_0 \leq m < M), \end{aligned} \tag{20}$$

having a tridiagonal matrix and the boundary conditions

$$\alpha_0 U_2 + \beta_0 U_1 + \gamma_0 U_0 = \delta_0, \qquad \alpha_M U_M + \beta_M U_{M-2} + \gamma_M U_{M-2} = \delta_M. \tag{21}$$

Here, U_m ($m = 0, 1, \ldots, M$) are to be found and $\alpha_m, \beta_m, \gamma_m, \delta_m$, and a_{m_0} are known. The solution is found by means of auxiliary unknowns A_m, B_m ($m = 0, 1, \ldots, M-1$), and Q_m ($m = 0, 1, \ldots, m_0 - 1$), such that

$$\begin{aligned} &U_m = A_m U_{m+1} + B_m + Q_m U_{m_0} \qquad (m = 0, 1, \ldots, m_0 - 1), \\ &U_m = A_m U_{m+1} + B_m \qquad (m = m_0, m_{0+1}, \ldots, M - 1). \end{aligned} \tag{22}$$

Difference Equations 20 place further conditions on these auxiliaries, which are satisfied if the recurrence relations

$$\begin{aligned} &A_{m-1} = -\frac{(\alpha_m + \beta_m A_m)}{\gamma_m A_m}; \qquad B_{m-1} = \frac{(\delta_m A_m + \alpha_m B_m)}{\gamma_m A_m} \quad (m = 1, 2, \ldots, M-1); \\ &Q_{m-1} = \frac{a_{m_0} A_m + \alpha_m Q_m}{\gamma_m A_m} \quad (m = 1, 2, \ldots, m_0 - 1); \quad \text{and } Q_{m_0 - 1} = 0 \end{aligned} \tag{23}$$

hold.

The second boundary condition Equation 21 is satisfied if

$$A_{M-2} = -\frac{(\alpha_M + \beta_M A_{M-1})}{\gamma_M A_{M-1}} \quad \text{and} \quad B_{M-2} = \frac{\delta_M A_{M-1} + \alpha_M B_{M-1}}{\gamma_M A_{M-1}}.$$

Together with Equation 23 for $m = M - 1$, these enable A_{M-1} and B_{M-1} to be determined, from which follow A_m, B_m, and Q_m for decreasing m, successively. If we now write

$$U_m = C_m + D_m U_{m_0} \qquad (m = 0, 1, \ldots, M),$$

then the first boundary condition Equation 21 and Equations 22 determine C_0 and D_0, from which follow C_m and D_m for increasing m, successively. Setting $U_{m_0} = C_{m_0}/(1 - D_{m_0})$ finally determines U_m.

The difference form of the joint system of Equations 1–4 and 1, 2*, 3, and 4 can be written in the same tridiagonal form as Equations 20 and 21, generalized to matrices α_m, β_m, γ_m, δ_m, a_{m_0}, A_m, B_m, Q_m, and four-dimensional vectors U_m.

Note that, at the upper boundary B_1C_1, only smoothness conditions are applied; the pressure distribution is unknown there, to be found as part of the solution of the joint system. The computer time needed for a region such as IV with the algorithm described is about 100 times less than that with the full Navier-Stokes equations at the same Reynolds number. The reason is that only a boundary-value problem can be posed for the steady Navier-Stokes equations; the methods of solution of such problems used nowadays are iterative, requiring many iterations especially at the large Reynolds numbers that arise in flows at supersonic speeds.

References

1. Pavlov B. M., 1968. Numerical investigation of supersonic flow around blunt bodies. Izv. Akad. Nauk SSSR Mekh. Zhidk. Gaza **3**.
2. Brailovsky, I. Yu. 1971. Near-wake flow. Dokl. Akad. Nauk. SSSR Ser. Gidromeh. **197**(3).

DYNAMICS OF VORTEX LINES IN SUPERFLUID ^{4}HELIUM*

W. I. Glaberson

Department of Physics
Rutgers University
New Brunswick, New Jersey 08903

At temperatures below about 2 K, liquid ^{4}He undergoes a phase transition to become a superfluid. A particularly interesting property of the superfluid state is the set of restrictions on the possible hydrodynamic flow patterns in the system. In particular, the superfluid velocity field must be curl-free:

$$\vec{\nabla} \times \vec{V}_s = 0. \tag{1}$$

Solid-body rotation, characteristic of a uniformly rotating ordinary fluid, is therefore clearly forbidden. It does not follow, however, that for superfluid contained within a cylindrical container rotating about the cylinder axis, the velocity itself is necessarily zero. Above some critical angular velocity (much smaller than the angular velocities investigated in the experiments discussed in this paper), the free energy of the system is decreased with the introduction of a filamentary discontinuity threading the fluid from container wall to container wall along the rotational axis. In the presence of this discontinuity (termed a vortex line), the superfluid executes azimuthal motion about the line, with a velocity given by

$$\vec{V}_s(r) = \frac{\hbar}{mr}\hat{1}_\theta, \tag{2}$$

where $\hbar$ is Planck's constant, m is the mass of a helium atom, and r is the distance from the line. Note that in this state, every helium atom has $\hbar$ angular momentum. The vortex velocity field clearly cannot persist to indefinitely small r, and it is assumed that the superfluid is somehow substantially altered within some core, which turns out to have a radius of the order of several angstroms. At higher angular velocities, additional vortex lines enter the fluid until, at very high angular velocities, the fluid contains a more or less uniform distribution of lines having a density that yields an average superfluid velocity equal to that expected for solid-body rotation.

In this paper, I shall discuss two recent experiments on the dynamics of vortex lines. The first deals with a situation where the interactions between vortex lines are crucial and where the system appears to be well described by an average hydrodynamic description. The second deals with the excitations of isolated vortex lines. These experiments have been discussed to some extent elsewhere.[1,2]

Thermorotation Effects

Anderson[3] has shown that the motion of vortex lines through superfluid helium is accompanied by a chemical potential gradient in the fluid. This prediction is based on a description of the superfluid in terms of a complex order parameter

* Supported in part by grant DMR-78-25409 from the National Science Foundation.

whose time rate of change of phase is proportional to the chemical potential. Anderson showed that the passage of a vortex line between two points in the fluid produces a change of 2π in the order-parameter phase difference between the points. Applying this idea to a situation in which a large number of vortex lines move uniformly through the superfluid leads to the result that, on a scale large compared to the vortex spacing, a chemical-potential gradient will be present and that this gradient is proportional to the line density and velocity and oriented in a direction perpendicular to the direction of motion. This simple situation, in which an array of vortex lines is moved through the superfluid in a controlled manner, has been closely approximated in the series of experiments discussed in this paper. The hydrodynamics of helium in rotation is an insufficiently studied hybrid between ordinary classical hydrodynamics and the familiar curl-free two-fluid hydrodynamics and is therefore of intrinsic interest. It may also have relevance to superconducting alternating-current generator cooling problems[4] and to heat transfer in rotating neutron stars.

Consider an array of straight parallel vortex lines in the presence of thermal counterflow transverse to the lines. One can easily show that chemical-potential gradients with components both along the counterflow as well as perpendicular to it are set up. One way of visualizing this effect is to consider a single vortex line. In the presence of transverse counterflow, the line will, in general, experience frictional forces both along the counterflow and perpendicular to it. The line will then move with velocity $\vec{V}_L$ at an angle with respect to the counterflow in such a way that the magnus force produced by the relative motion between the line and the superfluid exactly balances the frictional forces. A chemical-potential gradient is then set up perpendicular to $\vec{V}_L - \vec{V}_S$. An analogous effect was observed[5] in superconductors in which chemical-potential gradients were observed in a type-II superconducting film in a magnetic field in the presence of a heat current transverse to the flux lines.

Our experimental arrangement, shown schematically in FIGURE 1, involves a glass channel of rectangular cross section, closed at one end, with the other end open to a pumped helium bath. A resistive heater is placed near the closed end, and the channel is rotated about a vertical axis perpendicular to the heat current. The channels investigated were of large aspect ratio: the heights ranged from 0.18 to 2.6 mm, the widths were 1.4 cm, and the lengths were $5\frac{1}{2}$ cm. Temperature sensors were aluminum films, held in their superconducting transition regions ($T \sim 1.3$ K), evaporated onto one of the glass channel walls. The sensors were placed in the middle of the channel, approximately 1 cm apart, and separated on a line at an angle of about 45 degrees with respect to the channel axis. The sensitive thermometric technique required in these experiments, having a temperature-difference noise level of about 40 nanodegrees per $\sqrt{\text{Hz}}$, is described elsewhere.[6] Temperature differences were measured in both clockwise and counterclockwise rotations, and the components of the temperature gradients parallel and perpendicular to the heat current were obtained. Under the conditions of our experiment, both components were of similar magnitude.

Direct chemical-potential measurements were obtained by means of a differential pressure transducer. One side of the transducer was open to the bath, and the other side was connected through a superleak to an opening in one of the channel walls between the heater and the closed end of the channel. The transducer, using a stretched aluminized mylar sheet as one side of a capacitor in a tunnel-diode oscillator circuit,† had a noise level of about 10^{-3} dyn per cm^2—$\sqrt{\text{Hz}}$. The

† The circuit is similar to that described in Van Degrift.[7]

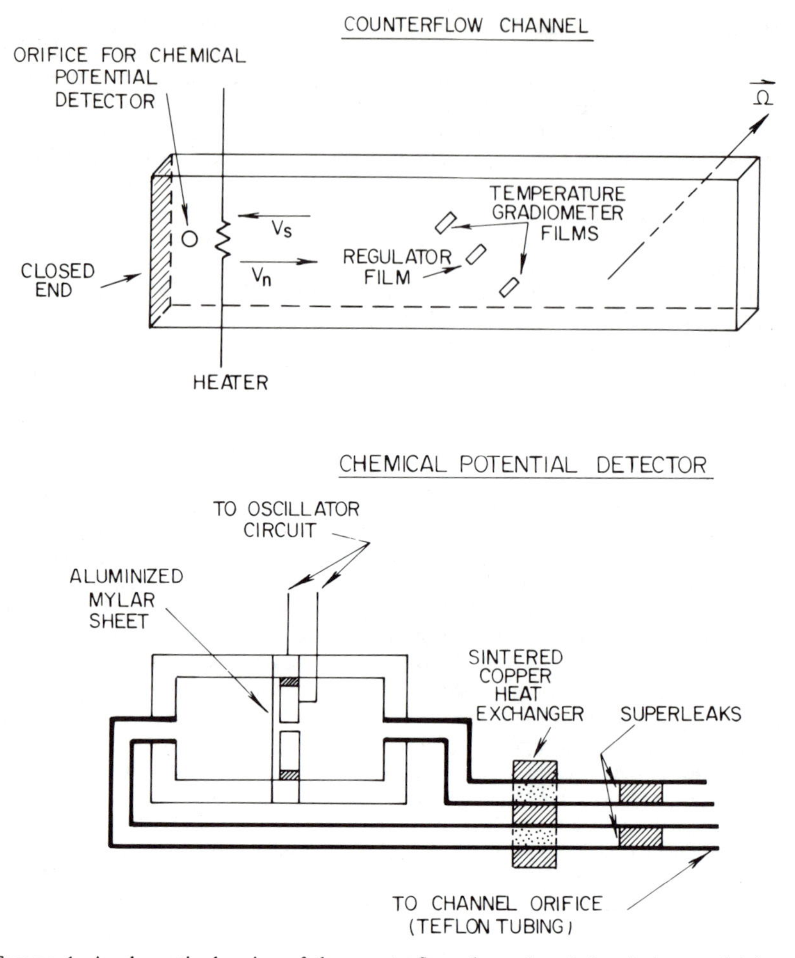

FIGURE 1. A schematic drawing of the counterflow channel and chemical-potential detector. Of the three aluminum films, the center one is used to regulate the ambient bath temperature, and the outer two are used in a bridge as temperature-difference detectors.

transducer was thermally isolated from the channel by the superleak, and thermal coupling between the two sides of the transducer was accomplished by means of a sintered-copper heat exchanger. The effect of thermal isolation from the channel and thermal coupling across the transducer is to convert chemical-potential differences to pure pressure differences detected by the transducer.

The parallel component of the temperature gradient in the 0.5-mm channel is shown in FIGURE 2 as a function of heater power for several rotational speeds. We note that (a) the temperature gradients in the linear (nonturbulent) regimes increase as the rotational speed is increased, (b) the critical heater power, Q_{c2}, associated with the onset of turbulent counterflow, is increased as the rotational speed is increased—apparently becoming proportional to $\sqrt{\Omega}$ as Ω gets large, (c) turbulent onset is characterized by a relatively smooth deviation from linearity when not

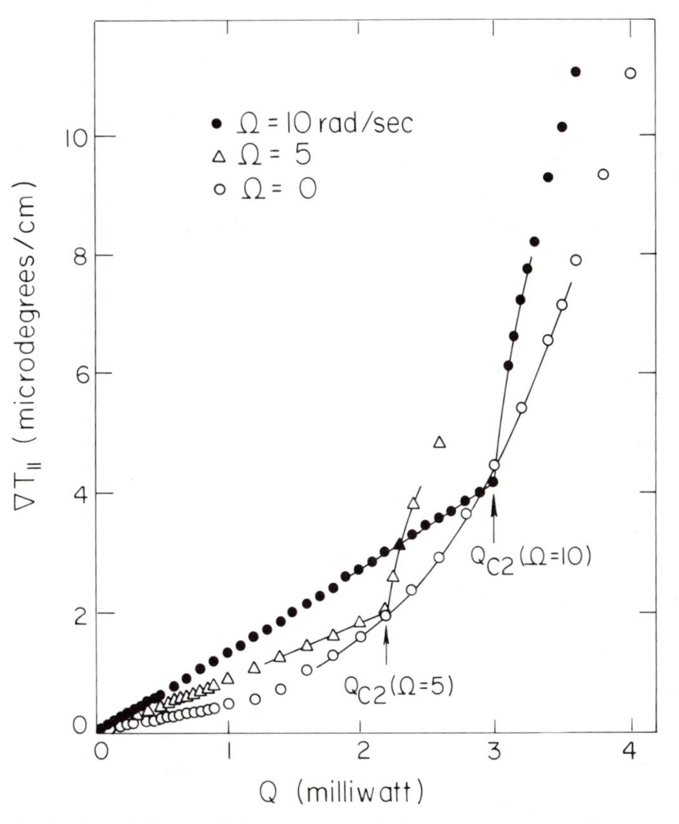

FIGURE 2. A plot of the parallel component of the temperature gradient versus heater power for several rotational speeds. The solid lines are an aid to the eye.

rotating but by a sharp change of slope when rotating, (d) turbulent onset occurs at a heater power close to the power where the laminar flow-temperature gradients would cross the nonrotating turbulent curve, and (e) the temperature gradient in the fully turbulent regime ($Q \gg Q_{c2}$) differs from that in the nonrotating situation by an amount that becomes Ω-independent as Ω gets large.

Hall's[8] macroscopic equations of motion for the normal and superfluid components of helium II form a generalization of the two-fluid model to include the effects of quantized vortex lines on large-scale flow phenomena. They are based on a detailed model of vortex-line dynamics in which momentum is transferred between the normal fluid and superfluid through the action of roton scattering from the vortex field. The phenomenologic equations of Bekharevich and Khalatnikov[9] are based on the assumption that the fluid internal energy contains a term proportional to the vorticity density. By identifying the parameters in their equations with the mutual friction parameters of Hall, the two sets of equations become nearly identical.

Since they are based on a microscopic model of vortex-line dynamics, Hall's equations allow the interpretation of experimental observations in terms of vortex-line motion that reveals their relationship to the concept of phase slippage proposed by Anderson.[3] In the following, the concept of phase slippage will be briefly reviewed, and the result will be applied to an idealization of this experiment in which Hall's microscopic equations yield simple results for the vortex-line motion. A more complete description of this experiment is then obtained by solving Hall's macroscopic equations in the case where the flow is confined between parallel plates.

In his 1966 paper, Anderson considers the superfluid to be described by a complex order parameter, the phase of which is related to macroscopic superfluid properties. In particular, it is shown that

$$-\frac{\hbar}{m}\frac{\mathrm{d}\phi}{\mathrm{d}t} = \mu, \tag{3}$$

where ϕ is the order-parameter phase, m is the mass of a helium atom, and μ is the chemical potential. By requiring that the order parameter be single-valued and noting that its phase need only be defined to within an integer times 2π, the quantization condition is obtained:

$$\oint \vec{\nabla}\phi \cdot \mathrm{d}\vec{l} = 2N\pi. \tag{4}$$

N is an integer and may be nonzero if the path of integration encloses a region of superfluid exclusion. Associating the superfluid velocity with the gradient of the order-parameter phase,

$$\vec{V}_s = (\hbar/m)\vec{\nabla}\phi, \tag{5}$$

we see that Equation 4 is a statement of the quantization of circulation. A vortex line passing between two points in the superfluid then leads to a phase "slippage" of 2π in the phase difference between the points. In the case where a large number of vortices move through the fluid, appropriate time derivatives may be defined in an average sense, so that

$$(\mu_2 - \mu_1)_{\mathrm{av}} = -\frac{\hbar}{m}\frac{\mathrm{d}}{\mathrm{d}t}(\Delta\phi) = -\frac{h}{m}\left\langle\frac{\mathrm{d}n}{\mathrm{d}t}\right\rangle, \tag{6}$$

where $\langle \mathrm{d}n/\mathrm{d}t\rangle$ is the average rate at which vortices cross a line connecting the two points of interest. The chemical-potential gradient is perpendicular to the circulation of the line $\vec{K} = (h/m)\hat{l}_k$ and to the velocity of the lines relative to the superfluid $\vec{V}_L - \vec{V}_s$, so that

$$\vec{\nabla}\mu_v = -\sigma_L \vec{K} \times (\vec{V}_L - \vec{V}_s), \tag{7}$$

where σ_L is the number density of vortex lines. (Note that $\rho_s \vec{\nabla}\mu_v$ is the magnus force per unit volume on the superfluid.) In the absence of superfluid acceleration, all of the chemical potential gradient in the system is associated with vortex motion:

$$\vec{\nabla}\mu = \frac{\vec{\nabla}p}{\rho} - s\vec{\nabla}T = \vec{\nabla}\mu_v; \tag{8}$$

whereas, if acceleration is allowed, the system can also respond to the chemical-potential gradient by accelerating:

$$\vec{\nabla}\mu = \vec{\nabla}\mu_v - \mathrm{d}\vec{v}_s/\mathrm{d}t. \tag{9}$$

Note that Equations 3 and 5 are valid locally (outside of vortex cores), whereas Equation 9 deals with an average over distances large compared with the intervortex line spacing.

This relation simply expresses the fact that a chemical-potential gradient gives rise to a time-varying phase gradient, which, in turn, can arise from either accelerating fluid($\dot{v}_s \propto \nabla\dot{\phi}$) or moving vortices. In a system in steady state while rotating at frequency Ω, this becomes

$$\vec{\nabla}\mu = \vec{\nabla}\mu_v - 2\vec{\Omega} \times \vec{v}_s, \tag{10}$$

where $\vec{v}_s$ is now measured in the rotating coordinate system.

To relate this last expression to measurable quantities, it remains to determine $\vec{v}_L$ in terms of the directly controllable normal and superfluid velocities. Following Hall, we write

$$-\rho_s\vec{K} \times (\vec{v}_L - \vec{v}_s) = D(\vec{v}_n - \vec{v}_L) + D''\hat{\omega} \times (\vec{v}_n - \vec{v}_L), \tag{11}$$

where $\vec{v}_n$ and $\vec{v}_s$ are the normal fluid velocities averaged over a region containing many vortex lines, $\vec{K}$ is the vortex circulation, $\hat{\omega}$ is the direction of local vorticity, and D and D'' are parameters related to roton-collision diameters for parallel and perpendicular momentum transfer to the vortex lines. The right-hand side of Equation 11 is a general expression for the frictional drag experienced by the vortex line as a consequence of a transverse normal fluid flow, and Equation 11 states that the net force per unit length on a vortex line (magnus force plus friction force) must be zero.

For the case of thermal counterflow in an infinite medium, $\vec{v}_n$ and $\vec{v}_s$ are given by

$$\vec{v}_n = v_o\hat{x}, \qquad \vec{v}_s = -(\rho_n/\rho_s)v_o\hat{x}, \tag{12}$$

where it has been assumed that a uniform heat-flux density given by ρSTv_o is applied in the x direction. In terms of the more convenient mutual friction coefficients B, B' (simply expressible in terms of the parameters D, D''), Equations 7 and 11 become

$$\vec{\nabla}\mu = \frac{\rho_n}{\rho_s}\Omega V_o(B\hat{x} + B'\hat{y}), \tag{13}$$

and Equation 10 yields

$$\vec{\nabla}\mu = \frac{\rho_n}{\rho_s}\Omega V_o(B\hat{x} + (2 - B')\hat{y}). \tag{14}$$

Naturally, the same result is obtained when the macroscopic equations are solved directly. The macroscopic equations, however, yield additional information. By including the effects on the normal fluid, one obtains the separate pressure- and temperature-gradient contributions to $\vec{\nabla}\mu$. In the absence of vortex curvature, the equations of motion in the rotating coordinate system are

$$\begin{aligned} \rho_s\frac{d\vec{v}_s}{dt} &= -\frac{\rho_s}{\rho}\vec{\nabla}p + \rho_s S\vec{\nabla}T - 2\rho_s\vec{\Omega} \times \vec{v}_s + \vec{F} \\ \rho_n\frac{d\vec{v}_n}{dt} &= -\frac{\rho_n}{\rho}\vec{\nabla}p - \rho_s S\vec{\nabla}T - 2\rho_n\vec{\Omega} \times \vec{v}_n + \eta\left[\nabla^2\vec{v}_n + \frac{1}{3}\vec{\nabla}(\vec{\nabla}\cdot\vec{v}_n)\right] - \vec{F}, \end{aligned} \tag{15}$$

where $\vec{F}$ is the mutual friction force given by

$$\vec{F} = (\rho_s\rho_n/2\rho)\{B'\hat{\omega} \times [\vec{\omega} \times (\vec{v}_s - \vec{v}_n)] + B'\vec{\omega} \times (\vec{v}_s - \vec{v}_n)\}, \tag{16}$$

and $\vec{\omega} = \vec{\nabla} \times \vec{v}_s + 2\vec{\Omega}$. The steady-state results for uniform counterflow transverse to the uniformly rotating superfluid are

$$\vec{\nabla} T = -(\rho_n \Omega v_o / \rho_s S)[B\hat{x} + (2 - B')\hat{y}], \qquad \vec{\nabla} p = 0. \tag{17}$$

The results in Equations 14 and 17 indicate roughly what is to be expected in a rotating-counterflow experiment if the counterflow channel is very large. The effects of the channel surfaces on vortex-line motion, however, are not included, and to deal with these effects, the macroscopic equations must include terms due to vortex curvature. Effects of curvature cannot be completely neglected in a channel of finite height because, in such a system, the normal fluid velocity is not constant throughout space and therefore will affect the lines differently at different points along their lengths. A solution of the full hydrodynamic equations for flow in a channel of finite height and including the effects of vortex line pinning on the channel surfaces has been obtained.

A plot of the experimental linear-regime slopes versus rotational speed for both parallel and perpendicular components of the temperature gradient are shown in FIGURE 3. The solid lines are fits to the data of the solutions to the full hydrodynamic equations. The values of the mutual friction coefficients obtained from this fit are in reasonable agreement with those previously measured.[10] The two-fluid equations for a channel are incomplete, and a unique solution can only be obtained by introducing a boundary condition on the tangential component of $\vec{V}_s$ or of curl $\vec{V}_s$. We chose to specify that the components of curl $\vec{V}_s$ perpendicular to the rotational axis are zero at the boundaries for the purpose of fitting the data. The boundary

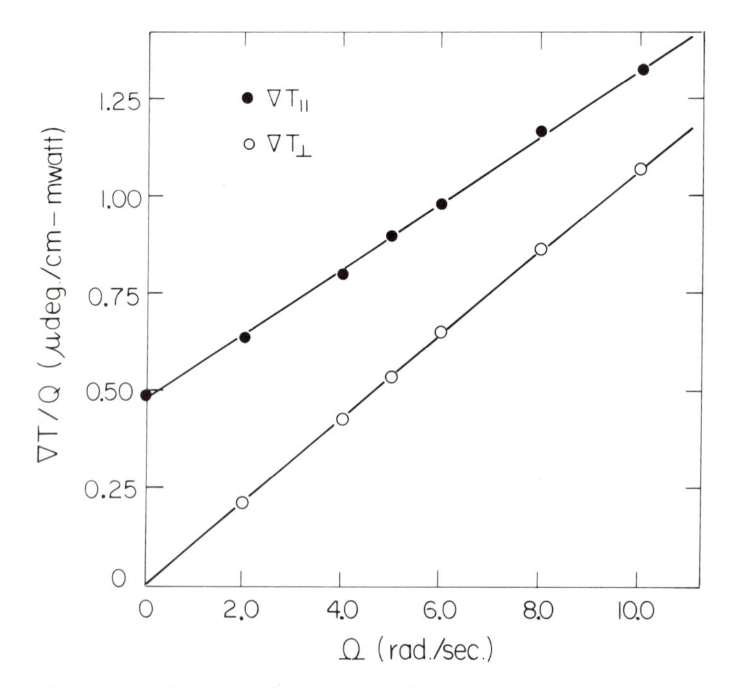

FIGURE 3. A plot of $\nabla T_{\parallel}/Q$ and $\nabla T_{\perp}/Q$ in the linear regime versus rotational speed. The solid lines are fits to the theory, as discussed in the text.

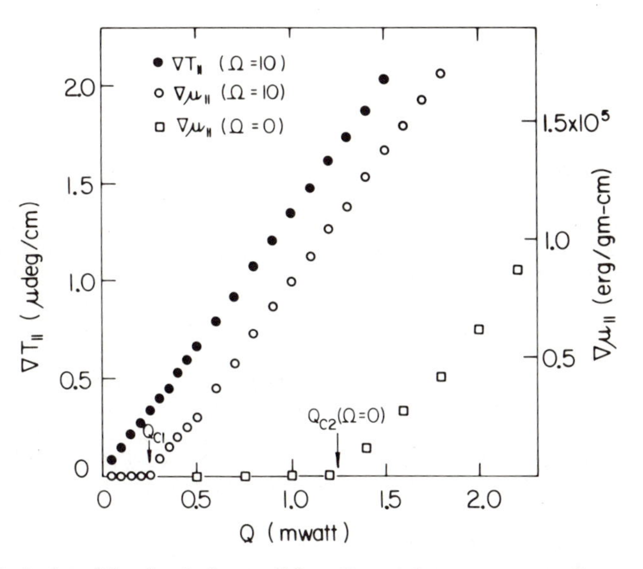

FIGURE 4. A plot of the chemical-potential gradient and temperature gradient versus heater power for $\Omega = 0$ and $\Omega = 10$ rad/sec.

conditions involved are macroscopic and should not be confused with microscopic boundary conditions involving isolated vortex lines intersecting walls.

Because we wished to observe, without ambiguity, the presence of a chemical-potential gradient associated with the vortex motion, and because we wished to extend the measurements to different temperatures, we introduced the chemical-potential detector discussed earlier. For reasons that we do not fully understand, noise problems prevented us from inserting the probe in the counterflowing region. Perhaps the orifice disturbed the flow in its vicinity sufficiently to cause vortices to nucleate or at least to hang up there in such a way as to cause large fluctuations in the chemical-potential difference across the orifice. In placing the probe behind the heater, we seem to have completely avoided this difficulty. We found that all of the effects [(a) to (e)] disussed above were observed with respect to chemical potential at all temperatures investigated (1.15 K $< T \lesssim$ 2.17 K).

A plot of the chemical-potential gradient $\vec{\nabla}\mu$ versus heater power Q for $\Omega = 0$ and $\Omega = 10$ rad per second is shown in FIGURE 4. Also shown for comparison is the temperature gradient. Note that there exists a critical heater power Q_{c1}, not observable in the temperature measurements, below which $\vec{\nabla}\mu$ is zero. We associate Q_{c1} with the power at which the vortex array "depins" and begins to move in response to the counterflow. Below Q_{c1}, the vortices are pinned to protuberances in the channel walls and accommodate the counterflow by deforming. If we adjust the boundary condition on curl $\vec{V}_s$ in the calculation so as to produce no longitudinal chemical-potential gradient, we find that, indeed, little influence on $\vec{\nabla}T$ is predicted: the pressure gradient does, of course, become large. We observed that Q_{c1} is weakly Ω dependent, becoming smaller as Ω is increased. We suggest that it is unlikely that individual vortices in the array will move without the others and that Q_{c1} is associated with some average pinning force. Clearly, the largest protuberances will be the first to be occupied by vortices, so that as the vortex density is increased, the average pinning force will decrease. We cannot, however, rule out an explanation of the Ω dependence of Q_{c1} in terms of mechanical vibration levels of the apparatus. Note that at Q_{c1}, at 1.3 K, the superfluid velocity is only 0.02 cm per second. A crude estimate based on this velocity gives a depinning force of about 10^{-7} dyn per line.

To test the suggestion that Q_{c1} is associated with vortex-line depinning, the experiment was repeated with a channel in which the walls were roughened with sand grains. As seen in FIGURE 5, the effect of roughening the surface is to increase Q_{c1}, as expected.

The actual flow pattern in the region below Q_{c2} is quite interesting. A solution of the Hall equations yields a secondary flow in which an element of normal fluid executes a counterclockwise spiral-like motion if in the upper half of the channel and a clockwise spiral-like motion if in the lower half of the channel. The superfluid behaves similarly, although in the opposite direction, of course. This sort of pattern is similar to the behavior of the flow of ordinary fluids in rotating channels.[11] An increase in the critical Reynolds number for ordinary fluids in a rotating channel has also been observed.[12] This increase for both ordinary fluids and the superfluid can probably be explained as a consequence of the Taylor–Proudman Theorem,[13] which tends to make a rotating fluid more stable with respect to three-dimensional perturbations.

The series of experiments I have discussed are the first to probe the large-scale macroscopic hydrodynamics of a rotating superfluid. They are, in a sense, complementary to experiments on the velocity and damping of sound in such systems. The latter are chiefly concerned with small deviations from equilibrium, whereas our experiments deal with large-scale flow phenomena and, of course, are at "zero" frequency. Our measurements constitute observations of large-scale vorticity flow, involving vortex-depinning phenomena and the transition to turbulence. They pro-

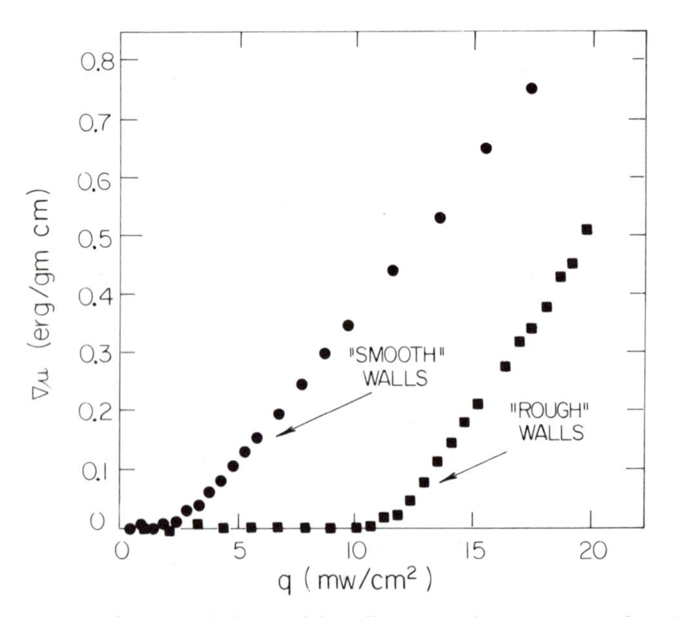

FIGURE 5. A plot of the chemical-potential gradient versus heater power at $\Omega = 10$ rad/sec for untreated-glass channel walls and for walls roughened with grains of sand.

vide convincing evidence of the correctness of the basic two-fluid equation for a rotating superfluid and of the phase-slippage arguments underlying them.

The transition to turbulence in nonrotating superfluid flow is neither well understood theoretically nor is there a clear experimental picture of the phenomenon. The experimental situation in rotating helium seems to be much better, in that the critical heat-current density approaches independence of channel size and wall composition and is simply proportional to the square root of the rotational speed at high rotational speeds. The phenomenon of vortex pinning is not well understood, and the techniques developed in this work seem particularly well suited for a detailed investigation of it.

Vortex Waves

Vortex waves are traveling helical deformations of a vortex line having a dispersion relationship[14]‡

$$\omega(k) = \frac{\hbar k^2}{2m}\left[\ln\left(\frac{1}{ka}\right) + 0.1159\right], \tag{18}$$

where a is the effective "radius" of the vortex core. These waves are polarized, so that an element of vortex line executes circular motion in a sense opposite to the circulation sense of the velocity field. Several experiments have been done in which vortex waves having a frequency of order 1 Hz have been observed.[16] I shall now discuss an experiment in which the vortex-wave dispersion relation is tested at radio frequencies.

Halley and Cheung,[17] and more recently Halley and Ostermeier,[18] have suggested that a radio-frequency (RF) electric field, transverse to a vortex line charged with ions moving along the line, would couple strongly to vortex waves under suitable conditions. It is reasonable to expect strong coupling, that is, "resonant" generation of vortex waves, when the following two conditions are satisfied:

$$\omega_{\mathrm{RF}} = \omega(k) - kv_{\mathrm{ion}}, \tag{19}$$

$$v_{\mathrm{ion}} = \frac{\partial\omega(k)}{\partial k}. \tag{20}$$

The first condition is simply that the vortex-wave frequency and sense of rotation, in the frame of reference of the moving ion, is the same as the radio frequency and sense of rotation. The second condition ensures that any energy pumped into the vortex wave, which, of course, moves with the group velocity of the vortex wave, remains in the vicinity of the ion. These two conditions determine a characteristic ionic velocity that depends on the radio frequency. If the longitudinal direct-current electric field, driving the ions along the vortex line, is measured as a function of ionic velocity, an anomaly should be observed at the characteristic velocity. The vortex waves produced near "resonant" conditions clearly have phase velocities in the same direction as the ionic velocity, so that the ionic drag should be enhanced.

The only propagating vortex-wave modes in our frequency range are those that are polarized in a sense opposite to the circulation sense of the vortex. Because the

‡ A general treatment of vortex oscillations is presented by Fetter.[15] Vortex oscillations having wavelengths comparable to the core radius or associated with deformations of the shape of the core itself are not relevant to the work described in this paper.

vortex-wave group velocity is larger than its phase velocity, the sign of ω_{RF} is opposite to that of $\omega(k)$, and Conditions 19 and 20 can only be satisfied simultaneously for a radio-frequency electric field polarization in the same sense as the vortex circulation. The ionic-velocity anomaly should therefore be observed in only one radio-frequency field polarization for a given sense of rotation of the apparatus.

The experiment involves the use of a rotating ^{3}He refrigerator in which a sample of ^{4}He can be cooled to 0.3 K while rotating at 10 rad per second. This rotational speed yields a uniform distribution of vortex lines oriented along the rotational axis with a density of 2×10^4 cm^{-2}. A schematic diagram of our experimental cell is shown in FIGURE 6 and, except for the radio-frequency electrodes, is similar to those used previously for ionic-mobility studies.[19] To properly observe the predicted anomaly, it was necessary to ensure that both the longitudinal constant electrical field and the transverse radio-frequency electrical field were reasonably uniform. This was accomplished by having the drift field defined by four stacks of electrodes, each of which consisted of eight electrodes stepped in potential. Radio-frequency potentials were capacitively coupled to the electrodes and applied in a circularly polarized mode. Circular polarization not only helped distinguish real from spurious effects (because of the intrinsic polarization of the vortex waves) but also helped achieve radio-frequency field uniformity over the cross section of the drift region.

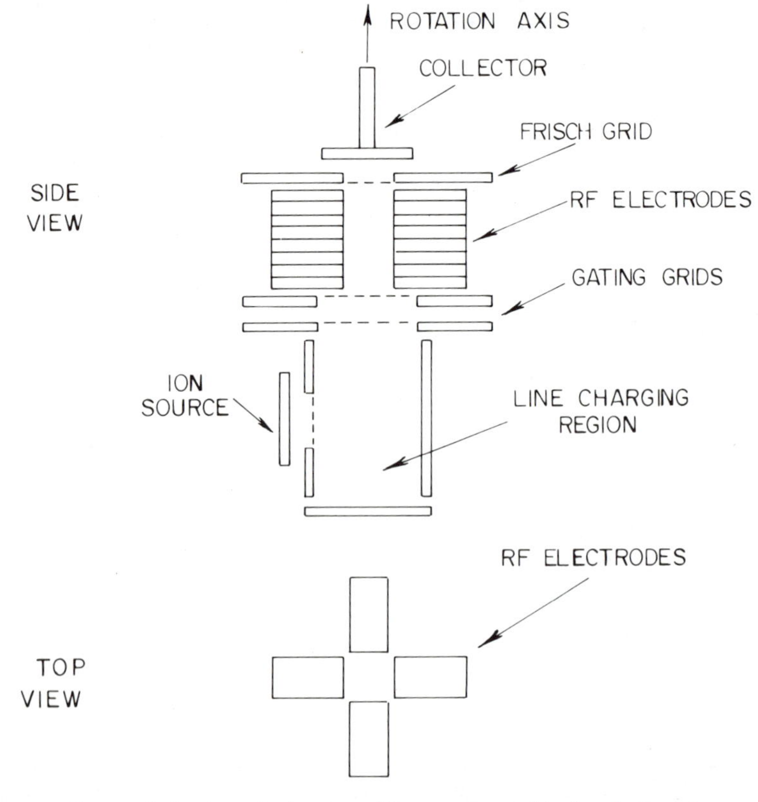

FIGURE 6. A schematic drawing of the experimental ionic-velocity cell.

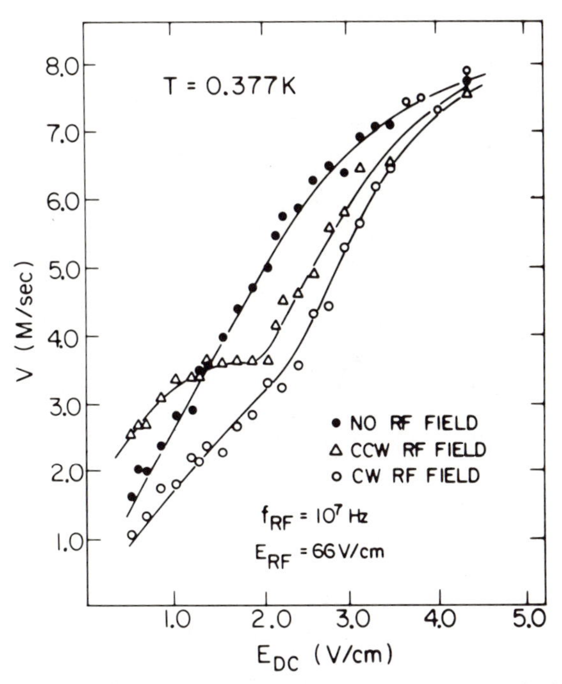

FIGURE 7. A plot of ionic velocity versus longitudinal electrical field for no radio-frequency field and for the radio-frequency field polarized in the clockwise and counterclockwise senses. The solid lines are freely drawn through the data points to guide the eye.

A plot of ionic velocity versus longitudinal electric field, for the cases of no radio-frequency field and for the radio-frequency field polarized in the clockwise and counterclockwise senses, is shown in FIGURE 7 (in this case, the ion cell was rotated in the counterclockwise sense). Note that the ionic velocity is anisotropic with respect to radio-frequency field polarization. Reversing the apparatus' rotational sense, and hence the vortex's circulation sense, reproduces the data with the roles of the two radio-frequency field polarizations reversed. An anomalous kink and plateau in the velocity versus longitudinal electric field curve, for counterclockwise radio-frequency field polarization, is observed near the characteristic velocity determined by Equations 19 and 20. A simultaneous solution of Equations 19 and 20 yields $v_{\text{ion}} = 3.0$ m per second for the radio frequency used. The small discrepancy between this value and the plateau velocity observed can be explained in terms of the radio-frequency field inhomogeneity: the ions move much faster near the top and bottom of the drift region where the radio-frequency field amplitude is small. At the characteristic velocity, the wavelength of the resonantly generated vortex waves (2000 Å) is two orders of magnitude larger than the ionic radius and three orders of magnitude larger than the vortex-core radius. The kink is not perfectly sharp, of course, because of finite vortex-wave damping as well as residual field inhomogeneities. This observation confirms the most important prediction of Halley *et al.*[17] and constitutes a measurement of the vortex-wave dispersion relation at radio frequencies.

Halley *et al.*[18] go beyond a calculation of the characteristic ionic velocity and, using the formalism of Ohmi and Usui[20] to describe the dynamics of the coupled

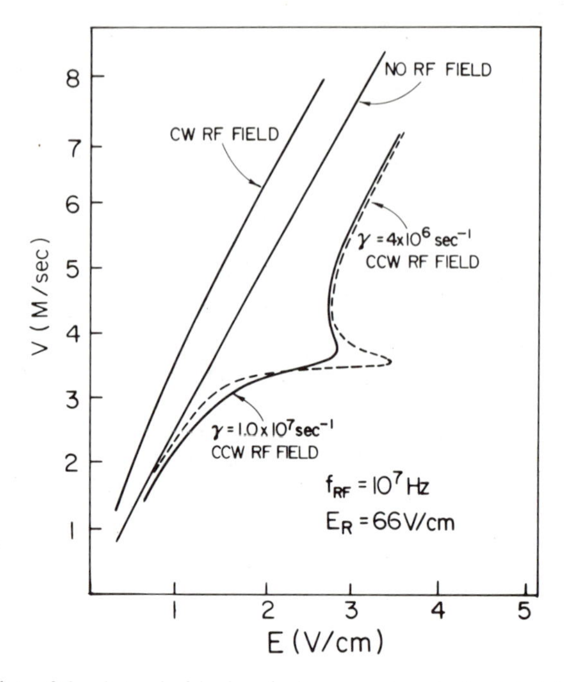

FIGURE 8. Plots of the theoretical ionic velocity versus longitudinal electric field relations with and without a transverse radio-frequency field. Curves for two different vortex-wave damping parameters in the case of resonant radio-frequency polarization are drawn.

ion–vortex-line system, they calculate the complete velocity versus electric field relation. A plot of their results is shown in FIGURE 8. Our results differ in some important ways from those of the calculation.§ For counterclockwise (i.e., "resonant") radio-frequency polarization and for ionic velocities smaller than the characteristic velocity, we observe that the effect of the radio-frequency field is to increase the ionic velocity substantially. One can show that the only vortex waves that should be generated under these circumstances propagate in the direction of the ionic velocity, thus increasing the ionic drag. Furthermore, the magnitude of the effect is much larger than predicted (at least for the expected damping constant). For clockwise (non-"resonant") polarization, the effect of the radio-frequency field is observed to be comparable in magnitude but also has the opposite sign from the calculated effect. Vortex waves propagating both along the direction of the ionic velocity and opposite to it should be generated under these circumstances. Assuming, quite reasonably, that the damping coefficient for vortex waves is not very dependent on wave number, the calculation shows that the net effect is to decrease the ionic drag rather than increase it, as observed.

These discrepancies should not be too surprising given that the Ohmi–Usui formalism fails by orders of magnitude to account for the observed ionic velocity in

§ In Reference 18, the authors erroneously ignored the intrinsic polarization of vortex waves and therefore included a larger class of vortex waves in their calculation than they should have. In this paragraph, we compare our results with their revised calculation.[21]

the absence of a radio-frequency field.[19] The nature of the failure, which disappears in the presence of a small concentration of ^{3}He impurities, is not at all clear but is probably related to a lack of sufficient damping experienced by the vortex line or of the ion–vortex system. Ohmi and Usui calculate the reflection coefficient for vortex waves by assuming that the ion is located on the axis of an otherwise perfectly straight vortex line. It is quite possible that the reflection coefficient would be much smaller for an ion not at rest and located off the vortex-line axis. A transverse radio-frequency field might enhance the damping if it properly polarizes and destabilizes the system in the opposite polarization. In any event, the existence of a plateau at the characteristic velocity should not be affected by these considerations.

In the course of studies of the mobility of ions along vortex lines in the absence of a radio-frequency field, Ostermeier and Glaberson[22] reported an interesting electric field dependence at large electrical fields. In weak fields, the ionic velocity increased linearly with field, characteristic of ordinary mobility behavior. As the electric field was increased further, the velocity saturated and became completely field independent over a range of more than two orders of magnitude in field. They suggested that the limiting velocity was associated with strong coupling between the ion and an appropriate vortex wave. Conditions identical to Equations 19 and 20 were proposed, with the radio frequency replaced by the natural frequency of the bound ion in its hydrodynamic potential well. The velocity plateau observed in the present experiment lends support to this interpretation of the limiting velocity. It is worth pointing out that the determination of the ion natural frequency based on this interpretation is the only measurement to date that probes the details of the structure of a vortex core.

The quality of our data is not as high as we had hoped and was principally limited by radio-frequency heating in the apparatus, which, in turn, severely limited the rate at which data could be acquired. We have, however, been able to observe vortex-wave generation and have demonstrated the feasibility of the technique for mapping out of the vortex-wave dispersion relation over a wide frequency range.

References

1. Yarmchuk, E. J. & W. I. Glaberson. 1978. Phys. Rev. Lett. **41**: 564.
2. Ashton, R. A. & W. I. Glaberson. 1979. Phys. Rev. Lett. **42**: 1062.
3. Anderson, P. W. 1969. Rev. Mod. Phys. **38**: 298.
4. Bald, W. B. & B. A. Hands. 1974. Cryogenics **14**: 179.
5. Sichel, E. K. & B. Serin. 1970. J. Low Temp. Phys. **3**: 635.
6. Yarmchuk, E. J. & W. I. Glaberson. 1978. Rev. Sci. Instrum. **49**: 460.
7. Van Degrift, C. T. 1974. Rev. Sci. Instrum. **45**: 1171.
8. Hall, H. E. 1963. *In* Liquid Helium, International School of Physics "Enrico Fermi" Course XXI. G. Careri, Ed.: 326–335. Academic Press. New York, N.Y.
9. Bekharevich, I. L. & I. M. Khalatnikov. 1961. Sov. Phys.-JETP **13**: 643.
10. Lucas, P. 1970. J. Phys. C: **3**: 1180.
11. Benton, G. S. & D. Boyer. 1966. J. Fluid Mech. **26**: 69; Bennetts, D. A. & L. M. Hocking. 1974. Phys. Fluids **17**: 1671.
12. Ito, H. & K. Nanbu. 1971. J. Basic Eng. **93**: 383.
13. Taylor, G. I. 1921. Proc. R. Soc. London Ser. A **100**: 114; Proudman, J. 1916. Proc. R. Soc. London Ser. A **92**: 408.
14. Thomson, W. 1880. Philos. Mag. **10**: 155; Raja Gopal, E. S. 1964. Ann. Phys. **29**: 350.
15. Fetter, A. L. 1969. Lect. Theor. Phys. **XI-B**: 321.
16. Andronikashvili, E. L. & J. S. Tsakadze. 1960. Sov. Phys.-JETP **10**: 227; Hall, H. E. 1960. Adv. Phys. **9**: 89.

17. Halley, J. W. & A. Cheung. 1968. Phys. Rev. **168**: 209; Erratum. 1970. Phys. Rev. A **1**: 1272.
18. Halley, J. W. & R. M. Ostermeier. 1977. J. Low Temp. Phys. **26**: 877.
19. Ostermeier, R. M. & W. I. Glaberson. 1976. J. Low Temp. Phys. **25**: 317.
20. Ohmi, T. & T. Usui. 1971. Prog. Theor. Phys. **45**: 1717.
21. Halley, J. W. & R. M. Ostermeier. 1978. Personal communication.
22. Ostermeier, R. M. & W. I. Glaberson. 1975. Phys. Rev. Lett. **35**: 241.

GROUP THEORY AND SUPERFLUID SYSTEMS

A. I. Solomon

Faculty of Mathematics
The Open University
Milton Keynes, England

Foreword

By its very nature, a meeting of physicists deals with abstractions. Mathematics, perhaps the most abstract language of all, is used to model the concrete physical processes we see in the real world. But not every state of the physical world is satisfactorily modelled this way; in particular, as Edmund Burke pointed out, "Abstract liberty, like other mere abstractions, is not to be found." The holding of this Third International Conference on Collective Phenomena bears eloquent testimony to the very real and concrete existence of a spirit of scientific liberty that no outside force can quench. I am grateful to my Soviet colleagues for having given me the opportunity to share with them the exhilirating joy of a free scientific exchange of ideas, and to experience at first hand their dedication to this highest expression of liberty.

Introduction

A general quantum mechanical problem is the following: solve the time-independent Schrödinger equation

$$H\Psi = E\Psi$$

for a given Hamiltonian H. Group theory can help solve the problem in essentially two different ways. First of all, there is a group associated with the Hamiltonian operator H consisting of all unitary operators U that leave H invariant, $UHU^{-1} = H$. Clearly, such a subset of the group of all unitary operators does, in fact, form a group; it is called the invariance group, or the symmetry group, associated with H. One value of the symmetry group is the following: if Ψ is a solution of the above Schrödinger equation, so, too, is $\Psi' = U\Psi$, which has the same eigenvalue E, and so the elements U can be used to generate solutions Ψ' from a given solution Ψ. However, a second, and less obvious, way in which group theory can enter is by considering operators U under which H is not left invariant! We may illustrate this in the following simple manner: suppose we take for our Hamiltonian H a 2×2 Hermitian matrix, which we shall also assume traceless. The set of all 2×2 unitary matrices U will maintain this property: the property of H being Hermitian and traceless; this set forms the group $U(2)$. However, the elements of this group will not in general leave H invariant, and it is just this property which is so valuable, since it enables us to use a group element to diagonalize H. For we may write $H = aX + bY + cZ$, where

$$X = \tfrac{1}{2}\begin{bmatrix} & 1 \\ 1 & \end{bmatrix} \qquad Y = \tfrac{1}{2}\begin{bmatrix} & -i \\ i & \end{bmatrix} \qquad Z = \begin{bmatrix} 1 & \\ & -1 \end{bmatrix},$$

so that

$$[X, Y] = iZ \qquad [Y, Z] = iX \qquad [Z, X] = iY.$$

We may think of H as being a 3-vector; diagonalization consists in rotating this 3-vector until it is along the Z axis. Thus,

$$UHU^{-1} = kZ.$$

This rotation does not change the values of the eigenvalues, and they are immediately obtained since the spectrum of Z is known. The value of k is given by the "metric length" of H, $k^2 = a^2 + b^2 + c^2$, so the eigenvalues are $\sqrt{a^2 + b^2 + c^2}$, $-\sqrt{a^2 + b^2 + c^2}$. This rather trivial example illustrates the general procedure that we shall adopt in our discussion of superfluid systems. In the general situation, our Hamiltonian will be an element of a Lie algebra g; in the above, we had $g = \mathrm{su}(2)$, the rotation algebra. This is called the spectrum-generating algebra, for obvious reasons. The diagonalization of H is effected by an automorphism ϕ of the algebra g $\phi: g \to g$ which sends H into a sum of l mutually commuting elements k_i

$$H \to \phi(H) = \sum_{i=1}^{l} c_i k_i .$$

In our applications, the coefficients c_i will be real; the integer l is called the rank of g. A basis for g that includes l such mutually commuting elements always exists. The automorphism ϕ may be implemented by a rotation U $H \to UHU^{-1} = \phi(H)$; this is the generalization of the Bogoliubov transformation, familiar in superfluid ^{4}He and superconductivity. The "metric length" of H corresponds, in the general case, to the (square root of the) Killing form, a useful invariant for Lie algebras, which, in simple cases, is sufficient to determine the energy spectrum.

What are the advantages of using this group theoretic approach, apart from its undoubted elegance? First of all, in reducing the solution of a wide class of problems to a single technique, one unifies and simplifies the problems. Further, the role of the Bogoliubov transformation is clarified, energy spectra are readily found, and coherent states emerge naturally as eigenstates of group elements. Finally, the power of group theory can be brought to bear, in particular, representation theory. One need not implement the diagonalizing automorphism in the (generally) large or infinite-dimensional realization of the Hamiltonian supplied by the physics; one may choose a smaller faithful representation in which to perform one's calculations, confident that the manipulations will be mirrored faithfully in all the representations.

The above ideas will be illustrated with reference to the superfluid ^{4}He and ^{3}He systems in the following text.

Superfluid Bose System

The first application of these ideas is to superfluid ^{4}He, modeled by a bose many-body system with two-body repulsive interactions. Such a system has Hamiltonian $H = \sum_k H_k$, where

$$H_k = E_k a_k^+ a_k + \tfrac{1}{2} \sum_{p,q} V_k a_{p+k}^+ a_{q-k}^+ a_p a_q .$$

Here, a_k (respectively, a_k^+) is the annihilation (respectively, creation) operator for a boson of momentum k; it obeys

$$[a_k, a_{k'}^+] = \delta_{kk'} .$$

E_k is the associated energy and V_k the (Fourier transform of the) interaction potential. The simplication to an exactly solvable model is effected by assuming macroscopic occupation of the $k = 0$ state at zero temperature and treating a_0 and a_0^+ as the c number $\sqrt{N}$, where N is the number density of $k = 0$ bosons. With this simplification, the Hamiltonian H_k reduces to

$$H_{\mathrm{k}}^{\mathrm{red}} = (E_{\mathrm{k}} + NV_{\mathrm{k}})(a_{\mathrm{k}}^+ a_{\mathrm{k}} + a_{-\mathrm{k}}^+ a_{-\mathrm{k}}) + NV_{\mathrm{k}}(a_{\mathrm{k}}^+ a_{-\mathrm{k}}^+ + a_{\mathrm{k}} a_{-\mathrm{k}}).$$

Introducing operators

$$X^{(k)} = -\tfrac{1}{2}(a_{\mathrm{k}}^+ a_{-\mathrm{k}}^+ + a_{\mathrm{k}} a_{-\mathrm{k}})$$

$$Y^{(\mathrm{k})} = i/2(a_{\mathrm{k}}^+ a_{-\mathrm{k}}^+ - a_{\mathrm{k}} a_{-\mathrm{k}})$$

$$Z^{(\mathrm{k})} = \tfrac{1}{2}(a_{\mathrm{k}}^+ a_{\mathrm{k}} + a_{-\mathrm{k}}^+ a_{-\mathrm{k}} + 1),$$

we find that they obey the commutation relations

$$[X, Y] = -iZ \qquad [Y, Z] = iX \qquad [Z, X] = iY$$

(on suppressing the subscript k). Apart from the negative sign in the first commutator, these relations are reminiscent of the su(2) rotation algebra of the introduction. In fact, they are the defining commutation relations of the noncompact "pseudorotation" algebra su(1, 1). This is the three-dimensional analog of Minkowski space. We may write the reduced Hamiltonian (for each k) as

$$H_{\mathrm{k}}^{\mathrm{red}} = a_{\mathrm{k}} X^{(\mathrm{k})} + c_{\mathrm{k}} Z^{(\mathrm{k})} \qquad (a_{\mathrm{k}} = -NV_{\mathrm{k}}, c_{\mathrm{k}} = NV_{\mathrm{k}} + E_{\mathrm{k}}),$$

in terms of the generators of su(1, 1). An explicit rotation about the "$Y^{(\mathrm{k})}$" direction in this case will send $H_{\mathrm{k}}^{\mathrm{red}}$ along the $Z^{(\mathrm{k})}$ direction. This corresponds precisely to the Bogoliubov transformation. However, there is no need to perform this rotation explicitly to obtain the spectrum. The "metric length" of $H_{\mathrm{k}}^{\mathrm{red}}$ in this case is $\sqrt{-a_{\mathrm{k}}^2 + c_{\mathrm{k}}^2} = \sqrt{E_{\mathrm{k}}^2 + 2NVE_{\mathrm{k}}}$; this factor times the spectrum of $Z^{(\mathrm{k})}$ gives the energy spectrum of the reduced Hamiltonian for each k (and, consequently, for the complete reduced Hamiltonian). It can be shown that the spectrum of $Z^{(\mathrm{k})}$, a compact element of a Hermitian representation of a noncompact Lie algebra, in this case consists essentially of the natural numbers. Thus, we obtain the well-known excitation spectrum for this model of superfluid ^{4}He. Note that the spectrum-generating algebra g for this case is $g = \sum_k g_{\mathrm{k}}$ in the sense of a direct sum of Lie algebras, where each g_{k} is isomorphic to su(1, 1).

Superfluid Fermion System

Superfluidity in fermion systems seems to be a consequence of pair formation in opposite momentum states; in recognition of that fact and without further ado, we write down a model Hamiltonian in which interactions other than those pairing interactions are ignored:

$$H = \sum_{k,\alpha} E_{\mathrm{k}} a_{\mathrm{k}\alpha}^+ a_{\mathrm{k}\alpha} + \tfrac{1}{2} \sum_{k,k',\alpha,\beta} V_{\mathrm{kk}'} a_{\mathrm{k}\alpha}^+ a_{-\mathrm{k}\beta}^+ a_{-\mathrm{k}'\beta} a_{\mathrm{k}'\alpha}.$$

The fermion annihilation and creation operators obey the anticommutation rules

$$[a_{\mathrm{k}\alpha}, a_{\mathrm{k}'\beta}^+]_+ = \delta_{\mathrm{kk}'} \delta_{\alpha\beta},$$

where k, k' are 3-momentum labels as before, and the additional α, β are spin labels that may be either up (↑) or down (↓). Reduction to an exactly solvable form is achieved by using the following linearization procedure: for any two operators A and B, we have the identity

$$AB = (A - \langle A \rangle)(B - \langle B \rangle) + A\langle B \rangle + B\langle A \rangle - \langle A \rangle\langle B \rangle,$$

where $\langle A \rangle$, $\langle B \rangle$ are the expectation values in some ground state. To the extent that we may ignore deviations from this ground state, we may approximate

$$AB \doteqdot A\langle B \rangle + B\langle A \rangle,$$

ignoring an additive c-number constant. Applying this process to our model Hamiltonian leads to the reduced form

$$H^{\mathrm{red}} = \sum_k H_{\mathrm{k}},$$

where

$$H_{\mathrm{k}} = E_{\mathrm{k}} a_{\mathrm{k}\alpha}^{+} a_{\mathrm{k}\alpha} + V(k, \alpha, \beta) a_{\mathrm{k}\alpha}^{+} a_{-\mathrm{k}\beta}^{+} + V^*(k, \alpha, \beta) a_{-\mathrm{k}\beta} a_{\mathrm{k}\alpha}$$

and

$$V(k, \alpha, \beta) = \sum_k \langle \tfrac{1}{2} V_{\mathrm{kk}'} a_{\mathrm{k}'\beta} a_{-\mathrm{k}'\alpha} \rangle.$$

Since our reduced Hamiltonian decouples into a sum of independent H_{k}'s (just as in the boson case previously), we may treat each H_{k} individually and suppress the k subscript for typographic convenience. As a consequence, the spectrum-generating algebra we obtain for H^{red} will be a direct sum of (isomorphic) algebras associated with H_{k}. To identify this algebra, we introduce the 4-vector

$$(A_1, A_2, A_3, A_4) = (a_{\uparrow}, a_{\downarrow}, a_{-\downarrow}^{+}, a_{-\uparrow}^{+})$$

and the matrices $X_{\mu\nu}$ $\mu, \nu = 1, 2, 3, 4$, where

$$(X_{\mu\nu})_{\mathrm{ij}} = \delta_{\mu}^{\mathrm{i}} \delta_{\nu}^{\mathrm{i}} A_{\mathrm{i}}^{+} A_{\mathrm{j}} \qquad i, j = 1, 2, 3, 4.$$

A little manipulation leads to the commutation relations,

$$[X_{\mu\nu}, X_{\mu'\nu'}] = \delta_{\nu}^{\mu'} X_{\mu\nu'} - \delta_{\nu'}^{\mu} X_{\mu'\nu}.$$

These are the defining relations of $gl(4, R)$, the Lie algebra of all real 4×4 matrices, as may be readily seen by choosing a basis of 16 different 4×4 matrices, each of which possesses precisely the single nonzero entry 1. The operators in the Hamiltonian actually only occur in Hermitian combinations of $gl(4, R)$ elements; so the algebra obtained is, in fact, u(4)—the Lie algebra of Hermitian 4×4 matrices. Thus, in this case, the spectrum is generated by the Lie algebra g, where

$$g = \sum_k g_{\mathrm{k}}$$

and each

$$g_{\mathrm{k}} \sim \mathrm{u}(4).$$

Since u(4) is a rank 4 Lie algebra, each H_{k} can be sent to a sum $c_1 k_1 + c_2 k_2 + c_3 k_3 + c_4 k_4 (c_{\mathrm{i}} \in R)$ by the general Bogoliubov transformation, where the mutually commuting k_{i} are given by

$$\{k_1, k_2, k_3, k_4\} = \{a_{\uparrow}^{+} a_{\uparrow}, a_{\downarrow}^{+} a_{\downarrow}, a_{-\downarrow} a_{-\downarrow}^{+}, a_{-\uparrow} a_{-\uparrow}^{+}\}$$

for each (suppressed) momentum subscript k.

The physics of this model can most easily be recovered by inspecting the four-dimensional representation of the reduced Hamiltonians H_k; in each case, H_k may be represented by the 4×4 Hermitian matrix M,

$$M = \begin{bmatrix} E & V \\ V^+ & -E \end{bmatrix},$$

with $E = \varepsilon I$, $V = \sum_{\mu=0}^{3} b_\mu \tau^\mu$ (suppressing subscript k). Here, the b_μ are complex numbers, in general, and the τ^μ are the Pauli spin matrices

$$I = \tau^0 = \begin{bmatrix} 1 & 0 \\ 0 & 1 \end{bmatrix} \quad \tau^1 = \begin{bmatrix} 0 & 1 \\ 1 & 0 \end{bmatrix} \quad \tau^2 = \begin{bmatrix} & -i \\ i & \end{bmatrix} \quad \tau^3 = \begin{bmatrix} 1 & \\ & -1 \end{bmatrix}.$$

Superconductor (BCS model)

We assume spin-0 pairing only, so that

$$b_1, b_2, b_3 = 0, \qquad b_0 = x + iy \neq 0.$$

In this case, M simplifies to

$$M = xT_1 - yT_2 + \varepsilon T_3,$$

where

$$\{T_1, T_2, T_3\} = \{\tau^1 \times I, \tau^2 \times I, \tau^3 \times I\}.$$

The T_i clearly generate su(2), and the energy spectrum is given by

$$\sqrt{\varepsilon^2 + x^2 + y^2} = \sqrt{\varepsilon^2 + |V|^2},$$

with $|V|$ the energy gap.

Superfluid Helium Three

We assume spin-1 pairing only, so that

$$b_0 = 0, \qquad b_i \neq 0 (i = 1, 2, 3).$$

Again, the energy spectrum is given by the eigenvalues of M. Since

$$M^2 = \begin{bmatrix} E^2 + VV^+ & 0 \\ 0 & E^2 + V^+V \end{bmatrix},$$

we obtain two energy gaps, which are given by the square roots of the eigenvalues of VV^+. We obtain a unique energy gap when V obeys

$$VV^+ = \Delta^2 I \qquad \text{(unitary state)}.$$

This occurs when all the b_i are real (B phase) or when $|V_{12}| = |V_{21}|$ (A phase).

Source Reading

So as not to interrupt the flow of this presentation, I have refrained from giving references in the body of the text. In this section, I shall supply the interested reader with additional sources from which he may possibly embark upon his own further study of the material.

Spectrum-generating algebras originally arose in elementary particle physics as a technique for generating elementary-particle spectra. Extensive use has been made of them in this context by Ne'eman[1] and his collaborators. A more formal description is given by Joseph,[2] who introduces the idea of a minimal realization. The first person to use these techniques in the context of superfluid systems, as outlined in this note, seems to be the present writer[3]; this paper is also available in a Russian translation by Perelomov.[4]

The introduction of a canonical transformation to effect the solution of the superfluid ^{4}He model is, of course, due to Bogoliubov; the usefulness of the generalized form (as outlined here) is not limited to superfluidity. Other exactly solvable many-body problems succumb to the same procedure; these models include the so-called XY and Ising lattice models.[5]

A comprehensive theoretical review of ^{3}He is given in the article by Leggett[6]; more details of the present treatment of the anisotropic fermi superfluid given here will be published elsewhere.

References

1. Ne'eman, Y. 1967. Algebraic Theory of Particle Physics. Chap. 10. W. A. Benjamin. New York, N.Y.
2. Joseph, A. 1974. Minimal realizations and spectrum generating algebras. Commun. Math. Phys. **36:** 325.
3. Solomon, A. I. 1971. Group theory of superfluidity. J. Math. Phys. **12:** 390.
4. Perelomov, A. M. 1977. Usp. Fiz. Nauk. **123:** 23 (in Russian); in English: Generalized coherent states and some of their applications. Sov. Phys.-Usp. **20**(9), Sect. C **2:** 718.
5. Solomon, A. I. & W. Montgomery. 1978. Generalised XY model. J. Phys. A (Math. Gen.): 1633.
6. Leggett, A. J. 1974. A theoretical description of the new phases of liquid ^{3}He. Rev. Mod. Phys. **47:** 331.

HEAT-RADIATION TRANSFER IN HETEROGENEOUS MEDIA

A. A. Men'*

Leningrad, U.S.S.R.

The problem of energy transfer in heterogeneous media, which has a great number of practical applications, includes many aspects; heat-radiation transfer is one of them. This problem has been extensively studied for several decades, both theoretically and experimentally. Nevertheless, there are some questions still to be answered.

The characteristic feature of radiative transfer in a heterogeneous medium is strong scattering. Since the shape and dimensions of nonhomogeneities in materials can vary in a large scale, the theoretical difficulties of the problem are mainly associated with scattering. Here, I shall consider heat-radiation transfer in a medium with arbitrary scattering characteristics. In the first section, I shall discuss the case of a small concentration of scatterers for which the geometric optics approach is valid; the mathematical treatment is based on the radiative transfer equation. In the second section, I shall describe a more general case of an arbitrary concentration of nonhomogeneities in terms of fluctuative electrodynamics.

Asymptotic Solution for an Infinite Medium

Throughout this article, I shall consider heat radiation in an emitting, absorbing, and scattering medium of nonuniform temperature. My goal is to calculate the flux of radiant energy.

Let the concentration of scatterers be small enough that their mutual influences can be neglected. Then, as is well known,[1,2] the spectral radiation flux $\mathscr{E}_\nu(\bar{r})$ can be expressed in terms of the spectral intensity I_ν, which is a function of space coordinates and the direction of a traveling ray. For simplicity, it is assumed that the temperature depends on the x coordinate only. Then, $I_\nu = I_\nu(x, \mu)$ ($\mu = \cos\varphi$, where φ is the angle made by a ray with the direction of the axis $0x$), and the value of the radiant flux can be written as follows:

$$\mathscr{E}_\nu(x) = 2\pi \int_{-1}^{1} I_\nu(x, \mu)\mu \, d\mu. \tag{1}$$

The governing equation for $I_\nu(x, \mu)$ is the radiative transfer equation (RTE):

$$\mu \frac{dI_\nu}{d\tau} = -I_\nu + j_\nu(T) + \frac{g}{4\pi}\int_{-1}^{1} I_\nu(\tau, \mu')\rho(\mu, \mu')\, d\mu', \tag{2}$$

where $\tau = (k_\nu + \gamma_\nu)x$ is a dimensionless coordinate (optical depth), $g = \gamma_\nu/(k_\nu + \gamma_\nu)$ is the albedo of scattering, $\rho(\mu, \mu')$ is the angular function of scattering, $j_\nu(T)$ is the volumetric spectral emissive power of the matter, and

$$j_\nu(T) = k_\nu n_\nu^2 D I_B(\nu, T) = (1 - g)Dn_\nu^2 I_B(\nu, T),$$

* *Present address*: Faculty of Mechanical Engineering, Techmon, Haifa, Israel.

where D is the diffraction factor, $I_B(\nu, T)$ is Planck's function, and k_ν, γ_ν, and n_ν are the spectral absorptivity, spectral scattering coefficient, and spectral refractive index, respectively (for further details, see References 1 and 3).

Concerning the angular function, $\rho(\mu, \mu')$, we assume only the constraint to be imposed; that is, $\rho(\mu, \mu')$ can be expanded in the Legendre polynomial series

$$\rho(\mu, \mu') = 2\pi \sum c_i P_i(\mu) P_i(\mu'). \tag{3}$$

To study the asymptotic behavior of $\mathscr{E}_\nu(\tau)$ for points far removed from the boundaries, $\tau \gg 1$, we use the same method as in Reference 3. Let $I_\nu(x, \mu)$ be the sum of two terms, $I_\nu^{(1)}$ and $I_\nu^{(2)}$. The first term obeys the equation

$$\mu \frac{dI_\nu^{(1)}}{d\tau} = -I_\nu^{(1)} + (1 - g)Dn^2 I_B(\nu, T), \tag{4}$$

and the second obeys the equation

$$\mu \frac{dI_\nu^{(2)}}{d\tau} = -I_\nu^{(2)} + \frac{g}{4\pi} \int_{-1}^{1} I_\nu^{(2)} \rho(\mu, \mu')\, d\mu' + \frac{g}{4\pi} \int_{-1}^{1} I_\nu^{(1)} \rho(\mu, \mu')\, d\mu'. \tag{5}$$

Boundary conditions for Equation 4 include all kinds of nonhomogeneity (external radiation, reflection on the surfaces of the body; see Reference 3). Equation 4 is of the same kind as RTE in homogeneous substances, and it can be easily treated. By solving it, one can calculate the last term in Equation 5, which acts as a collection of heat-radiation sources distributed in the medium.

For the case $\tau \gg 1$, the solution of Equation 4 takes the form

$$\left.\begin{aligned} I_+^{(1)} &= (1 - g)Dn^2 \int_{-\infty}^{\tau} I_B[\nu, T(\tau')] e^{-(\tau-\tau')/\mu} \frac{d\tau'}{\mu} \\ I_-^{(1)} &= (1 - g)Dn^2 \int_{\tau}^{\infty} I_B[\nu, T(\tau')] e^{-(\tau'-\tau)/\mu} \frac{d\tau'}{\mu} \end{aligned}\right\}, \tag{6}$$

where the first line is valid for $\mu > 0$, and the second one for $\mu < 0$. These equalities yield the following expression of $\mathscr{E}_\nu^{(1)}(\tau)$:

$$\mathscr{E}_\nu^{(1)}(\tau) = 2\pi\left(\int_0^1 I_+^{(1)} \mu\, d\mu - \int_0^1 I_-^{(1)} \mu\, d\mu\right) = 2\pi \int_{-\infty}^{\infty} \frac{dI_B}{d\tau} E_3 |\tau - \tau'|\, d\tau', \tag{7}$$

where $E_3(z)$ is the exponential integral function of the third order. Integration by parts results in

$$\mathscr{E}_\nu^{(1)}(\tau) = -\frac{4}{3}\pi(1 - g)Dn^2\left(\frac{dI_B}{d\tau} + \frac{3}{5}\frac{d^3 I_B}{d\tau^3} + \cdots\right). \tag{8}$$

Now we proceed to Equation 5. To solve it, Green's function of one-speed transport equation, expressed in terms of Case's eigenfunctions,[4] can be applied.

Since the asymptotic behavior of $\mathscr{E}_\nu(\tau)$ is of main interest, we use Equations 6 to calculate the source function in Equation 5 and Green's function of an infinite medium $G_\infty(\tau, \mu; \tau_1, \mu_1)$. Substitution of the expansion Equation 3 into Equation 5 yields the equation whose Green's function can be expressed as

$$G_\infty = \begin{cases} \dfrac{1}{2\pi} \displaystyle\sum_{j=1}^{M} \varphi_{j+}(\mu_1)\varphi_{j+}(\mu) \dfrac{e^{-(\tau-\tau_1)/\nu_{j+}}}{N_{j+}} \\ \qquad + \displaystyle\int_0^1 \varphi_{\nu+}(\mu)\varphi_{\nu+}(\mu_1) e^{-(\tau-\tau_1)/\nu} \dfrac{d\nu}{N(\nu)}; \qquad \tau > \tau_1 \\ -\dfrac{1}{2\pi} \displaystyle\sum_{j=1}^{M} \varphi_{j-}(\mu_1)\varphi_{j-}(\mu) \dfrac{e^{-(\tau_1-\tau)/\nu_{j-}}}{N_{j-}} \\ \qquad - \displaystyle\int_{-1}^0 \varphi_{\nu-}(\mu)\varphi_{\nu-}(\mu_1) e^{-(\tau_1-\tau)/\nu} \dfrac{d\nu}{N(\nu)}; \qquad \tau < \tau_1 \end{cases} \tag{9}$$

where $\nu_{j\pm}$ and $\varphi_{\nu\pm}(\mu)$ are eigenvalues and eigenfunctions of continuum, and $N_{j\pm}$ are the normalizing constants.[4] For the value $\mathscr{E}_\nu^{(2)}(\tau)$, we obtain

$$\mathscr{E}_\nu^{(2)}(\tau) = \pi g(1-g)Dn^2 \sum_{i=0}^{m} c_i \int_{-\infty}^{\infty} K_i(\tau, \tau_1) \left[\int_{-\infty}^{\tau_1} I_B(\tau'')F_{i+}(\tau_1 - \tau'')\, d\tau'' + \int_{\tau_1}^{\infty} I_B(\tau'')F_{i-}(\tau'' - \tau_1)\, d\tau'' \right] d\tau_1, \tag{10}$$

where

$$K_i(\tau, \tau_1) = \int_{-1}^{1} \int_{-1}^{1} \mu G_\infty(\tau, \mu; \tau_1, \mu_1) \cdot P_i(\mu)\, d\mu_1\, d\mu; \tag{11}$$

$$F_{i\pm}(z) = \int_0^1 \frac{P_i(\mu)}{\mu} e^{-z/\mu}\, d\mu. \tag{12}$$

Substitution of Equation 9 and integration by parts of the internal integrals in Equation 10 (see the Appendix) result in

$$\mathscr{E}_\nu(x) = \mathscr{E}_\nu^{(1)}(x) + \mathscr{E}_\nu^{(2)}(x) = -\left\{ \frac{4}{3}\pi(1-g)Dn^2 + \pi(1-g)gDn^2 \times \left[4J + \frac{2}{\pi}\left(1 + \frac{c_1}{3}(1-g)\right)S \right] \right\} \frac{1}{k_\nu + \gamma_\nu} \frac{\partial I_B}{\partial T} \frac{dT}{dx}, \tag{13}$$

where

$$J = \int_0^1 \frac{\nu^3}{N(\nu)}\, d\nu; \qquad S = \sum_j \frac{\nu_j^3}{N_{j+}},$$

and summation in S is done taking into account all discrete eigenvalues of Equation 5. Determination of all the values, ν_j, is not difficult, but it cannot be done analytically.[4] It should be emphasized that the terms with high-order derivatives of $\partial I_B/\partial T$ were neglected in Equation 13.

Since the proportionality factor of $\mathscr{E}_\nu$ and dT/dx is defined as the radiative conductivity, in our case we have

$$\lambda_{rad} = \frac{\pi n_\nu^2}{k_\nu + \gamma_\nu} \frac{\partial I_B}{\partial T} \left\{ \frac{4}{3} D(1-g) + Dg(1-g)\left[4J + \frac{2}{\pi}\left(1 + \frac{c_1}{3}(1-g)\right)S \right] \right\}. \tag{14}$$

It is also well known that for internal regions of optically thick media, the most widely used formula is Rosseland's expression of λ_{rad}:

$$\lambda_{rad_R} = \frac{4}{3}\pi \frac{n_v^2}{k_v + \gamma_v} \frac{\partial I_B}{\partial T}. \quad \textbf{(15)}$$

Comparison of Equations 14 and 15 shows that Rosseland's formula may not be valid for scattering materials.

Heat Radiation in Strong Scattering Materials

There are a lot of materials for which the assumptions of the previous section are not valid. Indeed, in the majority of scattering solids, the distances between inhomogeneities, as well as their dimensions, are comparable with the wavelength. In such cases, the geometric optics approach is not good enough. Violation of this approach mainly results from dependence of scattering patterns (angular function) on the distance of a scatterer (for small distances) and from diffraction of heat radiation on the boundary of a scatterer.[5]

To avoid the difficulties mentioned above, we pass to the more general physical model of the process where heat radiation is considered as a fluctuative electromagnetic field. To calculate the flux of radiation, it is necessary to solve Maxwell's equations with external sources of field. By such an approach, the scattering characteristics γ_v, $\rho(\mu, \mu')$, n_v are not introduced, and the only characteristic of materials is complex permittivity $\varepsilon(\bar{r})$ as a function of space coordinates. It can be shown[5] that the relation between nonuniform temperature in a medium and heat-radiation flux in a nonhomogeneous dielectric is based on application of the fluctuative-dissipative theorem and the electromagnetic theorem of reciprocity. This approach yields

$$\mathscr{E}_\omega(\bar{r}) = \int_{r'} \int_\omega \theta[\omega, T(\bar{r}')] K(\bar{r}, \bar{r}')\, d\omega\, d\bar{r}', \quad \textbf{(16)}$$

where

$$K(\bar{r}, \bar{r}') = \frac{ic\omega}{32\pi^3} [\varepsilon^*(\bar{r}') - \varepsilon(\bar{r})]\{[(\bar{E}_{ey}\bar{E}^*_{mz} - \bar{E}_{ez}\bar{E}^*_{my}) + \text{c.c.}] + [(\bar{E}_{ez}\bar{E}^*_{mx} - \bar{E}_{ex}\bar{E}^*_{mz}) + \text{c.c.}] + [(\bar{E}_{ex}\bar{E}^*_{my} - \bar{E}_{ey}\bar{E}^*_{mx}) + \text{c.c.}]\}, \quad \textbf{(17)}$$

where ω is the cyclic frequency, c is the speed of light, and $\bar{E}_{ex}$, $\bar{E}_{my}$, etc., are diffraction electrical fields of electric and magnetic dipoles directed along the $0x$ and $0y$ axes, respectively; c.c. denotes a complex conjugate value. The dipole fields act as a Green's function of the problem. They depend on the structure of the material but not on the temperature distribution. Since the scatterers are randomly distributed in the medium, it is necessary to average $K(\bar{r}, \bar{r}')$ with respect to the ensemble of structures.

In the simplest case of noninteracting scatterers, the electrical fields $\bar{E}_{ex}$, $\bar{E}_{my}$, etc., can be written for any inhomogeneity as if all the others were absent. The averaging in such a case implies the averaging with respect to shape and dimensions of a single scatterer. If we wish, for instance, to calculate the emissive power of a unique elementary volume δV, we can use Equation 15 with integration over volume δV substituted by summation over all the scatterers. Such a procedure enables us to take into account the effect of diffraction on the boundaries of scatterers. It can be shown[5] that the resulting expression is as follows (for spherical particles of R_i radius):

$$j_v = k_v n_v^r I_B(v, T) D(R_i, N_i), \tag{18}$$

where

$$D(R_i, N_i) = 1 + \sum N_i R_i \left(\frac{P_{vi}}{k_v n_v^2 I_B(v, T)} - \frac{4}{3} \pi R_i \right),$$

where N_i is the concentration of scatterers of type i, and P_{vi} is the heat radiation of a unit area of the scatterer surface. These values were obtained for a single sphere. $D(R_i, N_i)$ is the diffraction factor mentioned in the first section.

Appendix

We use the following properties of eigenfunctions and eigenvalues of the one-speed transport equation:[4]

$$\int_{-1}^{1} \varphi_v \, d\mu = 1;$$

$$\int_{-1}^{1} \varphi_{v+} \mu \, d\mu = - \int_{-1}^{1} \varphi_{v-} \mu \, d\mu; \qquad v_{j+} = -v_{j-i} \qquad N(v) = -N(-v).$$

Then, we obtain

$$\int_{-\infty}^{\infty} K_i\{\cdots\} \, d\tau_1 = \int_{-\infty}^{\tau} K_{i+}\{\cdots\} \, d\tau_1 - \int_{\tau}^{\infty} K_{i-}\{\cdots\} \, d\tau_1;$$

$$K_{i\pm} = \sum \frac{1}{2\pi N_{j\pm}} e^{-(|\tau-\tau_1|)/v_j} W_{ji\pm} W_{j1\pm} + \int_{0(-1)}^{1(0)} \frac{W_{vi\pm} W_{v1\pm}}{N_{\pm}(v)} e^{-(|\tau-\tau_1|)/v} \, dv;$$

$$W_{ji\pm} = \int_{-1}^{1} P_i(\mu) \varphi_{j\pm}(\mu) \, d\mu; \qquad W_{j1} = v_j(1-g); \qquad W_{v1} = v(1-g);$$

$$\{\cdots\}_{\tau_1} = I_B(\tau_1)(F_{i1+} + F_{i1-}) - \frac{dI_B}{d\tau}(F_{i2+} + F_{i2-}) + \cdots;$$

$$F_{ik\pm} = \int_{0}^{1} P_i(\pm\mu) \mu^{k-1} \, d\mu;$$

$$\int_{-\infty}^{\tau} e^{-(\tau-\tau')/v_j} \{\cdots\} \, d\tau' = v_j(F_{i1+} + F_{i1-}) I_B(\tau) - [v_j(F_{i2+} - F_{i2-}) + v_j^2(F_{i1+} + F_{i1-})] \frac{dI_B}{d\tau} + \cdots;$$

$$\int_{\tau}^{\infty} e^{-(\tau'-\tau)/v_j} \{\cdots\} \, d\tau' = v_j(F_{i1+} + F_{i1-}) I_B(\tau) - [v_j(F_{i2+} - F_{i2-}) - v_j^2(F_{i1+} + F_{i1-})] \frac{dI_B}{d\tau} + \cdots;$$

$$f_{\pm}(v_j) = v_j(F_{i2+} - F_{i2-}) \pm v_j^2(F_{i1+} + F_{i1-});$$

$$\mathscr{E}_{\nu}^{(2)} = -\pi g(1-g)Dn^2 \sum_{ij} C_i\left(\frac{W_{ji+}W_{j1+}}{2\pi N_{j+}}f_+(\nu_j) - \frac{W_{ji-}W_{j1-}}{2\pi N_{j-}}f_-(\nu_j) + \int_0^1 \frac{W_{\nu i+}W_{\nu+}}{N(\nu)}f_+(\nu)\,d\nu - \int_{-1}^0 \frac{W_{\nu i-}W_{\nu-}}{N(\nu)}f_-(\nu)\,d\nu\right)\frac{dI_B}{d\tau} + \cdots$$

$$= -\pi g(1-g)Dn^2 \sum_j \left[C_0\left(\frac{\nu_j^2 W_{j1}}{\pi N_{j+}} + 2\int_0^1 \frac{W_{\nu 1+}}{N(\nu)}\nu^2\,d\nu\right) + \frac{2}{3}C_1\frac{W_{01}^2\nu_j}{N_{j+}}\right]\frac{dI_B}{d\tau} + \cdots$$

REFERENCES

1. CHANDRASEKHAR, S. 1960. Radiative Transfer. Dover Publications. New York, N.Y.
2. VISKANTA, R. & E. E. ANDERSON. 1975. Adv. Heat Transfer **11**: 317.
3. MEN', A. A. & O. A. SERGEEV. 1974. *In* Proceedings of the 5th International Heat Transfer Conference, Tokyo. 1974: 67.
4. CASE, K. M. & D. F. ZWEIFEL. 1967. Linear Transport Theory. Addison-Wesley Publishing Company. New York, N.Y.
5. MEN', A. A. 1977. Teplofiz. Vys. Temp. **15**(6): 1212.

ACOUSTIC-WAVE INSTABILITIES IN ANISOTROPIC PIEZOSEMICONDUCTORS

I. S. Ravvin

Moscow, U.S.S.R.

It is known that, given a supersonic drift of charge carriers in piezosemiconductors, there is an amplification of acoustic waves, particularly of equilibrium thermal acoustic noise, because of acoustoelectric interactions. Since the sample has a finite size, all amplified wave energy or part of it is reflected from a crystal surface; the energy then interacts with the medium again and is re-reflected, and so on. If after multiple reflections the amplified acoustic flow is in the direction of the initial flow, so-called round-trip instabilities of acoustic noise take place. To return the flow, it is possible to use either samples of rather complicated geometry or various features of the medium itself and the laws of wave reflection.

This paper deals with acoustic instabilities that occur as a result of the elastic features of the medium and the anisotropic features of the acoustoelectronic interaction. Consider a sample of length L and width d (FIGURE 1), in which the direction of the drift field E_d coincides with the normal $\bar{n}$ to the plane-parallel surfaces of the sample. In general, we can observe three principal variants of wave propagation with quasitransverse and quasilongitudinal polarization[1]:

1. The waves propagate along the normal to the surface:

$$\vec{q}_f \| \vec{q}_r \| \vec{n} \| \vec{E}_d$$

2. Off-axis waves propagate without a polarization transformation when reflecting: T-T-T, L-L-L, where T and L correspond to transverse and longitudinal polarizations, respectively
3. Off-axis waves propagate with a polarization transformation when reflecting: T-L-T, L-T-L.

The carrier's drift enforces the amplification of acoustic noise in all directions where the electromechanical coupling factor does not equal zero and $(\vec{k}V_d/\omega - 1) > 0$, where $\vec{k}$ is the wave vector, V_d is the drift velocity, and ω is the sound frequency.[2,3] However, the round-trip amplification occurs only for directions in which the increment Γ_+ of the wave amplification by the drift exceeds the attenuation of the wave energy by propagation of the wave against the field (decrement Γ_-), by viscosity (Γ_{vis}), and by the reflection from the surface (R, $\tilde{R}$ are reflection quotients of power)

$$R\tilde{R} \exp\left(\frac{\Gamma_+ + \Gamma_{vis}}{\cos \psi_f} + \frac{\Gamma_- + \Gamma_{vis}}{\cos \psi_r}\right) d \underset{\bullet}{>} 1, \qquad (1)$$

where ψ_f and ψ_r are the angles between the group velocities of waves k_f and k_r and the direction n, respectively.

Specific conditions under which instability occurs (such as 1–3 above) are determined by the choice of the crystallographic axis with respect to the direction of drift. If the orientation is such that the curve of the electromechanical coupling coefficient $\eta(\varphi)$, which characterizes the value of electron–phonon interactions for the waves with a fixed polarization, is symmetric with respect to the normal to the

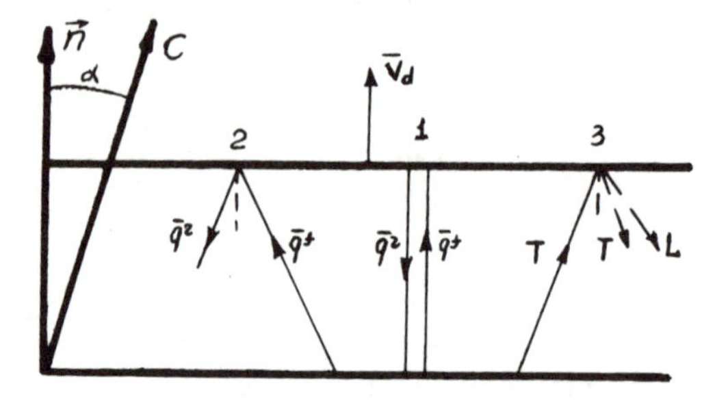

FIGURE 1. Schemes of wave reflection in a plate.

reflection surface, the waves amplified by the drift and the corresponding reflected waves propagate in the direction of the same $\eta_f = \eta_r$. In this case, there is dissipation of the reflected waves. When such an orientation is taken, the instabilities of acoustic noise are possible only because of asymmetric dependence of the electronic increment versus the drift field, when $|\Gamma_+| > |\Gamma_-|$. However, even under such conditions, Equation 1 is valid only for the waves from a narrow cone close to $\vec{q}_f \| \vec{q}_r \| V_d$ since for oblique waves of larger φ, the reflection coefficient decreases for the same polarization.[1] Absolute instability of the shear and longitudinal acoustic noise normal to the surface occurs, for instance, in AEO,[4] and off-axis acoustic-noise instabilities in such oriented samples has been considered in Reference 1. When the polarization is changed, convectional acoustic-noise instabilities occur when the wave is transformed into one with a small electromechanical coupling factor. Shapira *et al.*[5] discuss the development of acoustic-noise instabilities in the process T-L-T in the sample of GaAs when $\eta_T \gg \eta_L$.

Therefore, all the acoustic-noise instabilities that occur in a symmetric case are associated with large attenuation, which is produced either by the propagation of the wave against the field (Equations 1 and 2) or by reflection. Such instabilities can occur only under severe restrictions on the sound frequency ω, the crystal parameters (d, η), the sample geometry (d), and the field (E_d) imposed by the strong energy dissipation.

When a crystal has an arbitrary orientation, the conditions required for off-axis waves to be amplified are changed, since the electromechanical coupling factor is not the same for incident and reflected waves. In any specific case, the domain of acoustic instabilities is given by Equation 1. The most interesting case is an orientation in which most amplified waves are reflected in the direction where the electron–phonon interaction vanishes.[6,7]

FIGURE 2 shows the domain of acoustic-noise instabilities in a plate of CdSe with $d = 0.02$ cm, where the angle between the polar axis C_{6V} and $\vec{n}$ equals $\alpha = -15$ degrees. Notice that the increments Γ_+ and Γ_- are calculated according to White's linear theory when the acoustic-noise spectrum is replaced by a monochromatic wave $\omega_{max} = (\omega_c \omega_D)^{1/2}$, where $\omega_c = \sigma/\varepsilon$ and $\omega_D = V_s^2/D$ are the conductivity relaxation frequency and the free-charge diffusion frequency, respectively, according to Hutson–White.

If the field exceeds the threshold $\mu > dE_{th}/\cos\varphi$, such instabilities occur in domain 1 for the shear waves $\vec{q}_f \| \vec{q}_r \| \vec{n}$ and $\eta_f^2 = \eta_r^2 = 0.017$; in domains 2 and 3 for the shear waves $\vec{q}_f(\varphi = -15$ degrees), $\vec{q}_r(\varphi = +16$ degrees), $\eta_f^2 = \eta_{max}^2 = 0.027$, $\eta_r = \eta_{min} = 0$; in domain 4 for shear waves $\vec{q}_f(\varphi = +45$ degrees) and $\vec{q}_r(\varphi = -47$ degrees), $\eta_f^2 = \eta_{max}^2 = 0.027$, $\eta_r = \eta_{min} = 0$. In domains 5 and 6, such instabilities occur for the T-L-T process: for 5, $\vec{q}_f(\varphi = -15$ degrees), $R_{LT} = R_{TL} = 0.21$; for 6,

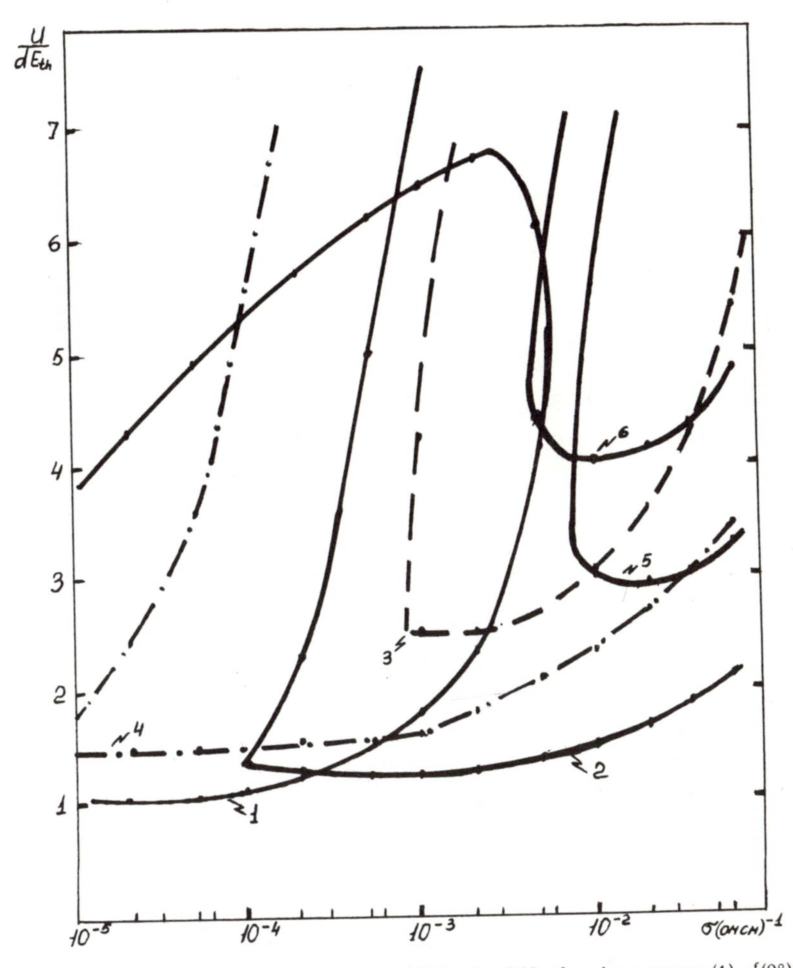

FIGURE 2. Domains of acoustic-noise instabilities in CdSe for shear waves (1) $q^{f}(0°)$, (2) and (3) $q^{f}(-15°)$, and (4) $q^{f}(45°)$. For a T-L-T process: (5) $q^{f}(-15°)$ and (6) $q^{f}(-30°)$, except for (3), where $\omega = (\omega_C \omega_D)^{1/2}$, $3 - \omega = 0.5(\omega_C \omega_D)^{1/2}$.

$\vec{q}_f(\varphi = -30$ degrees), $R_{TL} = R_{LT} = 0.45$. As is shown in FIGURE 2, in the case of such an orientation, acoustic-noise instabilities occur in many directions. However, for this orientation, the instabilities occur with $\vec{q}_f(\varphi = -15$ degrees, $+45$ degrees), $\eta_f^2 = \eta_{max}^2$, $\vec{q}_r(\varphi = +16$ degrees, -47 degrees), $\eta_r = \eta_{min} = 0$. For such waves, $|\Gamma_+| > |\Gamma_-|$, beginning with the drift velocities, which slightly exceed the threshold of this direction. At the same time, $R \simeq R \simeq 1$. Hence, the severe restrictions on the crystal parameters are no longer required, and the round-trip amplification occurs in a wide range of σ, while the round-trip cannot occur at all in a symmetrically oriented crystal. FIGURE 2 shows also that such instabilities do occur when the conductivity is large and the superthreshold field is small. Time increments of such waves also much exceed those of waves in other directions.

Off-axis instabilities of this type can also occur in materials that have small

TABLE 1

Media	Polarization	Axis of Counting	η_{max}	$\theta^f(\eta_{max})$	$\theta^r(\eta_{min})$	$\alpha = \angle \vec{n}\vec{C}$
ZnO			0.1		0°	15°
CdS	⊥	C_{6v}	0.056	31°	47°	16°
CdSe			0.027		16°	47°
	∥		0.017	0°	90°	45°
GaAs		[100]	0.0037			
GaP	⊥		0.0015	90°	55°	70.8°
InSb			0.001		0°	45°

electromechanical coupling factors because of the very weak attenuation mentioned above. TABLE 1 lists data on sample orientation and the directions where such instabilities occur for A^2B^6 and A^3B^5.

Absolute acoustic-wave instabilities occur if the group velocities of the incident and reflected waves are collinear indcpendent of the directions of their wave vectors (FIGURE 3):

$$\theta_g^r = \theta_g^f + \pi, \tag{2}$$

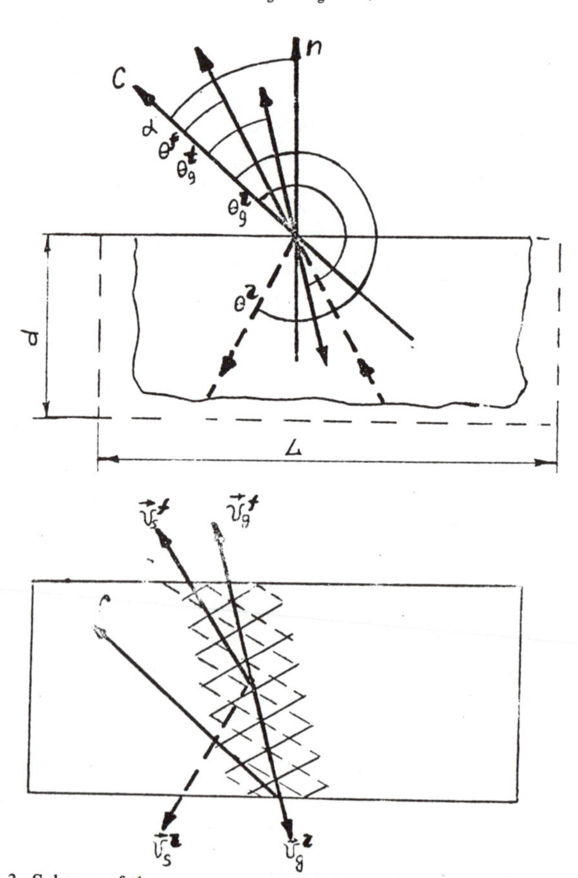

FIGURE 3. Scheme of the occurrence of absolute instability in anisotropic media.

where θ_g^f, θ_g^r, are the angles between axis C and the group velocities of the incident and the reflected waves, respectively. In this case, the wave package, after passing twice between the reflecting planes, does not move along the planes—it remains trapped in the plate.

Consider the phenomenon of collinearization of the group velocities in anisotropic media; for instance, the hexagonal crystal CdSe. FIGURE 4 shows the case of

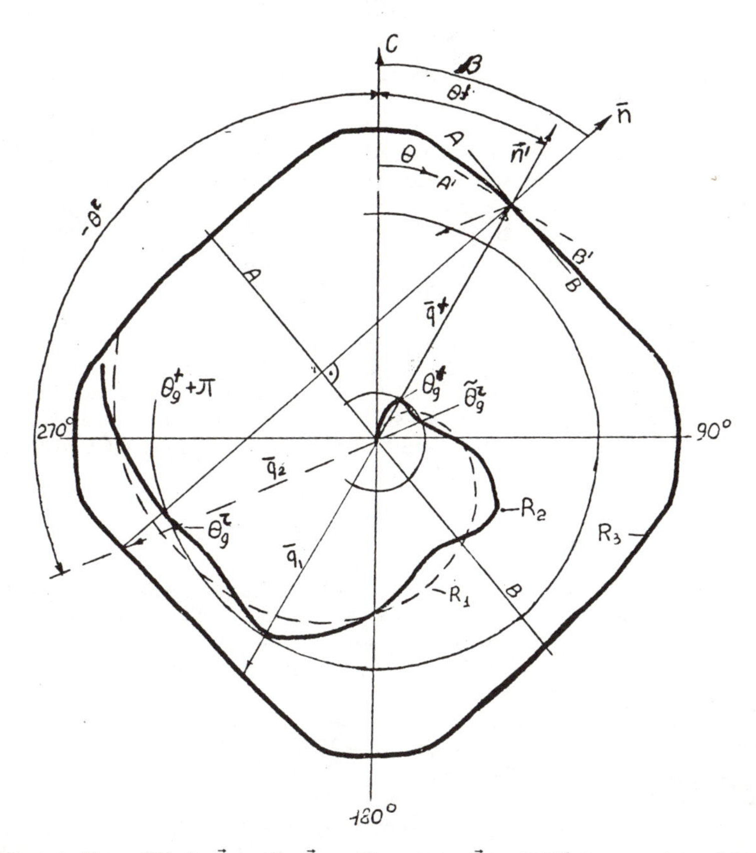

FIGURE 4. Plane (Olm). $\vec{R}_1 = \vec{q}\theta$, $\vec{R}_2 = \vec{q}\theta_q$, where $\vec{R}_3 = V_t/V_T^2$ is a section of inverse surface.

a wave propagating in a plane that contains axis C_6 (the plane (Olm)), and it shows the curves in polar coordinates of $\vec{R}_1 = \vec{q}\theta(\theta = \angle \vec{C}_6\vec{q})$, $\vec{R}_2 = \vec{q}\theta_g(\theta_g = \angle \vec{C}_6 V_g)$, where $\vec{R}_3 = V_T/V_T^2$ is a section of inverse surface for shear waves, and $\vec{q}$ is a unit vector, as above. It can be seen that for a fixed wave $k_0\vec{q}_f$, the group velocity direction θ_g^f can be collinear only to the waves $k_0\vec{q}_1$ and $k_0\vec{q}_2$, since only these waves satisfy Equation 2. Therefore, Equation 2 is valid for the incident wave $k_0\vec{q}_f$ if the reflected wave has either the direction $\vec{q}^r \| \vec{q}_1$ or the direction $\vec{q}^r \| q_2$. For the pair $k_0\vec{q}^f$ and $k_0\vec{q}_1$, the directions of the phase velocities are collinear, as are those of the group velocities. They satisfy Equation 2 when the plane A′B′ with the normal $\vec{n}_1 \| \vec{q}^{\rm f} \| \vec{q}^{\rm r} \| \vec{q}_1$ is a reflecting surface. To have a reflection of the wave in the direction $\vec{q}_2$, the wave must be reflected from plane AB, which is orthogonal to the straight line connecting the cross points of R_3 with $\vec{q}^{\rm f}$ and $\vec{q}_2$. In a general case, if Equation 2 is satisfied by the pair $q^{\rm f}(\theta^{\rm f})$ and $q^{\rm r}(\theta^{\rm r})$, they are transformed into each other when reflecting from the plane such that the angle between the normal $\vec{n}$ to this plane and the axis C is β,

$$\beta = \arctan \frac{\sin \theta^{\rm f} + a \sin \theta^{\rm r}}{\cos \theta^{\rm f} + a \cos \theta^{\rm r}}, \qquad a = \frac{V_{\rm f}}{V_{\rm r}}. \tag{3}$$

By use of FIGURE 4, we can graphically determine the wave directions and the orientation of the reflecting surface for a crystal of any symmetry as well as for the process T → L. If the directions $\vec{q}$ and $-\vec{q}$ are equivalent, it is sufficient to consider the curve in the first quadrant (FIGURE 5) to find the waves satisfying Equation 2. Such waves occur in piezosemiconductors when there is no carrier drift.[8] As in FIGURE 4, the solutions correspond to the cross points of the straight line $\theta_{\rm g}(\theta) =$ const., with the curve $\theta_{\rm g}(\theta)$. In general, each point of $\theta_{\rm g}(\theta)$ is a solution of Equation 2 for normally incident waves with $\vec{q}^{\rm f} \| \vec{q}^{\rm r} \| \vec{n}$ and group velocities slant to $\vec{n}$. The group velocities of the oblique waves are collinear for $\theta^{\rm f}$ and $\theta^{\rm r}$ when $\theta_1 < \theta^{\rm f} < \theta_2$ and $\theta_3 < \theta^{\rm r} < \theta_4$ or $\theta_2 < \theta^{\rm f} < \theta_3$ and $\theta_3 < \theta^{\rm r} < \theta_4$, each pair being uniquely determined by an orientation β given by Equation 3.

The anisotropy is not a sufficient condition for Equation 2 to hold. For instance, despite the strong elastic anisotropy in Te($\theta_{\rm g} - \theta \simeq 40$ degrees) no orientation provides Equation 2 for quasi-shear oblique waves. The same is true for longitudinal waves in all crystals considered. FIGURE 5 shows that Equation 2 is true only for waves propagating in the directions θ, where

$$\frac{\partial \theta_{\rm g}}{\partial \theta} < 0, \text{ or, which is the same, } \frac{1}{V}\frac{\partial^{\rm r} V}{\partial \theta^{\rm r}} < -1.$$

This condition is equivalent to the existence of a topological equivalent to a figure "8" on a section of the wave surface. According to this principle, the collinearization of group velocities can be observed for incident and reflected waves not only in CdSe but in materials with other symmetries, namely, hexagonal ZnO, CdS, Zn, and Be; cubical InSb, GaAs, GaP, $Bi_{12}GeO_{20}$, Ni, and Al; diamond-shaped C_6H_4 (COOH)(COOK); and so on. In a cubical crystal (InSb), there are two trapping regions (FIGURE 5): one is -27 degrees $< \theta < 13$ degrees, and the other is 63 degrees $< \theta < 103$ degrees.

Acoustic-noise instabilities have been well studied[4] for waves with $\vec{q}^{\rm f} \| \vec{q}^{\rm r} \| \vec{n}$. To be more precise, such instabilities may occur only when $\vec{n}$ is parallel to the direction of an extremum of the electromechanical coupling factor. When the orientation is arbitrary, Equation 2 is true for waves close to the normal but not coinciding with it, because the acoustoelectric anisotropy influences the value of the phase velocity and, respectively, the value and the direction of the group velocity.[8]

Consider a more general case of instability for the waves of noncollinear vectors $\vec{q}^{\rm f} \| \vec{q}^{\rm r}$. A particular case has been studied.[9] FIGURE 5 shows that the specific anisotropy of the elastic and acoustoelectric features of these crystals admits that a pair of such waves simultaneously satisfy the condition

$$\eta_{\rm f} = \eta_{\max} \gg \eta_{\rm r} \simeq \eta_{\min} \simeq 0.$$

In CdSe, Equation 2 is satisfied for waves $\vec{q}^{\rm f}(\theta = 31$ degrees), $\eta^2 = 0.027$, and $\vec{q}^{\rm r}(\theta = 64$ degrees), $\eta = 5 \cdot 10^{-4}$; whereas, in InSb, for $\vec{q}^{\rm f}(\theta = 90$ degrees), $\eta_{\rm f}^2 = \eta_{\max} = 10^{-3}$ and $\vec{q}^{\rm r}(\theta = 116$ degrees), $\eta_{\rm r}^2 = 2 \cdot 10^{-4}$ when $\beta = 13$ degrees (θ is

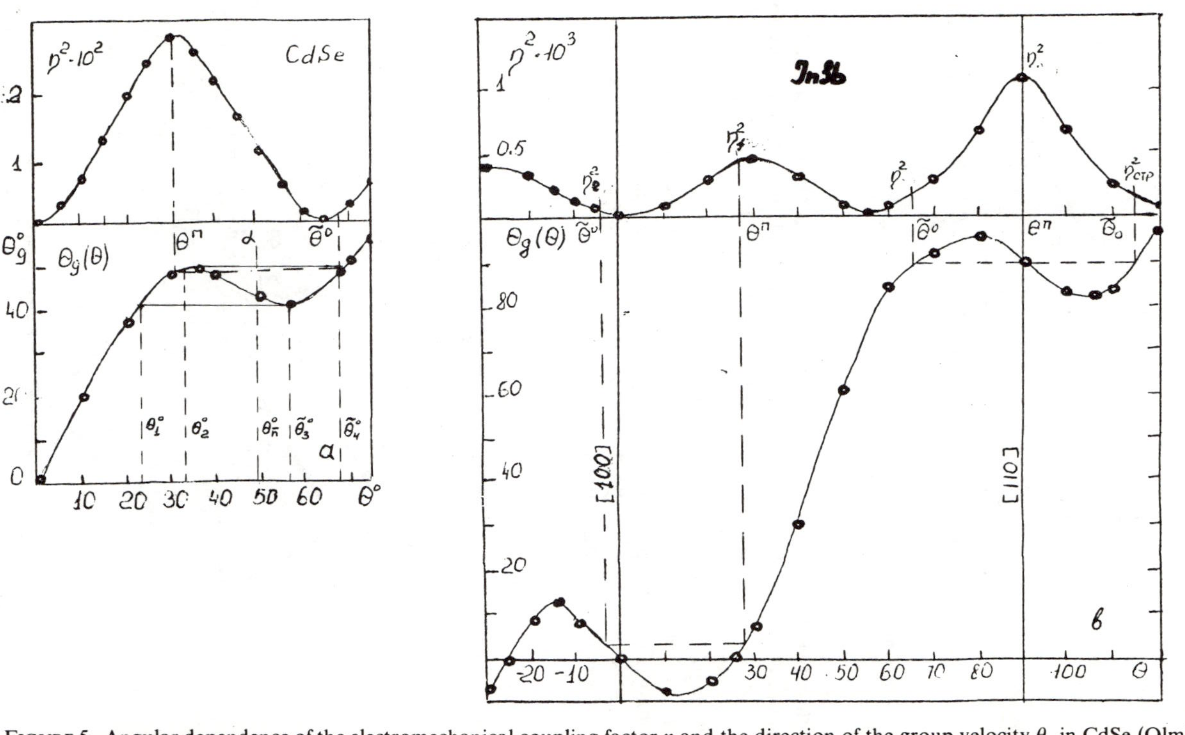

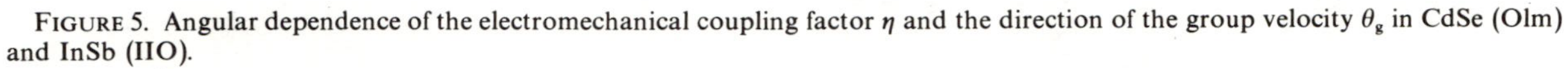

FIGURE 5. Angular dependence of the electromechanical coupling factor η and the direction of the group velocity θ_g in CdSe (Olm) and InSb (IIO).

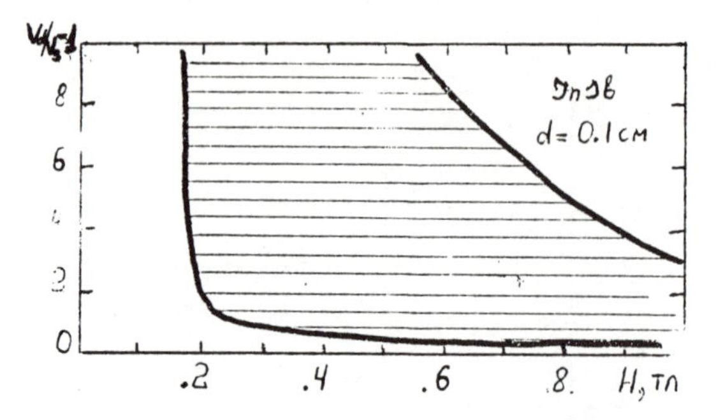

FIGURE 6. Domain of absolute instability in InSb for waves $\vec{q}^{f}(90°)$ propagating in the (IIO) lane.

the angle between the axis [100] and $\vec{q}$ in the plane (100)). As already described, there is an instability in such directions. Calculations reveal that acoustic-wave instabilities also occur for other pairs of waves. Consider the existence of such instabilities in the case shown in FIGURE 4 for waves $\vec{q}^{f}(\theta = 31$ degrees) and $\vec{q}^{r}(\theta = 64$ degrees), $\alpha = 49$ degrees. This case differs from the others in two ways. First of all, the difference of the electromechanical coupling constant is maximal; hence, the maximal round-trip gain is possible. Secondly, for these directions, acoustoelectric anisotropy has no influence on the directions of the group velocities. Also, $\vec{n} \| \vec{V}^{f}_{g} \| \vec{V}^{r}_{g}$, which makes the considerations above clearer.

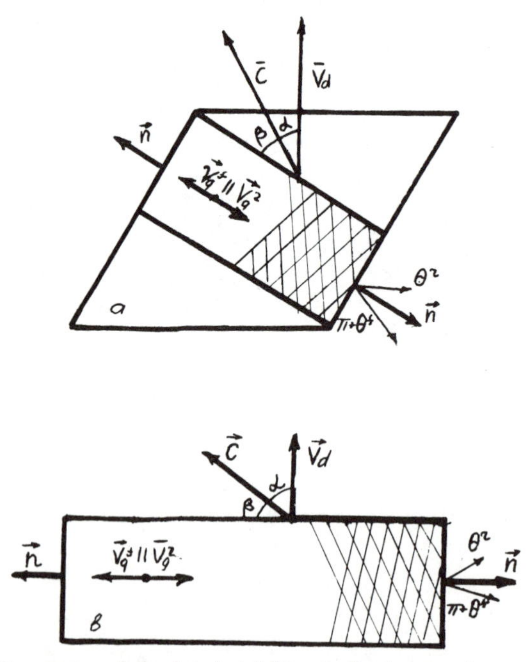

FIGURE 7. Occurrence of absolute instability of off-axis waves for an arbitrary orientation (a) and the special case $\vec{n}$ V_{d} (b).

It should be emphasized that for off-axis absolute acoustic-noise instabilities to occur, the carrier's drift direction with respect to n must not be strongly conditioned. The only requirement is that angle β for any given $\vec{q}^f$ and $\vec{q}^r$ satisfy Equation 3. If the reflecting plane is also an electrical-contact surface for the case in question, $\vec{n} \| E_d$ and we have the situation of FIGURE 3, except for $V_g \| n$. Actually, the domain of such absolute instabilities for this orientation coincides with the domain of convective instability (FIGURE 2, curves 2 and 3) of off-axis noise in a sample of the same parameters. FIGURE 6 shows the domain of such an instability in the sample of InSb with $d = 0.1$ cm, $\mu = 6 \cdot 10^5$ cm^2/V sec; $n = 10^{14}$ cm^{-3} for the case of an optimal gain $\vec{q}^f$(90 degrees), $\vec{q}^r$(116 degrees), $\alpha = 13$ degrees. Absolute instabilities with $\vec{q} \| \vec{n}$ have been observed in InSb[10] when a large crystal and electrical and magnetic fields of large magnitude were used.

For an arbitrary crystal orientation, absolute instabilities occur in samples shaped like a parallelepiped, as is shown in FIGURE 7. This case requires the condition $d > L \cos(\alpha + \beta)/2$. The most intriguing possibility is to amplify acoustic waves by a transverse carrier drift (FIGURE 7b). For the pair of directions $\vec{q}^f$(31 degrees) and $\vec{q}^r(\theta = 64$ degrees), such a possibility appears when $\alpha = 41$ degrees. By use of such samples, amplification of acoustic waves can be achieved for small absolute values of voltage $u > E_{th} d/\cos 72$ degrees. Moreover, in the direction of wave propagation, there is no nonuniform distribution of the field.

DISCUSSION

FIGURE 8 compares domains of absolute instabilities calculated from Equation 1 together with the data on the existence of acoustoelectric current in a sample of CdSe with $\alpha = 15$ degrees and $d = 0.02$ cm. The occurrence of acoustoelectric cur-

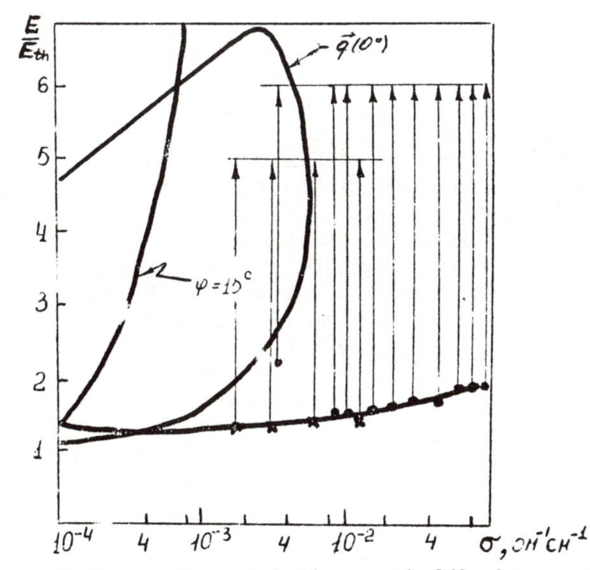

FIGURE 8. Domains of acoustoelectric current in CdSe plates, $\alpha = 15°$.

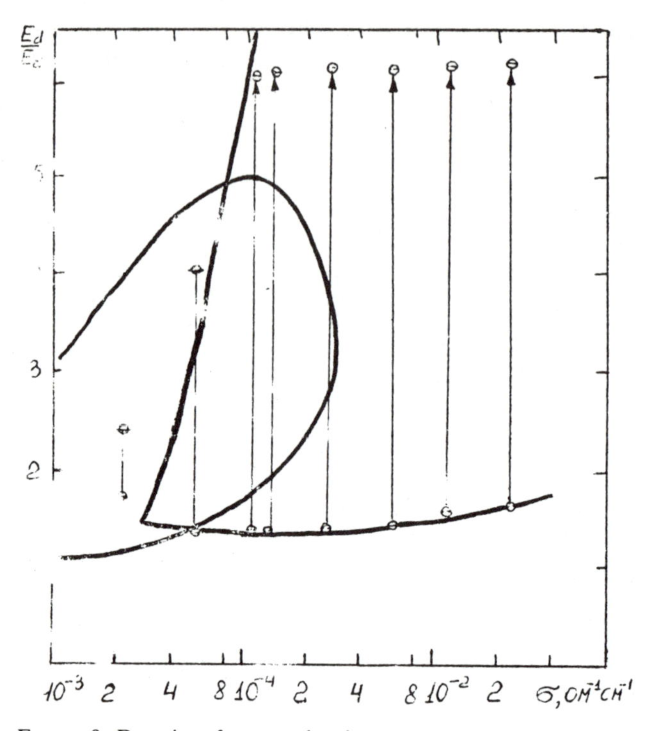

FIGURE 9. Domains of acoustoelectric current in CdSe plates, $\alpha = 49°$.

rent in a sample with conductivity $\sigma > 5 \cdot 10^{-3}\ \Omega^{-1}\ \mathrm{cm}^{-1}$, where no acoustic noise normal to the surface could be amplified, proves that this phenomenon is associated with a round-trip gain of off-axis acoustic noise. The same conclusion can be drawn by comparing the analogous parameters in FIGURE 9, which were obtained from crystal CdSe oriented at an angle of 49 degrees. As mentioned above, this angle permits the development of off-axis acoustic-noise instability $V_g \| \vec{n} \| \vec{E}$ (FIGURE 3). To determine the space-power distribution of the amplified acoustic noise, we can take advantage of the fact that the amplification direction provides not only the magnitude of amplification but also the maximal time it takes to pass through the crystal. Therefore, in dynamic conditions, when the power of sound flow is small and there are no nonlinear effects, we can determine the angular distribution of acoustic-noise intensities from the increase of acoustoelectric current. The results of such measurements[7] have shown that a round-trip gain occurs in samples CdSe, $\alpha = 15$ degrees, in directions $(\varphi) = (-6$ degrees–20 degrees$)$ and $\varphi = +45$ degrees. The domains of acoustic noise are close to those calculated in the first section. A rapid increase of acoustoelectric current has been observed, which is associated with an acoustic-noise gain in optimal directions. In the samples with $\alpha = 49$ degrees, the situation is simplified since there is no instability of longitudinal waves, and intensive flows of quasi-shear phonons appear either in the directions $\vec{q}^{f} \| \vec{q}^{r} \| \vec{n}$ or $\vec{V}_g^{f} \| \vec{V}_g^{r} \| \vec{n}$. However, the round-trip amplification of off-axis phonons is much greater, and the incubation time t_i is thus enforced by their amplification for almost all conductivities studied.

If the group velocities of an amplified incident wave and an attenuated reflected wave are at an angle Ω with respect to each other,

$$\Omega = |\theta_g^f - \theta_g^r - \pi|,$$

the time it takes for a wave to pass along the reflecting surfaces of length L is

$$t = 2L \cos(\theta_g - \alpha)/V_g \sin \Omega.$$

Therefore, by measuring t_i experimentally, we can estimate the maximal value of the group-velocity noncollinearity, which is determined as $\Omega \leq 2L/V_g t_i$ for a given orientation of the crystal. If the length L of the crystal is decreased, a more precise determination of the noncollinearity Ω is obtained. FIGURE 10 shows measurements of t_i versus E_d for a sample of CdSe with $L = 0.02$ cm, for which the sensitivity equals $\Omega < 2.1 \cdot 10^{-7}$(rad sec). FIGURE 10 shows that in weak fields, when the increments Γ_+ are small for all $\vec{q}$, acoustoelectric current is induced only by off-axis acoustic-noise. Therefore, for the experimental magnitudes $t_i = 100$ mks, the noncollinearity is of $\Omega = |\theta_g^f - \theta_g^r - \pi| < 2.1 \cdot 10^{-3}$ rad $\simeq 7'$. The limit angle of noncollinearity that can be determined by this technique increases when the voltage increases. For the minimal experimental value $t_i = 1.5$ mks, angle Ω has to be less than 4 degrees.

The curves of t_i versus E in FIGURE 10 have clear minima with values that are conditioned by the sample conductivity. This phenomenon is due to the fact that acoustic-noise instabilities occur in only one direction. According to the linear theory, the amplification attains its maximum when

$$\gamma_{\text{opt}} = (\bar{k}\vec{V}\alpha/\omega - 1) = (\omega_D/\omega + \omega/\omega_D)$$

for the wave frequency ω. Therefore, by use of the experimental data of FIGURE 10, frequencies are obtained from the latter formula showing that absolute instabilities

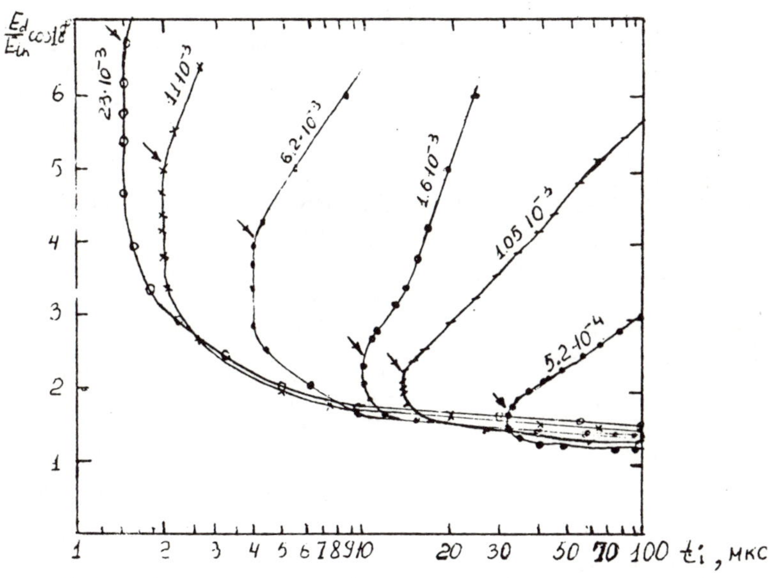

FIGURE 10. Dependence of incubation time t_i versus relative drift field in CdSe plate $0.02 \times 0.08 \times 0.02$ cm^3, $\alpha = 49°$.

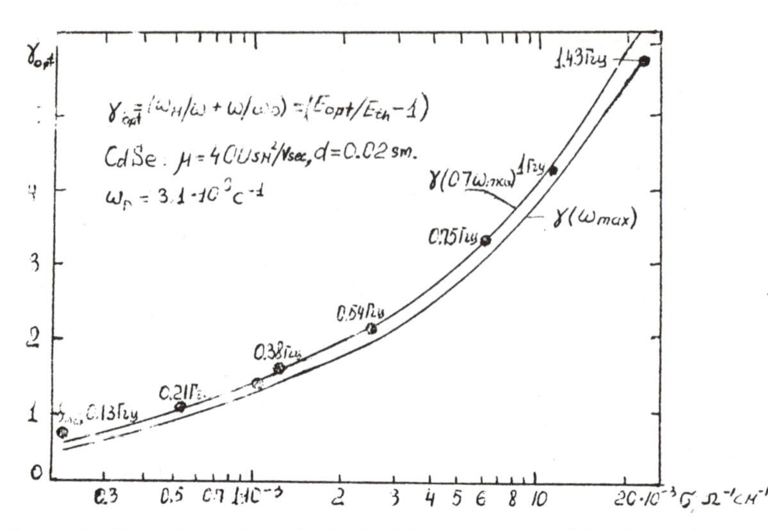

FIGURE 11. Dependence of maximal gain field versus conductivity. Calculations are based on the data from FIGURE 10.

exist. FIGURE 11 shows the results of these calculations for all experimental conductivity values together with the curve of the optimal field as a function of corresponding conductivity for frequencies of maximal gain. When comparing the experimental data with those calculated for $\omega_{\max} = (\omega_C \omega_D)^{1/2}$, we come to the following conclusion: absolute acoustic-noise instabilities occur in the range $\omega \simeq \omega_{\max}(0.7\text{–}1)$.

SUMMARY

The condition for off-axis acoustic-wave instabilities appearing in cut-off samples are developed. I describe the orientation of the samples that provides round-trip amplification of acoustic noise with large time increments. In piezosemiconductors, it is shown that absolute instabilities exist for waves with noncollinear wave vectors. I show that it may be possible to amplify the acoustic waves and that absolute instabilities of acoustic noise may occur when amplifying by use of a transverse electrical field. The experimental results demonstrated that such instabilities occur in a wide range of conductivities for frequencies = $(0.7\text{–}1)(\omega_c \omega_D)^{1/2}$.

REFERENCES

1. VRBA, I. & R. R. HAERING. 1975. J. Acoust. Soc. Am. **57:** 116; 1972. Can. J. Phys. **51:** 1350.
2. KLEIN, R. 1968. Phys. Lett. **28A:** 428.
3. MOORE, A. R., R. W. SMITH & R. WORCHESTER. 1969. IBM J. Res. Dev. **13:** 503; MOORE, A. R. & J. W. DAVENPORT. 1972. J. Appl. Phys. **43:** 4513.
4. MAINES, I. D. & E. G. S. PAIGE. 1969. J. Phys. C. **2:** 175.
5. SHAPIRA, Y., B. FISHER, B. PRATT & A. MANY. 1972. Solid State Commun. **10:** 761.
6. CHERNOZATONSKY, L. A. & V. M. LEVIN. 1972. Fiz. Tverd. Tela (Leningrad) **14:** 2069.
7. RAVVIN, I. S. & L. A. CHERNOZATONSKY. 1976. Pis'ma Zh. Eksp. Teor. Fiz. **2:** 404.
8. KELLER, O. 1973. Phys. Status Solidi (a) **16:** 87.
9. RAVVIN, I. S. & L. A. CHERNOZATONSKY. 1976. Reports of IV All-Union Symposium on A_2B_6 Semiconductors, Odessa: 169; PUSTOVOIT, V. I., I. S. RAVVIN & L. A. CHERNOZATONSKY. 1977. JETP Lett. **25:** 144.
10. OGAWA, Y., T. AOKI & T. ARIZUMI. 1972. J. Appl. Phys. **43:** 4804.

STRUCTURE OF THE ELECTROMAGNETIC FIELD IN A MAGNETOACTIVE PLASMA

Jacov L. Al'pert

Moscow, U.S.S.R.

This report briefly summarizes some results on electric dipole emission in a homogeneous magnetoactive multicomponent plasma in the frequency range $0 \leq \omega \lesssim \omega_0$, where ω_0 is the Langmuir frequency. An important feature of electromagnetic waves propagating in such media is the anisotropic distribution of their energy flux with respect to the direction of the constant magnetic field $\bar{H}_0$. This property of waves leads, for example, to the guiding and trapping of electron and ion whistlers along the magnetic-field lines. The main peculiarities of this process have been qualitatively established on the basis of a detailed analysis of the behavior of the group-velocity vector in the whole frequency range in a multicomponent magnetoactive plasma.[1,2] It was shown that the electromagnetic emission in the range of frequencies that includes plasma-resonance oscillations is concentrated in two cones, one inserted into the other, their axes being along the magnetic field $\bar{H}_0$. The boundary of one cone is determined by the angle $\alpha_1 = \alpha_\infty$, at which the resonance cutoff of waves occurs ($n \to \infty$). The boundary of the other is determined by the angle $\alpha = \alpha_M$, which is the maximum angle between the group velocity $\partial\omega/\partial\bar{k}$ and the vector $\bar{H}_0$. Angle α_M is known as the Storey angle. Storey was the first to determine its value for whistlers.[3] However, study of the complete waves of this process is a difficult task, particularly because of the complicated expressions for the plasma-refractive index n. Some aspects of this problem, namely, frequencies $\omega \ll \Omega_{HS}$ (Ω_{HS} is the ionic gyrofrequency), or, in practical terms, the properties of Alfven waves, have been considered,[4,5] but not quite consistently. The same problem was studied previously in the general form by many authors.[6,7] However, the field around the direction of the magnetic field $\bar{H}_0$ was not studied carefully. Some of the results in those papers are therefore incorrect. Moreover, the second approximation was not calculated, and the applicability of the asymptotic formulas in some of these papers were, therefore, not analyzed. In addition, numerical results were given only for $\omega \gtrsim \omega_0$.[6] I have analyzed all these problems[8] and have given numerical results for whistler-mode waves in the frequency range $\omega_L < \omega \lesssim \omega_H$, where ω_L and ω_H are the low-hybrid and electron gyrofrequencies, respectively.

Formulation of the Problem: The Main Formulas

The emission by an electric-point dipole, the dipole moment of which is equal to $\bar{I}_0$ and whose angular frequency is ω, is analyzed in a homogeneous magnetically active plasma where $\bar{I}_0 \| \bar{H}_0$. Maxwell's equations are written in a dimensionless cylindrical coordinate system, $\rho = (\omega/c)r = 2\pi(I/\lambda_0)$, φ, $\zeta = (\omega/c)z = 2\pi(z/\lambda_0)$, in which a simpler and physically adequate solution of the wave equations can be found, namely:

$$\left.\begin{aligned}
-\frac{\partial^2 E_\rho}{\partial \zeta^2}+\frac{\partial^2 E_\zeta}{\partial \zeta\,\partial \rho} &= \mathscr{E}_1 E_\rho - i\mathscr{E}_2 E_\varphi,\\
-\frac{\partial^2 E_\varphi}{\partial \zeta^2}-\frac{\partial}{\partial \rho}\left[\frac{1}{\rho}\frac{\partial}{\partial \rho}(\rho E_\varphi)\right] &= i\mathscr{E}_2 E_\rho + \mathscr{E}_1 E_\varphi,\\
\frac{1}{\rho}\frac{\partial}{\partial \rho}\left[\rho\left(\frac{\partial E_\rho}{\partial \zeta}-\frac{\partial E_\zeta}{\partial \rho}\right)\right] &= \mathscr{E}_3 E_\zeta + 4I\frac{\delta(\rho)\delta\zeta}{\rho},
\end{aligned}\right\} \quad (1)$$

where $\zeta \| \bar{H}_0$, $\bar{H}_0 = \text{const.}$; $E_{\rho,\varphi,\zeta}$ are the space components of the electrical field, which are proportional to $\exp(i\omega t)$; $\mathscr{E}_1$, $\mathscr{E}_2$, and $\mathscr{E}_3$ are the components of the dielectric coefficient tensor; $\delta(\rho)$, $\delta(\zeta)$ are delta functions; and $I = I_0(\omega^3/c^3)$.

The solution of the differential Equations 1 is obtained in the following way. The Fourier–Bessel and Fourier transformations with respect to the variables ρ and ζ, respectively, reduce Equations 1 to a system of algebraic equations of the Fourier components $\tilde{E}_{\rho,\varphi,\zeta}$ of the electrical field. These equations determine the values

$$\tilde{E}_\rho = K(n_\|, n_\perp)\tilde{E}_\varphi, \qquad \tilde{E}_\zeta = B(n_\|, n_\perp)\tilde{E}_\varphi, \qquad \tilde{E}_\varphi = 2I\frac{(-\mathscr{E}_2)n_\| n_\perp}{D(n_\|, n_\perp)}, \quad (1a)$$

where

$$\left.\begin{aligned}
K(n_\|, n_\perp) &= i\frac{n_\|^2 + n_\perp^2 - \mathscr{E}_1}{(-\mathscr{E}_2)}, \qquad B(n_\|, n_\perp) = \frac{(n_\|^2 + n_\perp^2 - \mathscr{E}_1)(\mathscr{E}_1 - n_\|^2) + \mathscr{E}_2^2}{(-\mathscr{E}_2)n_\| n_\perp},\\
D &= \mathscr{E}_3 n_\|^4 + n_\|^2[(\mathscr{E}_1 + \mathscr{E}_3)n_\perp^2 - 2\mathscr{E}_1\mathscr{E}_3]\\
&\quad + \{\mathscr{E}_1 n_\perp^4 + n_\perp^2[\mathscr{E}_2^2 - \mathscr{E}_1(\mathscr{E}_1 + \mathscr{E}_3) + \mathscr{E}_3(\mathscr{E}_1^2 - \mathscr{E}_2^2)]\},
\end{aligned}\right\} \quad (1b)$$

and $n_\|$, $n_\perp$ (see below). Then, the reverse Fourier transformation of $\tilde{E}_{\rho,\varphi,\zeta}$ determines the components of the electrical field $E_{\rho,\varphi,\zeta}$ in the space coordinate system. They are expressed by a twofold integral of the variables $n_\perp$ and $n_\|$. These integrals have poles $n_\|(n_\perp)$ obtained from the dispersion Equation 1b, $D = 0$. But only the real branches that determine unattenuated waves are of interest.

This paper deals only with the resonance branches of $n_\|(n_\perp)$ of these waves, which have the polarization sense of electron-whistler and ion-whistler modes (the sign $\langle + \rangle$ in $n_\|$ (see Equation 3c) $\mathscr{E}_3 < 0$). Finally, the solution of interest to us is:

$$\begin{Bmatrix} E_\rho \\ E_\varphi \\ E_\zeta \end{Bmatrix} = 2\pi \int_0^\infty dn_\perp\, n_\perp^2\, F(n_\perp) \begin{Bmatrix} K(n_\perp)\tau_1(n_\perp \rho) \\ \tau_1(n_\perp \rho) \\ B(n_\perp)\tau_0(n_\perp \rho) \end{Bmatrix} e^{-in_\|(n_\perp)\zeta} \quad (2)$$

However, this formula is difficult for calculation purposes, and it does not clearly reveal its physical essence. The substitution in Equation 2 of the Bessel functions by their integral representations and the analysis of the stationary points of the integrals result in the following solution of the problem:

$$\begin{Bmatrix} E_\rho \\ E_\varphi \\ E_\zeta \end{Bmatrix} = (I_1^{(\mathscr{J})} + I_{\mathscr{J}}^{(\mathscr{J})}) + (I_1^{(\mathscr{E})} + I_2^{(\mathscr{E})}). \quad (3)$$

It is not difficult to see that $I_{1,2}^{(\mathscr{J})}$ and $I_{1,2}^{(\mathscr{E})}$ in Equation 3 determine the structure of the electric field in the shape of two cones: indices $\langle 1 \rangle$ and $\langle 2 \rangle$. We shall show below that the greater part of the field is "trapped" in the cavities of these cones at sufficiently large distances. The second important peculiarity of Equation 3 is that these equations immediately allow us to obtain asymptotic formulas for large ρ and ζ. In Equation 3, the symbol $(\mathscr{J})$ is used to describe the field inside and the symbol $(\mathscr{E})$, the field outside the cones, and

$$I_{1,2}^{(\mathscr{J})} = \int_{\delta}^{\infty} dn_{\perp} \int_{\Delta}^{\pi-\Delta} dx n_{\perp}^2 F(n_{\perp}) e^{-i\omega_{1,2}(n_{\perp}, x, \rho, \zeta)} \begin{Bmatrix} K(n_{\perp})e^{\mp ix} \\ e^{\mp ix} \\ B(n_{\perp}) \end{Bmatrix},$$

$$I_{1,2}^{(\varepsilon)} = \int_{0}^{\Delta} dx \int_{0}^{\delta} dn_{\perp} A(x, n_{\perp}) e^{-i\psi_{1,2}} + \int_{\delta}^{\infty} dn_{\perp} \int_{0}^{\Delta} dx\, A(x, n_{\perp}) e^{-i\psi_{1,2}}, \tag{3a}$$

$$A(x, n_{\perp}) = n_{\perp}^2 F(n_{\perp}) \begin{Bmatrix} \mp iK(n_{\perp}) \sin x \\ \mp i \sin x \\ B(n_{\perp}) \end{Bmatrix},$$

where

$$\left.\begin{aligned} F(n_{\perp}) &= i\frac{\mathscr{E}_2}{2\pi} \frac{I}{2\mathscr{E}_3 n_{\|}^2(n_{\perp}) + (\mathscr{E}_1 + \mathscr{E}_3)n_{\perp}^2 - 2\mathscr{E}_1\mathscr{E}_3}, \\ \psi(n_{\perp}, x, \rho, \zeta)_{1,2} &= \mp n_{\perp}\rho \sin x + n_{\|}(n_{\perp})\zeta, \end{aligned}\right\} \tag{3b}$$

$$\left.\begin{aligned} n_{\|}(n_{\perp}) &= \left\{-\frac{1}{2}\left[\left(1 + \frac{\mathscr{E}_1}{\mathscr{E}_3}\right)n_{\perp}^2 - 2\mathscr{E}_1\right] \pm \left[\frac{n_{\perp}^4}{4}\left(1 - \frac{\mathscr{E}_1}{\mathscr{E}_3}\right)^2 - \frac{\mathscr{E}_2^2}{\mathscr{E}_3}n_{\perp}^2 + \mathscr{E}_2^2\right]^{1/2}\right\}^{1/2}, \\ \mathscr{E}_1 &= 1 + \frac{\omega_0^2}{\omega_H^2 - \omega^2} + \sum_S \frac{\Omega_{OS}^2}{\Omega_{HS}^2 - \omega^2}, \\ \mathscr{E}_2 &= -\frac{\omega_H \omega_0^2}{\omega_H^2 - \omega^2} + \sum_S \frac{\Omega_{HS}}{\omega} \frac{\Omega_{OS}^2}{\Omega_{HS}^2 - \omega^2}, \\ \mathscr{E}_3 &= 1 - \frac{\omega_0^2}{\omega^2} - \sum_S \frac{\Omega_{OS}^2}{\omega^2}, \end{aligned}\right\} \tag{3c}$$

$n_{\|} = n \cos \theta$, $n_{\perp} = n \sin \theta$, n is the refractive index,* θ is the angle between the wave vector $\bar{x}$ and the static magnetic field $\bar{H}_0$, (ω_0, Ω_{OS}), (ω_H, Ω_{HS}) are the plasma and gyrofrequencies, respectively, and J_0 and J_1 are Bessel's functions. The upper and lower signs in Equations 3a and 3b correspond to the indices $\langle 1 \rangle$ and $\langle 2 \rangle$, and small values Δ, $\delta > 0$ of the limits of the integrals are chosen so that various asymptotic expressions of Equation 3a are singled out. The equations of the stationary points are:

$$\frac{\partial\psi}{dn_{\perp}} = \mp \frac{\rho}{\zeta} \sin x + \frac{dn_{\|}(n_{\perp})}{\partial n_{\perp}} = 0, \tag{3d}$$

$$\frac{\partial\psi}{\partial x} = \pm n_{\perp}\rho \cos x = 0, \tag{3e}$$

* In a collisional plasma, other formulas for $\mathscr{E}_1$, $\mathscr{E}_2$, $\mathscr{E}_3$ should be used.[9]

or

$$\frac{dn_{\parallel}(n_{\perp})}{dn_{\perp}} = \pm\frac{\rho}{\zeta}\sin x, \qquad n_{\perp}\cos x = 0,$$

where the upper and lower signs in Equations 3d and 3e determine the integrals $I_1^{(\mathcal{J})}$ and $I_2^{(\mathcal{J})}$. Equations 3e determine, when $n_{\perp} \neq 0$, the value $x = \pi/2$ and one stationary point $(n_{\perp 1}, x = \pi/2)$ of $I_1^{(\mathcal{J})}$ and two stationary points $(n_{\perp 2, 3}, x = \pi/2)$ of $I_2^{(\mathcal{J})}$. These points are real values when

$$\frac{\rho}{\zeta} < \tan\alpha_1 = \left(-\frac{\mathscr{E}_1}{\mathscr{E}_3}\right)^{1/2}, \qquad \frac{\rho}{\zeta} < \tan\alpha_2 = -\left[\frac{dn_{\parallel}(n_{\perp})}{dn_{\perp}}\right]_{n_{\perp}(2)}, \tag{3f}$$

where $\alpha_1 = \alpha_{\infty}$, $\alpha_2 = \alpha_M$ (see the introduction); $n_{\perp}^{(2)}$ is the point where

$$\left[\frac{d^2 n_{\parallel}(n_{\perp})}{dn_{\perp}^2}\right]_{n_{\perp}(2)} = 0, \qquad \left[\frac{dn_{\parallel}(n_{\perp})}{dn_{\perp}}\right]_{n_{\perp}(2)} < 0.$$

The properties described for Equation 3 show that the field is composed of two emission cones, one inserted into the other: the resonance cone (cone $\langle 1\rangle$) and Storey's cone (cone $\langle 2\rangle$). Both cones have a common symmetry axis: the direction of the magnetic field $\bar{H}_0$. All the wave modes of the dispersion branch propagate inside these cones. However, the asymptotic approximations of these formulas are violated near the symmetry axis ($\rho/\zeta \to 0$) and on the surfaces of the cones ($\rho/\zeta = \tan\alpha_{1,2}$), where, as we shall see, the field intensity may increase considerably. The main point of this problem is investigation of the field in these regions; this requires the use of different mathematical methods and taking into account the disposition of the stationary points. I have used corresponding methods in the next section. Moreover, the second approximations in the numerical calculations (see RESULTS OF NUMERICAL CALCULATIONS) were used to determine the applicability of the formulas used.

We shall consider the structure of the field in these regions in the following two cases:

(a) $\dfrac{d^2 n_{\parallel}(0)}{dn_{\perp}^2} < 0$ and $|\mathscr{E}_3| \gg \mathscr{E}_1, |\mathscr{E}_2|$ cone $\langle 2\rangle$ exists (α_2 is real),

(b) $\dfrac{d^2 n_{\parallel}(0)}{dn_{\perp}^2} > 0$ and $|\mathscr{E}_3| \lesssim \mathscr{E}_1, |\mathscr{E}_2|$ cone $\langle 2\rangle$ does not exist (α_2 is imaginary).

(4)

It is important to note that in analyzing the stationary points of the integrals in Equation 3 and deriving the asymptotic formulas given below, we also used the following limitations:

$$\mathscr{E}_3 < 0, (\omega < \omega_0), \mathscr{E}_1 < 0.$$

ASYMPTOTIC FORMULAS

The Field in the Vicinity of the Axis of Symmetry ($\rho/\zeta \to 0$)

Parameter ρ is not large, and Equation 3e of the stationary points is not true in this region. However, Equation 3d is found only for large ζ. Therefore, Equation 3d is true for determining of the stationary points when $\rho \to 0$, but as $n_{\perp} = n_{\perp}(x)$, only the integral on the variable $n_\perp$ is calculated asymptotically. In contrast to the situation encountered when ρ is large enough, when $\rho \to 0$ the integral on variable x should be calculated exactly. It should also be noted that one of the stationary points is placed near the boundary of the contour of integration. Therefore, integrals $I_{1,2}$ are divided into two parts:

$$\int_0^\pi \mathrm{d}x \int_0^\delta \mathrm{d}n_\perp(\cdots) + \int_0^\pi \mathrm{d}x \int_\delta^\infty \mathrm{d}n_\perp(\cdots),$$

which have different asymptotic formulas.[10] For Case 4a, if $|\mathscr{E}_2| \gg \mathscr{E}_1\sqrt{-\mathscr{E}_1/\mathscr{E}_3}$, we obtain

$$\begin{Bmatrix} E_\rho \\ E_\varphi \\ E_\zeta \end{Bmatrix} = \frac{\mathscr{E}_2 I}{\mathscr{E}_3\sqrt{\zeta}} \frac{\exp\{-i[3\pi/4 + n_\parallel(n_\perp^{(1)})\zeta]\}}{\sqrt{2\pi\left|\dfrac{\mathrm{d}^2 n_\parallel(n_1^{(1)})}{\mathrm{d}n_\perp^2}\right|}} \begin{Bmatrix} i \operatorname{sign} \mathscr{E}_2\sqrt{-\dfrac{\mathscr{E}_3}{\mathscr{E}_1}}\mathscr{J}_1(n_\perp^{(1)}\rho) \\ -\mathscr{J}_1(n_\perp^{(1)}\rho) \\ -B(n_\perp^{(1)})\mathscr{J}_0(n_\perp^{(1)}\rho) \end{Bmatrix}, \quad \textbf{(5a)}$$

and for Case 4b,

$$\begin{Bmatrix} E_\rho \\ E_\varphi \\ E_\zeta \end{Bmatrix} = i\frac{I}{2\mathscr{E}_3}\frac{\rho}{\zeta^2\left|\dfrac{\mathrm{d}^2 n_\parallel(0)}{\mathrm{d}n_\perp^2}\right|^2} \exp i\left[\frac{\rho^2}{2\zeta\left|\dfrac{\mathrm{d}^2 n_\parallel(0)}{\mathrm{d}n_\perp^2}\right|} - n_\parallel(0)\zeta\right] \quad \textbf{(5b)}$$

$$\times\begin{Bmatrix} i \\ \operatorname{sign} \mathscr{E}_2 \\ \dfrac{n_\parallel(0)}{\mathscr{E}_3}\dfrac{1}{\zeta\left|\dfrac{\mathrm{d}^2 n_\parallel(0)}{\mathrm{d}n_\perp^2}\right|}\left(\dfrac{2\zeta}{\rho}\dfrac{\mathrm{d}^2 n_\parallel(0)}{\mathrm{d}n_\perp^2} + i\rho\right) \end{Bmatrix},$$

where

$$\left.\begin{aligned} &n_\perp^{(1)}\left(|\mathscr{E}_2|\sqrt{-\frac{\mathscr{E}_3}{\mathscr{E}_1}}\right)^{1/2}, \\ &\left(\frac{\mathrm{d}^2 n_\parallel(n_\perp)}{\mathrm{d}n_\perp^2}\right)_{n_\perp = n_\perp^{(1)}} = -4\frac{\sqrt{\mathscr{E}_1}}{\mathscr{E}_3 f(\mathscr{E})}, \\ &\left(\frac{\mathrm{d}^2 n_\parallel(n_\perp)}{\mathrm{d}n_\perp^2}\right)_{n_\perp = 0} = \frac{-(\mathscr{E}_1 + |\mathscr{E}_2| + \mathscr{E}_3|}{2\mathscr{E}_3\sqrt{\mathscr{E}_1 + |\mathscr{E}_2|}}, \end{aligned}\right\} \quad \textbf{(6)}$$

and the conditions of applicability of Equation 5a are:

$$\left.\begin{aligned} &\frac{\rho}{\zeta} \ll \frac{\sqrt{\mathscr{E}_2}(\mathscr{E}_1)^{1/4}}{\pi(-\mathscr{E}_3)^{3/4} f(\mathscr{E})}, \quad \frac{\rho^2}{\zeta} \ll \frac{4}{\pi}\frac{\sqrt{\mathscr{E}_1}}{|\mathscr{E}_3| f(\mathscr{E})}, \\ &\zeta \gg \frac{\sqrt{-\mathscr{E}_3}}{|\mathscr{E}_2|} f(\mathscr{E}), f(\mathscr{E}) = 1 + \frac{\mathscr{E}_2}{\sqrt{-\mathscr{E}_1\mathscr{E}_3}}. \end{aligned}\right\} \quad \textbf{(6a)}$$

The electric field strength in the vicinity of $\rho/\zeta \to 0$ has a maximum (see Equation 5a) whose modulus $|\bar{E}|_{0,\max}$ and position are determined by the following formulas:

$$|\bar{E}|_{0,\max} = 0.3\frac{I|\mathscr{E}_2|\mathscr{E}_1^{-3/4}}{\sqrt{2\pi\zeta}}[f(\mathscr{E})]^{1/2}, \qquad \rho_{0,\max} = \frac{2}{\left(-\dfrac{\mathscr{E}_3}{\mathscr{E}_1}\right)^{1/4}\sqrt{|\mathscr{E}_2|}}. \tag{5c}$$

The Field in the Vicinity of the Cone 1: ($\rho/\zeta = \tan\alpha_1$)

In the plasma resonance region ($n \to \infty$, $\rho/\zeta \to \tan\alpha_1$), the refractive index can be limited by introducing a small attenuation term that is proportional to the collision frequency ν. The main contribution to the field (see Equation 3a) is made by the stationary point ($n_{\perp 1}$, $x = \pi/2$) of the integral $I_1^{(\mathscr{I})}$. By use of the pass method, we obtain from Equation 3a:

$$\begin{Bmatrix} E_\rho \\ E_\varphi \\ E_\zeta \end{Bmatrix} = \frac{I e^{-i\psi_1(n_{\perp 1})}}{\sqrt{2}\zeta^4\sqrt{\beta^2+\gamma^2}(\mathscr{E}_1-\mathscr{E}_3)^4\sqrt{-\dfrac{\mathscr{E}_1}{\mathscr{E}_3}}}\begin{Bmatrix} \left(1-\dfrac{\mathscr{E}_1}{\mathscr{E}_3}\right)n_{\perp 1}^2 \\ i\mathscr{E}_2 \\ -\sqrt{-\dfrac{\mathscr{E}_1}{\mathscr{E}_3}}\left(1-\dfrac{\mathscr{E}_1}{\mathscr{E}_3}\right)n_{\perp 1}^2 \end{Bmatrix}, \tag{7}$$

where

$$\psi_1(n_{\perp 1}) = \zeta\cdot\mathrm{Re}\{n_{\perp 1}\}\frac{(\beta+\sqrt{\beta^2+\gamma^2})^2+\gamma^2+2i\gamma\sqrt{\beta^2+\gamma^2}}{\beta+\sqrt{\beta^2+\gamma^2}},$$

$$n_{\perp 1} = \frac{\sqrt{\mathscr{H}}}{\sqrt[4]{\beta^2+\gamma^2}}\exp(i/2[\mathrm{arctg}(-\gamma/\beta)+\pi\theta(-\beta)]),$$

$$\beta = \sqrt{-\frac{\mathscr{E}_1}{\mathscr{E}_3}\frac{\rho}{\zeta}}, \qquad \gamma = \frac{1}{2}\sqrt{-\frac{\mathscr{E}_1}{\mathscr{E}_3}}\left(\frac{\mathrm{Im}\,\mathscr{E}_1}{\mathscr{E}_1}-\frac{\mathrm{Im}\,\mathscr{E}_3}{\mathscr{E}_3}\right) < 0,$$

$$\mathscr{H} = \frac{\mathscr{E}_1(\mathscr{E}_1-\mathscr{E}_3)+\mathscr{E}_2^2}{2\sqrt{-\dfrac{\mathscr{E}_1}{\mathscr{E}_3}}(\mathscr{E}_1-\mathscr{E}_3)}, \qquad \theta(\cdots) \text{ is the Heaviside function.}$$

The use of Equation 7 is possible when the following conditions are fulfilled:

$$\left.\begin{aligned} &\frac{1}{\sqrt{\mathscr{H}|\gamma|}} \ll \zeta \ll \frac{3}{|\gamma|\sqrt{M}}, \quad |\gamma|, \beta \ll \frac{\mathscr{H}}{M}, \\ &M = \max\left\{\mathscr{E}_1, -\mathscr{E}_3\left[1+\frac{\mathscr{E}_2^2}{\mathscr{E}_1(\mathscr{E}_1-\mathscr{E}_3)}\right], \frac{|\mathscr{E}_2|}{\sqrt{-\dfrac{\mathscr{E}_1}{\mathscr{E}_3}\left(1-\dfrac{\mathscr{E}_1}{\mathscr{E}_3}\right)}}\right\}. \end{aligned}\right\} \tag{8}$$

At distances ($\zeta \ll 3/|\gamma|\sqrt{M}$), the amplitude of the field strength in Equation 7 has a maximum†

$$|\bar{E}|_{1,\max} \simeq \frac{\left(\frac{3}{e}\right)^3 I \sqrt{1 - \frac{\mathscr{E}_1}{\mathscr{E}_3}}}{\sqrt{\mathscr{E}_1 + \frac{\mathscr{E}_2^2}{\mathscr{E}_1 - \mathscr{E}_3}}\, |\mathscr{E}_3||\gamma|^3\zeta^4}. \tag{7a}$$

The position of this maximum is determined by the formula:

$$(\tan\alpha_1 - \tan\alpha_{1,\max}) = \frac{\mathscr{H}}{9}\gamma^2\zeta^2, \quad (\alpha_1 \geq \alpha_{1,\max}), \tag{7b}$$

where in Equation 3f the real parts of $\mathscr{E}_1$ and $\mathscr{E}_3$ are used since the attenuation of the waves is small enough.†

It follows from Equation 7a that in the vicinity of the plasma resonance (the surface of the cone ⟨1⟩), the field strength decreases rapidly with distance, namely, $\sim\zeta^{-4}$. Its maximum is shifted inside the cone. It can be shown that the power absorbed on a length unit along the boundary of the cone is as follows:

$$Q \simeq \frac{3I^2\sqrt{-\frac{\mathscr{E}_1}{\mathscr{E}_3}}}{16|\mathscr{E}_3||\gamma|^3\zeta^4}.$$

The Field in the Vicinity of the Surface of the Cone ⟨2⟩ ($\rho/\zeta = \tan\alpha_2$)

In this vicinity, integral $I_2^{(\mathscr{I})}$ has two converging stationary points ($n_{\perp 2} \to n_{\perp 3}$, $x = \pi/2$) and makes the main contribution to the field. By employing the method of the "standard integral,"[10] we obtain from Equation 3a that, in Case 4a, integral $I_1^{(\mathscr{I})}$ is exponentially small, and integral $I_2^{(\mathscr{I})}$ is

$$\begin{Bmatrix} E_\rho \\ E_\varphi \\ E_\zeta \end{Bmatrix} = -i\frac{Ie^{i[\pi/4 - n_\perp(2)\rho - n_\parallel(n_\perp(2))\zeta]}}{\mathscr{E}_3\rho^{1/2}\zeta^{1/3}}\sqrt[3]{n_\perp^{(2)}\cdot n_\parallel(n_\perp^{(2)})}\cdot(n_\perp^{(2)})^{3/2}v\left[-G\left(\frac{\rho}{\zeta}\right)\cdot\zeta^{2/3}\right]$$

$$\times\begin{Bmatrix} \frac{5}{4} \\ i\,\mathrm{sign}\,\mathscr{E}_2 \\ \frac{-|\mathscr{E}_2|}{16n_\perp^{(2)}n_\parallel(n_\perp^{(2)})} \end{Bmatrix}, \tag{9}$$

† When $\omega \gtrsim 0.7\omega_H$, Equation 7a is not precise enough since condition $\zeta \ll 3(|\gamma|M)^{-1}$ (see Equation 8) is not fulfilled well.

where

$$n_{\perp}^{(2)} \simeq \sqrt{\frac{|\mathscr{E}_2|}{2} + \frac{2}{5}\frac{\mathscr{E}_2^2}{\mathscr{E}_3}}, \qquad n_{\parallel}(n_{\perp}^{(2)}) \simeq \sqrt{\mathscr{E}_1 + \frac{3}{4}|\mathscr{E}_2| - \frac{2}{5}\frac{\mathscr{E}_2^2}{\mathscr{E}_3}},$$

$$G\left(\frac{\rho}{\zeta}\right) = \sqrt[3]{2n_{\perp}^{(2)} \cdot n_{\parallel}(n_{\perp}^{(2)})} \cdot \left(\tan\alpha_2 - \frac{\rho}{\zeta}\right), \qquad v(t) = \frac{1}{2\sqrt{\pi}}\int_{-\infty}^{\infty} e^{i(tx + x^3/3)}\,dx, \quad \textbf{(9a)}$$

$$\tan\alpha_2 = \frac{3}{8}\left(1 + 4\frac{\mathscr{E}_1}{\mathscr{E}_3}\right)\frac{n_{\perp}^{(2)}}{n_{\parallel}(n_{\perp}^{(2)})},$$

and $v(t)$ is the Airy function. The conditions of the applicability of Equation 9 are:

$$\zeta \gg 2\frac{n_{\parallel}^2(n_{\perp}^{(2)})}{(n_{\perp}^{(2)})^2}, \qquad \left|\tan\alpha_2 - \frac{\rho}{\zeta}\right| \ll \frac{n_{\perp}^{(2)}}{2^5 n_{\parallel}(n_{\perp}^{(2)})}. \quad \textbf{(10)}$$

The amplitude of the field in the region under consideration has a maximum

$$|\bar{E}|_{2,\,\max} = 2.2I\,\frac{[n_{\parallel}(n_{\perp}^{(2)})]^{5/6}[n_{\perp}^{(2)}]^{4/3}}{|\mathscr{E}_3|\zeta^{5/6}\sqrt{1 + 4\dfrac{\mathscr{E}_1}{\mathscr{E}_3}}}\sqrt{1 + \frac{\mathscr{E}_2^2}{656[n_{\parallel}(n_{\perp}^{(2)})]^2[n_{\perp}^{(2)}]^2}}, \quad \textbf{(9b)}$$

and its position is determined by

$$(\tan\alpha_2 - \tan\alpha_{2,\,\max}) = \frac{\zeta^{-2/3}}{\sqrt[3]{2n_{\perp}^{(2)}n_{\parallel}(n_{\perp}^{(2)})}}, \; \alpha_2 \gtrless \alpha_{2,\,\max}. \quad \textbf{(9c)}$$

It is evident from Equation 9b that the field decreases as $\zeta^{-5/6}$, unlike the dependence ζ^{-1} inside the cone at a sufficient distance from the maximum; that is, focusing of the field near the surface of cone $\langle 2 \rangle$ takes place.

Formulas of the amplitude of the field in the regions inside and outside of the cones are not given here (see Reference 8).

The amplitude of the field in the space inside one of the cones, but close enough to the boundary outside the other, at first decreases exponentially (see Equations 7 and 9) and then as $(\rho\zeta)^{-1/2}$.

In the space outside both of the cones $[\alpha > \max(\alpha_1, \alpha_2)]$, the sum $\sum_{1,2} I_{1,2}^{(\mathscr{I})}$ is exponentially small, and the field is expressed through the integrals $I_{1,2}^{(\mathscr{E})}$ (see Equation 3). Their asymptotic values are determined by the contribution of the stationary point lying near the boundary of the integration contour.

Outside the cones, the amplitudes of the wave components have different distance dependences, namely, E_ρ, $E_\varphi \sim 1/\rho^2\sqrt{\zeta}$, $E_\zeta \simeq 1/\rho^2\zeta^{3/2}$ or $1/\rho\zeta^{5/4}$. At a fixed distance ζ, the amplitude of the field rapidly decreases as the distance from the surface of the outer cone increases. It is interesting that the phase velocity of these waves is directed along the magnetic field, that is, along axis ζ, and that the energy of these waves is transferred at the angle to each wave, which is determined by:

$$\tan(\zeta\bar{H}_0) = \begin{cases} \dfrac{n_{\parallel}(0)}{2|\mathscr{E}_3|\left|\dfrac{d^2n_{\parallel}(0)}{dn_{\perp}^2}\right|}\cdot\dfrac{\rho}{\zeta}, & \text{if sign } \dfrac{d^2n_{\parallel}(0)}{dn_{\perp}^2} = 1, \\[2ex] \left(\dfrac{6}{\dfrac{d^4n_{\parallel}(0)}{dn_{\perp}^4}}\right)^{1/4}\dfrac{\Gamma\left(\dfrac{1}{4}\right)n_{\parallel}(0)\rho}{2\sqrt{\pi}|\mathscr{E}_3|\left|\dfrac{d^2n_{\parallel}(0)}{dn_{\perp}^2}\right|^{1/2}}\dfrac{1}{\zeta^{3/4}}, & \text{if sign } \dfrac{d^2n_{\parallel}(0)}{dn_{\perp}^2} = -1. \end{cases}$$

Results of Numerical Calculations

Some results of the numerical calculations given in this section for the whistler mode $\omega_L \lesssim \omega < \omega_H$ (ω_L, ω_H are the low-hybrid and electron gyrofrequencies, respectively) illustrate several properties of the field in the outer and inner regions of both radiation cones. Calculations were made for fixed frequencies:

$$\omega = 4 \cdot 10^{-2}\omega_H \gtrsim \omega_L, \qquad \omega = (0.1, 0.3, 0.7, \text{ and } 0.9)\omega_H,$$

with the following parameters of plasma consisting of electrons and ions of oxygen O^+:

$$\left\{ \frac{\omega_0}{\omega_H} = 2, \quad \begin{matrix} \omega_H = 7.5 \cdot 10^6 \text{ cm/sec}; & \lambda_H = \dfrac{2\pi c}{\omega_H} = 8\pi \cdot 10^3 \text{ cm}; \\ \dfrac{m}{M(O^+)} = 3.42 \cdot 10^{-5}; & \nu = 10^2 \text{ sec}^{-1}. \end{matrix} \right\}$$

The values of the tensor elements used in the calculations are given in Table 1.

Asymptotic formulas were used, and the regions of their applicability with respect to ρ, ζ, ω, and the plasma parameters were estimated from the second approximations of these formulas.

The main peculiarity of the field structure is its oscillation character, which depends on ρ and the maxima of the amplitude of the field strength. In the

Table 1

Values of the Tensor Elements

$\frac{\omega_0}{\omega_H} = 2$, $\omega_H = 7.5 \cdot 10^6$ cm/sec, $\frac{m}{M(O^+)} = 3.42 \cdot 10^{-5}$, $\nu = 100$ sec^{-1}

$\frac{\omega}{\omega_H}$	$4 \cdot 10^{-2}$	0.1	0.3	0.7	0.9
$\mathscr{E}_1$	4.9	5.0	5.4	9	22
Im $\mathscr{E}_1$	$-1.3 \cdot 10^{-3}$	$-5.5 \cdot 10^{-4}$	$-2.3 \cdot 10^{-4}$	$-4.4 \cdot 10^{-4}$	$-3.0 \cdot 10^{-3}$
$\mathscr{E}_2$	−100	−40	−15	−11	−23
$\mathscr{E}_3$	−2500	−400	−43	−7	−4
Im $\mathscr{E}_3$	−0.83	$-5.3 \cdot 10^{-2}$	$-1.97 \cdot 10^{-3}$	$-1.55 \cdot 10^{-4}$	$-7.32 \cdot 10^{-5}$
The values of various magnitudes at point $\rho_{0.\,max}$					
$n_\perp$	47	19	6.4	—	—
$n_\parallel$	3.7	3.7	4.0	—	—
α_0^0	0.02	0.06	0.18	—	—
θ_0^0	86	79	58	—	—
The same at point $\rho_{1.\,max}$					
$n_\perp$	2310	880	255	43	7.4
$n_\parallel$	102	99	90	48	17
α_1^0	2.5	6.4	20	48	67
θ_1^0	89	84	71	42	24
The same at point $\rho_{2,\,max}$					
$n_\perp$	7.0	4.3	2.3	—	—
$n_\parallel$	9.0	6.1	4.3	—	—
α_2^0	17	16	6.0	—	—
θ_2^0	39	37	28	—	—

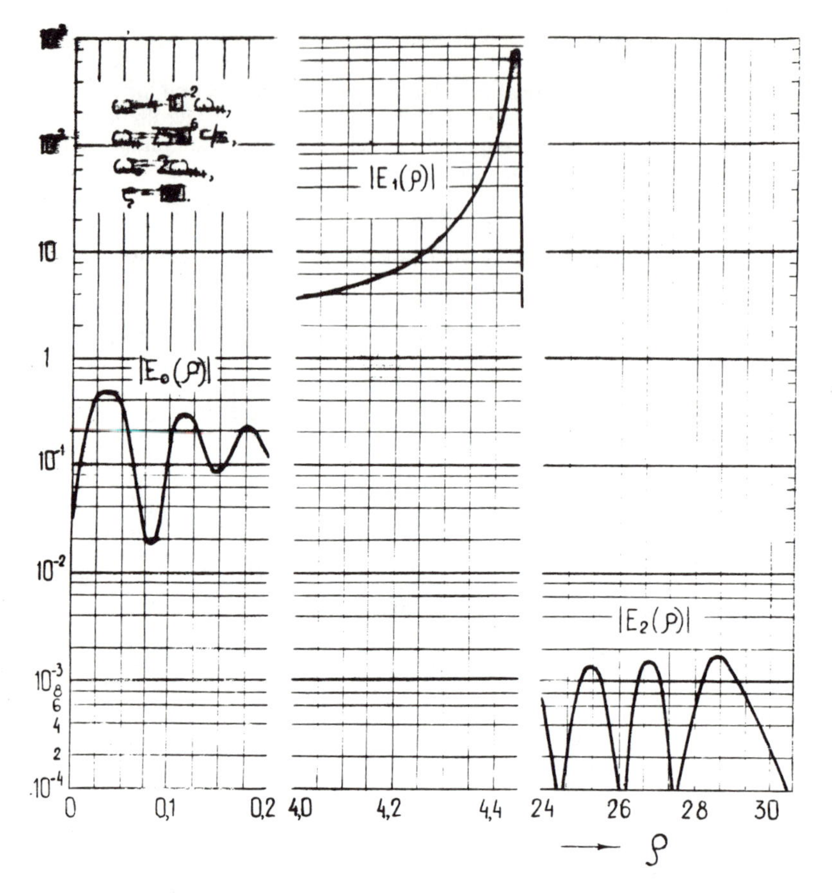

FIGURE 1.

frequency range $\omega_L \lesssim \omega < \omega_H/2$, the amplitude of the field $|\bar{E}| = \{|E_\rho|^2 + |E_\varphi|^2 + |E_\zeta|^2\}^{1/2}$ has three maxima: $|\bar{E}|_{0,\max}$, $|\bar{E}|_{1,\max}$, and $|\bar{E}|_{2,\max}$, which are close to the direction of the magnetic field $\bar{H}_0$ near the surface of the resonance cone $\langle 1 \rangle$ (angle α_1) and near the surface of the Storey cone $\langle 2 \rangle$ (angle α_2), respectively (FIGURE 1). These maxima are sharply pronounced, especially $|\bar{E}|_{1,\max}$. In the frequency range $\omega_H/2 \lesssim \omega < \omega_H$, the amplitude of the field has only one less pronounced maximum, $|\bar{E}|_{1,\max}$, which is near the resonance cone $\langle 1 \rangle$ (TABLE 2). The vanishing of the amplification of the field for the frequencies $\omega \gtrsim \omega_H/2$ near the direction $\bar{H}_0$ corresponds to the transition from Case 4a to Case 4b (see above). Maximum $|\bar{E}|_{2,\max}$ vanishes at these frequencies, since the Storey cone vanishes: $\alpha(\theta)$ has no maximum α_M when $\omega > \omega_H/2$ and $\omega_0^2 \gg \omega_H^2$.

As ω increases, the values of the maxima of $|E|$ decrease (FIGURE 2, TABLE 2). A particularly large amplification of the field near the surface of the resonance cone occurs (FIGURE 2) near the low-hybrid frequency, or when $\omega \gtrsim \omega_L$. However, maximum $|E|_{1,\max}$ is the highest only within a certain range of distances ζ from the emitter. With increasing distance, the value of $|\bar{E}|_{1,\max}$ gradually decreases and approaches the value of $|\bar{E}|_{0,\max}$; the main maximum of the field then becomes

TABLE 2

MAGNITUDES OF DIFFERENT VALUES AT THE RELATIVE DISTANCE FROM THE EMITTER $\zeta_H = 2.5 \cdot 10^3$

Relative Units	$\lvert\bar{E}\rvert(\rho = 0)$ **(5a), (5b)**	$\lvert\bar{E}\rvert_{0,\,max}$ **(5c)**	$\rho_{0,\,max}$ **(5c)**	$\lvert\bar{E}\rvert_{1,\,max}$ (7), **(7a)**	α_1^0	$\lvert\bar{E}\rvert_{2,\,max}$ **(9b)**	α_2^0 **(9a)**	$\lvert\bar{E}\rvert^{(\mathscr{E})}$ **(13)**
$\omega = 4 \cdot 10^{-2}\omega_H$ $\zeta = 100$	$3.5 \cdot 10^{-2}$	$4.8 \cdot 10^{-1}$	$3.9 \cdot 10^{-2}$	742	2.5	$1.6 \cdot 10^{-3}$	16.9	$1.1 \cdot 10^{-7}$
$\omega = 0.1\omega_H$ $\zeta = 250$	$2.2 \cdot 10^{-2}$	$1.2 \cdot 10^{-2}$	$9.7 \cdot 10^{-1}$	114	6.4	$1.8 \cdot 10^{-3}$	15.6	$6.7 \cdot 10^{-8}$
$\omega = 0.3\omega_H$ $\zeta = 750$	$1.4 \cdot 10^{-2}$	$2.4 \cdot 10^{-2}$	$2.8 \cdot 10^{-1}$	8.6	20.6	$3.0 \cdot 10^{-3}$	6.0	$1.3 \cdot 10^{-7}$
$\omega = 0.7\omega_H$ $\zeta = 1750$	$7.1 \cdot 10^{-7}$	—	—	$1.2 \cdot 10^{-1}$	48.0	—	—	$1.1 \cdot 10^{-9}$
$\omega = 0.9\omega_H$ $\zeta = 2250$	$1.4 \cdot 10^{-8}$	—	—	$9.2 \cdot 10^{-4}$	67.1	—	—	$5.8 \cdot 10^{-10}$

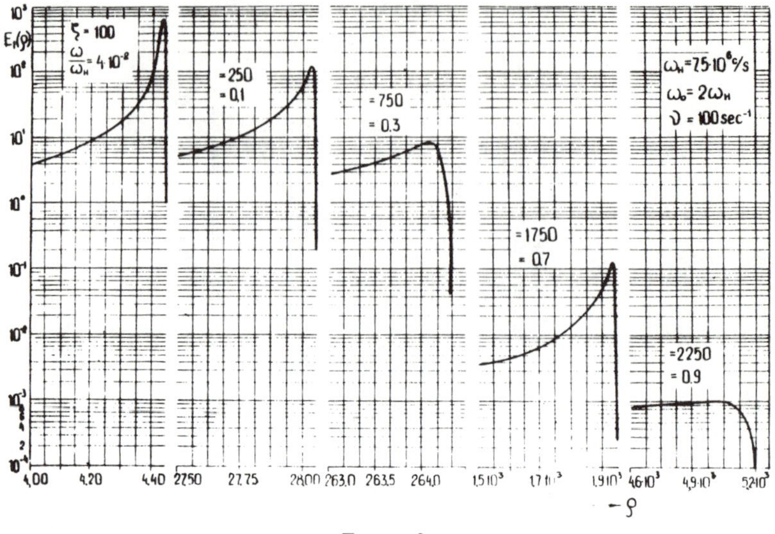

FIGURE 2.

$|\bar{E}|_{0,\max}$ (FIGURE 3). Within the entire distance range and in the whole frequency range, maximum $|\bar{E}|_{2,\max}$ is the lowest.

TABLE 1 lists the values of components $n_\perp$ and $n_\parallel$ of the refractive index n and the values of angles θ and α between the wave vector $\bar{k}$ and the directions of the maxima of the field strength and the magnetic field $\bar{H}_0$. It is worth noting that near the regions of the amplification in the vicinity of the axis $\bar{H}_0$ and the surface of the resonance cone $\langle 1 \rangle$, the wave vectors $\bar{k}$ are oriented across and not along the field $\bar{H}_0(\pi/4 < \theta_{0,1} \rightarrow \pi/2)$ (see TABLE 1). The wave vectors near the Storey cone $\langle 2 \rangle$ are oriented predominantly closer to the direction of H_0, namely, $\theta_2 < \pi/4$. The value of the field along $\bar{H}_0$ ($\rho = 0$) is an order of magnitude (or more) lower than the values of maximum $|\bar{E}|_{0,\max}$, whereas inside and outside the radiation cones, that is, between the maxima of the amplitude of the field strength and outside them, the amplitude of the field is many orders of magnitude lower than the maximum values of $|\bar{E}|_{1,\max}$ and $|\bar{E}|_{2,\max}$ (see TABLE 2 and FIGURES 1–3).

In a vacuum, the modulus of the electrical field of electric dipole I is determined by the expression $|E_0| = I \sin 2\theta / 2\zeta$. In the cases considered here in the vicinity of the maximum, $|\bar{E}|_{1,\max}$, the following values of the field $|\bar{E}_0|$ in a vacuum are obtained:

$\frac{\omega}{\omega_H}$	$4 \cdot 10^{-2}$	0.1	0.3	0.7	0.9
$\alpha_1 = \theta$	2.5	6.4	20.6	48.0	67.1
$\lvert\bar{E}_0\rvert$	$4.3 \cdot 10^{-4}$	$4.4 \cdot 10^{-4}$	$4.2 \cdot 10^{-4}$	$2.8 \cdot 10^{-4}$	$1.6 \cdot 10^{-4}$

Comparison of TABLE 2 and FIGURES 1–3 shows that the amplification of the field near the surface of the resonance cone (angle α_1) changes in relation to the field in a vacuum, in the case considered here, approximately $(2 \cdot 10^1–10^6)$ times.

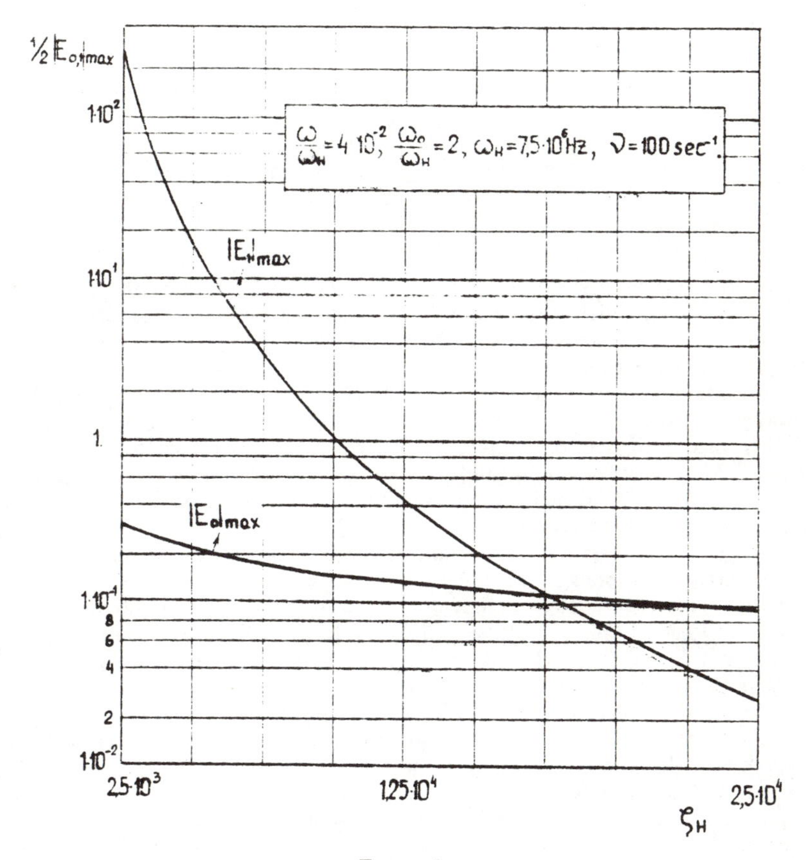

FIGURE 3.

Apparently, this effect plays a prominent part in various processes that occur in plasma. Outside the cones, near their surface, the field $|E|^{(\mathscr{E})}$ of the outer cone is approximately $3(10^2–10^4)$ times smaller than the field in a vacuum.

REFERENCES

1. AL'PERT, YA. L. 1979. J. Atmos. Terr. Phys. In press.
2. MOISEYEV, B. S. 1977. Izv. Vyssh. Uchebn. Zaved. Radiofiz. **20:** 1623.
3. STOREY, L. R. O. 1951. Philos. Trans. Roy. Soc. London Ser. A **246:** 113.
4. VANYAN, L. L. & V. A. YUDOVICH. 1969. Geomagn. Aeron. (USSR) **9:** 917.
5. BELLYUSTIN, N. S. & V. P. DOKUCHAYEV. 1975. Izv. Vyssh. Uchebn. Zaved. Radiofiz. **18:** 17.
6. BUNKIN, V. F. 1957. Sov. Phys.–JETF **32:** 338 (in Russian).
7. ARBEL, E. & L. B. FELSEN. 1963. Electromagnetic Theory and Antennas. E. C. Jordan, Ed. Part I: 391.
8. AL'PERT, YA. L. & B. S. MOISEYEV. 1979. J. Atmos. Terr. Phys. In press.
9. ALPERT, YA. L. 1979. J. Atmos. Terr. Phys. In press.
10. FEDORYUK, M. F. 1977. The Pass Method. Nauka. Moscow.

NUMERICAL METHODS IN SOLIDIFICATION THEORY*†

Jeffrey B. Smith and J. S. Langer

Physics Department
Carnegie–Mellon University
Pittsburgh, Pennsylvania 15213

Solidification processes often produce intricate nonequilibrium shapes and patterns. The most familiar examples are snowflakes and the related dendritic microstructures of metallic alloys. The problem of understanding how these shapes are generated is of great practical interest to metallurgists and physicists and is also of interest to mathematicians as a special example of a free-boundary problem. At the previous Moscow Conference, one of us (J. S. L.) presented preliminary results of a stability theory of dendritic growth. That work has continued to develop successfully. Much of it has been published,[1] and the analytic approach described in those papers should be understood as being complementary to the more directly numerical approach described here.

In order to deal with pattern-forming processes in solidification, we must generalize a conventional Stefan problem[2] by including capillary forces. To see what this generalization involves, let us confine our attention to the relatively simple situation shown in FIGURE 1. We consider a sample of a pure fluid, supercooled initially to a temperature T_0 that is less than the melting temperature T_M. The walls of the vessel are held constantly at temperature T_0. Solidification is started by the introduction of a solid seed at the center of the container. This seed grows into the surrounding supercooled fluid; its rate of growth is controlled by the rate at which the latent heat released at the moving solidification front can be conducted through the fluid and removed at the walls.

We now make several special simplifying assumptions. First, we assume that heat flow is purely diffusive, so that the temperature T in the fluid satisfies a diffusion equation of the form

$$\frac{\partial T}{\partial t} = D\nabla^2 T, \tag{1}$$

where D is the thermal diffusion coefficient. In what follows, we shall present results for only the special case in which the thermal diffusivity is the same in the solid as in the fluid; thus, T satisfies Equation 1 throughout both phases in the system. Note that we are neglecting convective heat flow in the fluid and that we are assuming that there is no diffusive anisotropy in the solid. The second ingredient of the model is heat conservation at the interface between fluid and solid. This conservation means that

$$Lv_n = Dc\hat{n} \cdot [(\vec{\nabla} T)_{\text{solid}} - (\vec{\nabla} T)_{\text{fluid}}], \tag{2}$$

* Supported by a grant from the National Science Foundation to the Center for the Joining of Materials, Carnegie–Mellon University.

† This lecture was presented by J. S. Langer. It is based largely on numerical analysis by J. Smith, and the collaborative project of which this work is a part also involves G. Fix and R. F. Sekerka of the Mathematics and Metallurgy Departments, respectively, Carnegie–Mellon University.

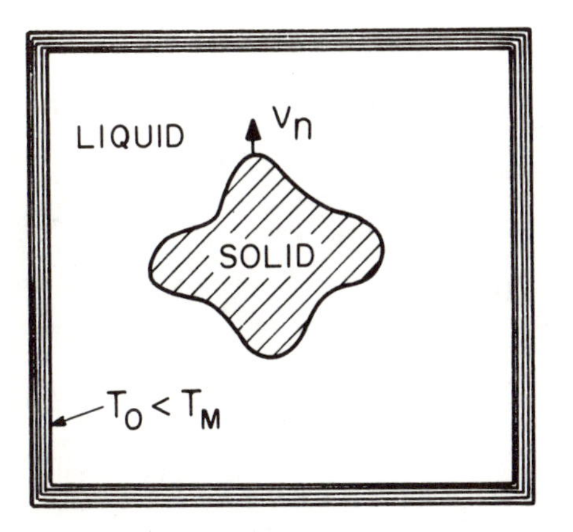

FIGURE 1. Schematic illustration of a solidifying system.

where L is the latent heat, c is the specific heat (assumed to be the same for both phases), v_n is the normal velocity of the interface, and $\hat{n}$ is the unit normal to the interface directed into the fluid.

Finally, we need a thermodynamic boundary condition at the interface. In the usual Stefan problem, one simply assumes that $T = T_M$ at all points on the surface where fluid and solid are in contact with each other. When the solid is growing into an undercooled fluid, however, this isothermal boundary condition produces a model in which the interface is unstable against deformations at indefinitely small length scales. This is a kinetic instability that occurs, roughly speaking, because the latent heat is dissipated most efficiently in configurations with large surface area. It is this instability that is responsible for pattern formation during solidification; however, a finite length scale for stable patterns requires the presence of capillary forces. We therefore include the Gibbs–Thomson term in the thermodynamic boundary condition:

$$T(\text{interface}) = T_M(1 - \Gamma \mathscr{K}), \quad (3)$$

where $\mathscr{K}$ is the curvature of the interface and Γ is a capillary length equal to the ratio of the surface tension to the latent heat per unit volume. Note that we are again neglecting crystalline anisotropies. We shall use an orientation-independent Γ and shall not consider any effects, such as attachment kinetics, that might lead to faceting or other specifically crystalline phenomena.

During recent years, a great deal has been learned analytically about the stability of simple shapes (e.g., spheres and cylinders) growing under the conditions described above. These are generally linear theories that take one just to the verge of pattern formation but not beyond. Direct numerical approaches in which one tries to solve the above equations by finite-difference or finite-element methods have been tried only for intrinsically stable versions of this problem, for example, the case in which the solid freezes in from the walls of the container; thus, these calculations do not consider the problem of pattern formation. It would be extremely desirable to be able to perform direct numerical calculations, both as a guide to more fundamental analytic work and as a means for simulating technologically interesting processes, such as welding. Moreover, the continuing rapid development of computers and computational algorithms has made the numerical approach seem more feasible now than it did just a few years ago. It is in this spirit that we have carried out the calculations that will be described briefly here. A more complete report of this work has been submitted for publication elsewhere.[3]

A useful simplification of the two-phase problem can be achieved by introducing an enthalpy-density function H:

$$H = \begin{cases} c(T - T_M) & \text{(fluid)} \\ c(T - T_M) - L & \text{(solid).} \end{cases} \tag{4}$$

The diffusion and conservation equations (Equations 1 and 2) can be combined into a single nonlinear equation of the form:

$$\frac{dH}{dt} = Dc\nabla^2 T. \tag{5}$$

In the case where T is always greater than T_M in the fluid, less than T_M in the solid, and exactly equal to T_M at the interface, the definition of H as a single-valued function of T can be completed by letting it jump sharply but continuously from its solid to fluid values in a narrow neighborhood around T_M. Equation 5 is then a complete description of the entire system. It can be solved directly by finite-difference or finite-element methods, and the position of the interface can be deduced at any time by locating the surface along which T passes through T_M. This technique appears to be quite efficient for stable solidification or melting problems where the interface is always isothermal and neither phase is ever supercooled or superheated.

The situations of interest here obviously require that H be a two-valued function of T near T_M and that we self-consistently determine which branch of H, or which phase, is appropriate at each point in the system at the same time that we are computing the motion of the interface. To do this, we have had to retreat from the simple H-function method by explicitly identifying each point in our system as fluid, solid, or interface. In effect, we have introduced a second space- and time-dependent function, a phase function, that must be computed simultaneously with the temperature T.

In our actual computations, we have considered only a two-dimensional system, specifically, a square of side $N\,\Delta x$ containing N^2 square cells. Our discretization procedure is based on the integral version of Equation 5:

$$\int_{\mathscr{S}} (Hn_t - \vec{Q} \cdot \vec{n}_x)\, dS = 0, \tag{6}$$

where $\mathscr{S}$ is a space-time cell of volume $(\Delta x)^2\,\Delta t$, dS is the element of surface area of that cell, $\vec{n}_x$ and n_t are the space and time components of the outward normal, and $\vec{Q}$ is the heat flux, $Dc\vec{\nabla}T$. Equation 6 may immediately be translated into an implicit finite-difference equation relating averages of H and Q over the various faces of the cell. Because this equation automatically conserves enthalpy, it is stable against round-off errors and even provides a certain robustness in the face of inaccuracies of our more detailed approximation.

The crux of our calculation is our treatment of interface cells, that is, the cells that contain pieces of the fluid–solid interface. We have found it important to specify the position and orientation of the interface to an accuracy greater than what we should have obtained had we simply identified "interface" cells and assigned to them some average "interface" properties. Specifically, we have approximated the interface by a series of straight-line segments and have located these segments within the interface cells self-consistently by computing enthalpy content, local heat fluxes, and so on. Each such interface segment is an isotherm whose temperature is set by the Gibbs–Thomson condition (Equation 3); the curvature is determined by finding the circle that passes through the center of the interface line and its two nearest neighbors. Finally, this procedure must be supplemented by computational rules that determine how an interface segment passes out of one cell and into another. Details will be found in Reference 3.

Numerical results typical of those that we have obtained by these procedures are shown in FIGURES 2–4. FIGURE 2 illustrates a method that we have used to check the accuracy of our calculations. Here, we have started with an initial solid seed with a slight fourfold asymmetry:

$$R(\phi) = R_1 + \delta \cos 4\phi, \tag{7}$$

where $R(\phi)$ is the interfacial position in polar coordinates; moreover, we have imposed the boundary condition $T = T_o < T_M$ on the fixed outer circle $R = R_o$. FIGURE 2 shows several successive positions of the interface as the solid grows unstably into the undercooled fluid. The advantage of this situation is that, for $\delta \ll R_1$, we can obtain accurate analytic approximations to compare with our numerical results. Specifically, we can carry out a quasistationary analysis of the

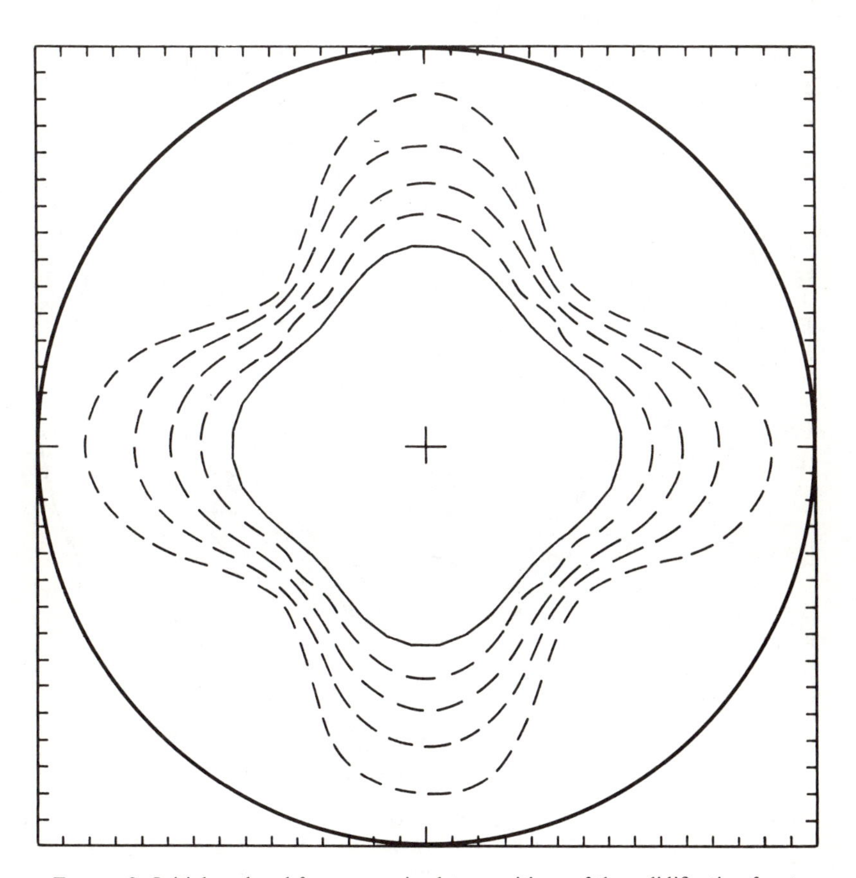

FIGURE 2. Initial seed and four successive later positions of the solidification front.

kind performed by Mullins and Sekerka[4] and can deduce from this analysis both the average growth rate, dR_1/dt, and the amplification rate for small deformations, $(d/dt)(\ln \delta)$. By looking at other asymmetries (e.g., twofold and threefold), and by varying the orientation of the seed, we can also check for grid-dependent effects in our approximation scheme. Solidification patterns of the kind shown in FIGURE 2 seem to pass all of these tests satisfactorily for grid sizes of the order $N = 40$ or larger.

FIGURE 3 is included as an illustration of the kind of detailed information that can be obtained by these methods. Shown here is a set of isotherms associated with the third in the series of interfaces in FIGURE 2. Note that the outward-pointing bulges are slightly cool and that the regions of inward curvature are relatively warm, consistent with the Gibbs–Thomson condition (Equation 3). For relatively simple situations such as this one, comparisons with analytically obtained thermal fields provide additional checks for the numerical procedure.

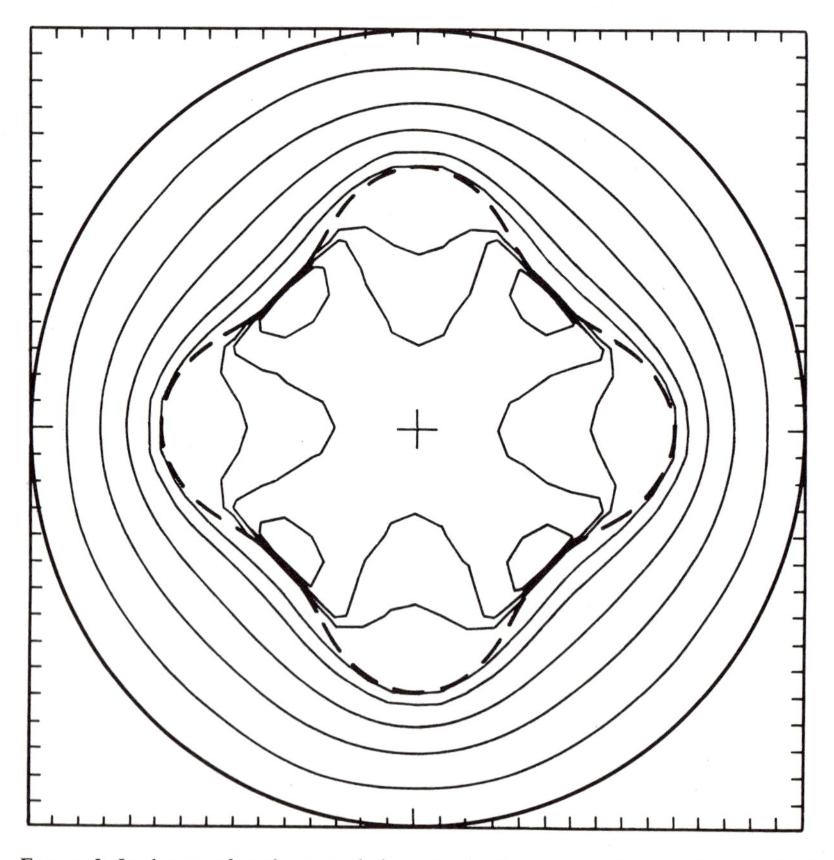

FIGURE 3. Isotherms after the second time step in the process shown in FIGURE 2. The dashed line is the solidification front. Successive temperatures, reading from the inside toward the bounding circle and in units of the undercooling $T_M - T_o$, are: 0.0, −0.01, −0.02, −0.1, −0.4, −0.6, −0.8, and −1.0.

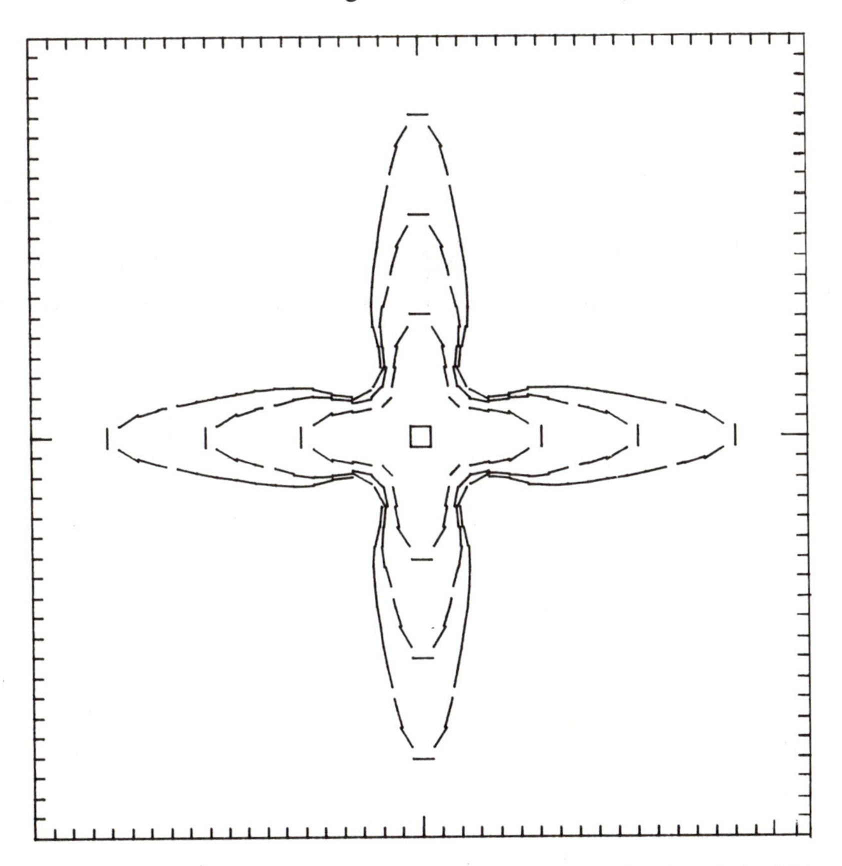

FIGURE 4. Dendritelike growth pattern developed from a single grid-cell seed of solid in a uniformly supercooled liquid.

Finally, FIGURE 4 has been selected to illustrate the kind of strongly unstable behavior that one can examine via the direct numerical method but that is largely inaccessible to analytic investigation. Here, the initial seed is a single cell at the center of the grid, and the outer boundary held at T_0 is the perimeter of the square. No attempt has been made to simulate either the shape or the thermal field associated with a smoothly growing solid. In effect, we have given the system an initial perturbation (the single grid-cell seed creates its own fourfold asymmetry) and have chosen the capillary parameter Γ to be small enough that the shape does not restabilize as it grows. The four arms grow out rapidly, producing the beginnings of a square "snowflake." There is even a hint of sidebranches, but, unfortunately, the tips of the main arms hit the boundaries long before these arms can develop into freely growing dendrites.

In summary, the numerical methods that we are developing do seem to be capable of simulating nontrivial pattern-forming processes in solidification. They should be immediately useful, for example, in studying how systems that are marginally stable respond to finite deformations—an intrinsically nonlinear and time-dependent problem that has an important bearing on the onset of instabilities in real systems. In the longer term, we expect that these methods may be useful for direct simulation of complex materials processes.

References

1. Langer, J. S. 1980. Instabilities and pattern formation in crystal growth. Rev. Mod. Phys. **52:** 1.
2. Rubinstein, L. 1971. The Stefan Problem (translated by A. D. Solomon). American Mathematical Society. Providence, R.I.
3. Smith, J. B. 1980. J. Comput. Phys. In press.
4. Mullins, W. W. & R. F. Sekerka. 1963. J. Appl. Phys. **34:** 323.

RECENT RIGOROUS RESULTS IN PERCOLATION THEORY

Hervé Kunz

Laboratoire de Physique Théorique
E.P.F.L.
CH-1001 Lausanne, Switzerland

Bernard Souillard

Centre de Physique Théorique
Ecole Polytechnique
F-91128 Palaiseau, France

Definition of Percolation Problems

Percolation Models ("*independent percolation*")

Let us consider a graph G, in order to fix ideas, the hypercubic d-dimensional lattice $\mathbb{Z}^d$. Let each site of the lattice be occupied with probability p and empty with probability $q = 1 - p$, independently of the other sites. We have then defined the *site-percolation model.* On the other hand, we could have considered the lines of the graph to be occupied, or opened, with probability p and empty, or closed, with probability $q = 1 - p$, independently of each other, instead of the sites. This would have defined the *bond-percolation model.* A transformation of the graph into its covering graph makes the bond-percolation model appear to be a special case of the site-percolation model, so we shall concentrate on the latter.

Clusters

In the site-percolation model, we have random configurations of occupied sites. We look then to the maximal connected components (connected through the bonds of the lattice) of these configurations, and we call such a maximal connected component a *cluster.*

Figure 1a shows a one-point cluster: one point is occupied and its four neighbors are all empty. In Figure 1b, we have a four-point cluster: each point is connected through the bond of the lattice, and their neighbors are all empty. In Figure 1c, we have an 18-point cluster. Note that the two sites in the hole inside the cluster can be occupied or empty; they do not belong to the cluster, since they are disconnected from the occupied points of the cluster.

Infinite Clusters and Percolation

We consider the configurations for which the origin of the lattice is occupied. Now, according to the configuration, the origin can sit in a finite cluster, but also in an infinite cluster. Let us then define $P_\infty(p)$ as the probability that the origin belongs to an infinite cluster of occupied sites. If $P_\infty(p)$ is strictly positive, we say that *percolation* occurs; $P_\infty(p)$ is the *percolation probability.*

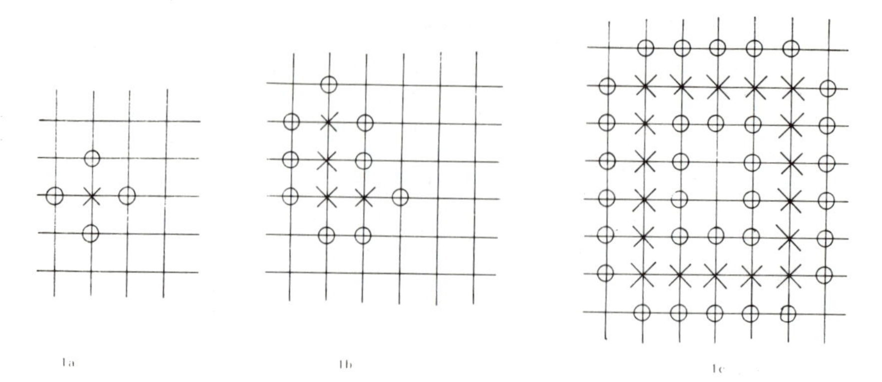

FIGURE 1. Various clusters on the $\mathbb{Z}^2$ lattice. ×, occupied site; ○, empty site.

For small concentrations, that is, when p is small, the clusters hopefully will be, in general, small isolated islands of occupied sites among an ocean of empty sites. When the concentration increases, the islands will grow and join together; for large concentrations, we will have an infinite continent of occupied sites, with only small lakes of empty sites.

Physical Motivations

The problem was introduced, in the form of the bond-percolation problem, by Flory in 1941[1] to describe the sol-gel transition of polymers.* For the sake of simplicity, the monomers are assumed to be fixed at the sites of the cubic lattice. Each monomer can establish chemical bonds with its neighbors. At first approximation, the occurrence of chemical bonds is random, independent from one monomer to the other, and the probability of such bonds will depend on the experimental conditions. Hence, the occurrence or absence of percolation when the experimental conditions are varied mimics the sol-gel transition, in which a large polymer can spread throughout the whole system or disappear, leaving only small polymers.

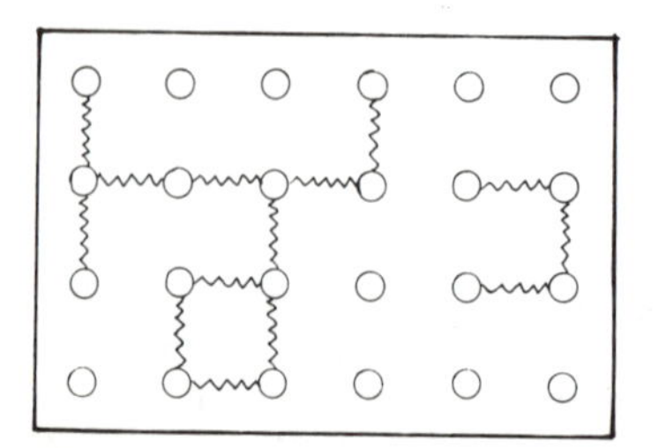

FIGURE 2. Flory model of polymer gelation. ○, Monomer; ⁓⁓, chemical bounds.

* For a similar but more sophisticated model of polymer gelation, see: CONIGLIO, A., H. E. STANLEY & W. KLEIN. 1979.[1]

The problem was then reintroduced in 1957 by Broadbent and Hammersley,[2] who obtained the first rigorous results. Their goal was to create a model for the percolation of a liquid through a porous medium: if there are only a few bubbles, or defects, in the solid, the liquid will never filtrate through it. Conversely, if many bubbles are present, there will be an "infinite path" drawing the liquid through the medium.

Many applications of these models have since been proposed; for example, to study dilute ferromagnets or binary alloys or to determine whether some disease in an orchard will turn into an epidemic!

Existence of the Percolation Transition

It is easy to see that, for a one-dimensional lattice, percolation never occurs, unless, of course, $p = 1$. However, as soon as $d \geq 2$, there exists a *percolation threshold* (p_c): $0 < p_c < 1$, such that:

$$p < p_c \Rightarrow P_\infty(p) = 0$$

$$p > p_c \Rightarrow P_\infty(p) > 0.$$

The first result is easily obtained by noting that the percolation probability is less than the probability for the origin to belong to a chain of ℓ occupied sites for all ℓ. Summing, then, over all possible chains of length ℓ, and letting ℓ go to infinity, proves that $P_\infty(p) = 0$ when p is small enough.

The second result also is obtained easily by showing that, for $p \simeq 1$, the possibility that the origin belongs to a finite cluster $P_f(p)$ can be made small, and by noting that $P_\infty(p) = p - P_f(p)$.

FIGURE 3 shows a typical shape of the curve $P_\infty(p)$ as it can be obtained from numerical computations. It should be mentioned that, for planar lattices, Harris[3] showed by means of a beautiful and difficult proof that $p_c \geq \frac{1}{2}$; hence, in these lattices, infinite clusters of occupied sites and infinite clusters of empty sites cannot coexist. In contrast, for dimensions greater or equal to 3, numerical estimates show that $p_c < \frac{1}{2}$, and so there is in this case a region of values of p for which infinite clusters of occupied and of empty sites will coexist.

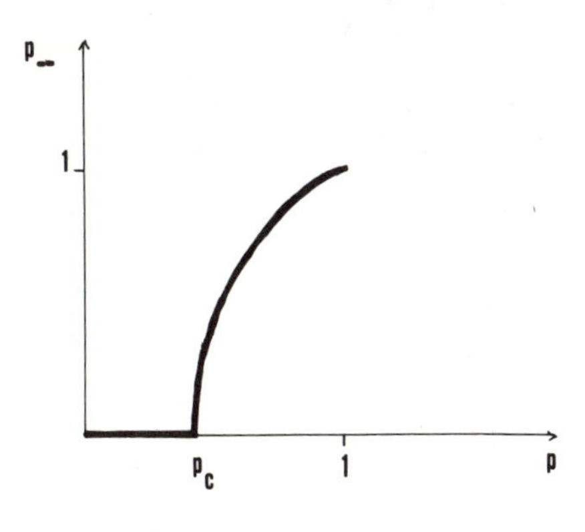

FIGURE 3. Percolation probability as a function of p. A typical case.

Generalizations: Percolation Problems

In the previous models, the sites or the bonds had only two possible states, occupied or empty, and the realization of each possibility was independent from one site or bond to the other. Hence, extensions of these problems are possible in two directions.

In the first extension, more than two states are accessible at each site or bond; that is, at each site or bond, we associate a random variable, which will represent, for example, the time necessary to travel from one end of a bond to the other end. One is then interested in problems like determining the minimal time to go from one site to another. This problem is the so-called *first-passage percolation*, introduced by Hammersley.[4] We shall not discuss this aspect further (see Smythe and Wierman[5] for more details).

In the second extension, the independence between the sites or the bonds is removed. For example, the distribution of the occupied sites in a site problem will be given by a probability law, which is not a product law. Of great interest is to choose a Gibbsian equilibrium distribution of particles on the lattice. That is, the occupied sites of our problem will be the particles of an interacting gas that we have made discrete. Analogously, the occupied sites can be associated to the positive spins of a model of ferromagnets. The most interesting problem in this regard is then to determine whether a link exists between percolation and phase transition. In this connection, it was proved[6] that, for the two-dimensional ferromagnetic Ising model with nearest-neighbor interactions and zero magnetic field, the region where percolation occurs exactly coincides with the region where spontaneous magnetization occurs. In higher dimensions, it is still true that the existence of spontaneous magnetization, of long-range order, implies percolation, but the converse is no longer true.

In addition to these extensions, one can introduce constraints in the percolation models. For example, an occupied point will be counted in a cluster only if it has more than, for example, m occupied neighbors (for more details, see Reference 7).

In the following section, we will discuss only the independent-site percolation model; the bond problem can be treated in a similar way. Results for the interacting-site model will be given at the end of the discussion.

Description of the Clusters

We now know that there is a qualitative change at p_c in the independent-site percolation model when we observe the appearance of an infinite cluster. Our true interest lies in a description of the clusters, their distribution according to size, their effective and real volumes, and their surface characteristics. In particular, we would like to know if the percolation transition will induce qualitative changes in the behavior of these quantities. To answer these questions, we need some definitions.

Definitions and Preliminary Results

The *cluster-size distribution function* is defined as $P_n(p)$, the probability that the origin belongs to a cluster of exactly n occupied sites. It will be useful to introduce the *cluster-size generating function*:

$$f_p(h) = \sum_{n\geq 1} \frac{1}{n} P_n(p)e^{-hn} = \sum_{n\geq 1} \frac{1}{n} P_n(p)z^n = g_p(z), \qquad (1)$$

which is a convergent series for Re $h \geq 0$, since $\sum_n P_n(p) \leq 1$. Note that

$$\left.\frac{\mathrm{d}^k}{\mathrm{d}h^k} g_p(z)\right|_{z=0} = (k-1)!\, P_k(p)$$

$$(-1)^k \left.\frac{d^k}{dh^k} f_p(h)\right|_{h=0} = \langle n^{k-1} \rangle,$$

where the brackets $\langle\ \rangle$ denote the average value on the configurations. Hence, $\langle n^{k-1} \rangle$ is the moment of order $k-1$ of the cluster size. In particular, $f_p(0)$ is the mean number of clusters per site, $-f_p^{(1)}(0)$ is the probability that the origin belongs to a finite cluster, and $f_p^{(2)}(0) = \langle n \rangle$ is the average size of finite clusters.

We first state a *preliminary remark* that derives mainly from the positivity of the series coefficients in Equation 1: $f_p(h)$ is analytic with respect to h at $h = 0 \Leftrightarrow g_p(z)$ is analytic in z for $|z| < \rho$, $\rho > 1 \Leftrightarrow P_n(p)$ decays exponentially with $n \Leftrightarrow \langle n^k \rangle$ is bounded by $c^k k!$

Low-Concentration Results

When p is small, it is easy to see that $P_n(p)$ *behaves exponentially with n.* This follows since the number of clusters of n sites that contain the origin is less than K^n for some K. On the other hand, a lower bound is obtained easily by looking at chains. Hence,

$$\exp(-\alpha_1 n) \leq P_n(p) \leq \exp(-\alpha_2 n) \qquad \text{for small } p.$$

Situation in the Percolative Region

The situation is much more interesting over the percolation threshold p_c. According to mean-field-type theories, or by analogy with the explicitly solvable case where the graph is a homogeneous tree, that is, a graph without closed loops, one could expect an exponential decay of $P_n(p)$ for every p, except at p_c; in fact, one can prove[8] that $P_n(p)$ *never decays exponentially at* p_c. Hence, for this question, one is really interested only in the case when $p \neq p_c$. On the contrary, an analogy with the conjectured singularity in the free energy of the Ising model at the phase transition suggests a nonexponential decay. Finally, various analyses of numerical computations have given divergent results. The singularity was predicted by Stauffer[9] and Flamang[9] for two- and three-dimensional cases, respectively.

We shall now summarize the solution[8] of this problem (for a complete discussion and other results, see Kunz and Souillard[8]). We shall see that a conceptually new phenomenon appears: *finite clusters do have an effective volume when* $p > p_c$, in contrast to the case where $p < p_c$.

We shall study the moments $\langle |C|^\ell \rangle$ of the size distribution $P_n(p)$:

$$\langle |C|^{\ell+1} \rangle = \sum_{C_f \ni 0} |C|^{\ell+1} p^{|c|} q^{|\partial c|} = \sum_{C_s} |C|^\ell p^{|c|} q^{|\partial c|}, \tag{2}$$

where the first sum runs over all finite clusters containing the origin and the second sum runs over all possible shapes of finite clusters. Let us now consider the contribution of all clusters with a cubic external boundary. All such clusters have (see FIGURE 4) the set $\partial\gamma$ empty, the set $\Delta\gamma$ occupied and a subset E of points of $\theta(\gamma)$

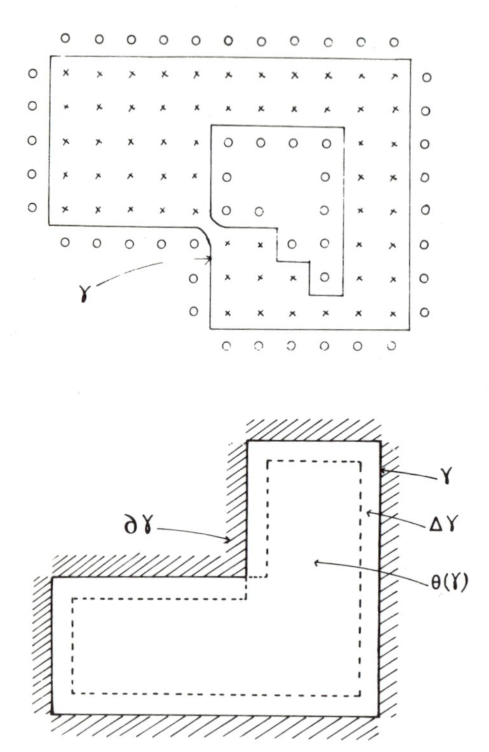

FIGURE 4. *Top*, contour γ associate to a cluster; *bottom*, sets $\partial\gamma$, $\Delta\gamma$, $\theta(\gamma)$ associate to a contour γ.

occupied, E satisfying the condition that any of its points is connected to the boundary $\Delta\gamma$, in order that $E \cup \Delta\gamma$ can form a cluster. Hence, we can write:

$$\langle |C|^{\ell+1} \rangle \geq \sum_{\substack{\text{cubic} \\ \text{contours } \gamma}} p^{\Delta\gamma} q^{\partial\gamma} \sum_{\substack{E \subset \theta(\gamma) \\ E c/\Delta\gamma}} (|\Delta\gamma| + |E|)^{\ell} p^{|E|} q^{|\partial E \cap \theta(\gamma)|}, \tag{3}$$

where the sum over E runs over all subsets of $\theta(\gamma)$ connected to $\Delta\gamma$, as described above, and where $|X|$ denotes the number of points of the set X. But one can transform Equation 3 and rewrite it as:

$$\langle |C|^{\ell+1} \rangle \geq \sum_{\substack{\text{cubic} \\ \text{contours } \gamma}} p^{\Delta\gamma} q^{\partial\gamma} \left\langle \left(|\Delta\gamma| + \sum_{x \in \theta(\gamma)} \chi_x^{\Delta\gamma} \right)^{\ell} \right\rangle_{\theta(\gamma)}, \tag{4}$$

where the average value runs over all configurations in $\theta\langle\gamma\rangle$ and where $\chi_x^{\Delta\gamma}$ is the characteristic function of the event: point x is occupied and connected to $\Delta\gamma$ through occupied sites.

We shall now use the F.K.G. inequality,[10] which applies here and tells us the following: let f and g be two increasing functions over the configurations; that is, if one adds occupied points to a given configuration, the values of f and g for this new configuration are larger than for the old one. Then, $\langle fg \rangle \geq \langle f \rangle \langle g \rangle$. A typical increasing function over the configuration is certainly $\sum_{x \in \theta(\gamma)} \chi_x^{\Delta\gamma}$. On the other hand, we have $\langle \sum_{x \in \theta(\gamma)} \chi_x^{\Delta\gamma} \rangle \geq P_\infty(p) \cdot |\theta(\gamma)|$, since the probability for a point x to be connected to $\Delta\gamma$ is certainly greater than the probability for it to be connected to infinity. By use of the F.K.G. inequality and the preceding remark, we then obtain:

$$\langle |C|^{\ell+1} \rangle \geq \sum_{\substack{\text{cubic} \\ \text{contours } \gamma}} p^{|\Delta\gamma|} q^{|\partial\gamma|} [\,|\Delta\gamma| + P_\infty(p)\,|\theta(\gamma)|\,]^\ell. \tag{5}$$

Finally, we will choose in Equation 3 the contribution of the cube whose side has precisely the length ℓ. Hence,

$$\langle |C|^{\ell+1} \rangle \geq A^\ell [\ell!]^{d/(d-1)} \tag{6}$$

whenever $P_\infty(p) > 0$.

Hence, in view of the preliminary remark above, we have proved that *for any* p *larger than* p_c, *the function* $f_p(h)$ *has a singularity at* $h = 0$, *and* $P_n(p)$ *does not decay exponentially*. As mentioned previously, we see that the term $P_\infty(p)\theta(\gamma)$ is responsible for the singularity over p_c and that it expresses that, over the percolation threshold, the clusters have an effective volume.

We shall now discuss in greater detail the distribution of the clusters according to size and shall describe their exact behavior. First of all, however, let us describe the surface of clusters.

Surface of Clusters

The cluster-distribution function satisfies[8] the following *upper-multiplicativity property*:

$$\frac{P_{n+m}}{n+m} \geq \frac{P_n}{n} \cdot \frac{P_m}{m}. \tag{7}$$

Such a property can be derived by making nearest neighbors in an appropriate way clusters of respective sizes n and m to build clusters of size $n + m$.

It is well known that such a property ensures that there exists a limit for $(1/n)\log(P_n/n)$. From the singularity proved above, that limit is then zero for $p > p_c$.

Let us now consider the function $g_n(p) = (P_n(p)/n) \cdot 1/p^n$. One can see that $(d/d \log q)g_n(p) = \langle |\partial C| \rangle_n / n$, where $\langle |\partial C| \rangle_n$ is the average value for clusters of size n of the size of their boundary. Moreover, the functions $g_n(p)$ are convex with respect to log q, since

$$\frac{d^2}{(d \log q)^2} g_n(p) = \frac{1}{n} \langle (|\partial C| - \langle |\partial C| \rangle_n)^2 \rangle_n \geq 0.$$

Furthermore, their limit is $-\log p$, which is derivable with respect to log q. Then, limit and derivations can be permuted, and finally[8,11] we obtain:

$$\lim_{n\to\infty} \frac{\langle |\partial C| \rangle_n}{n} = \frac{1-p}{p}, \qquad \text{for any } p > p_c.$$

Cluster-Size Distribution Function

As shown above, for small concentrations, $P_n(p)$ behaves exponentially with n. In contrast, when p is larger than p_c, $P_n(p)$ does not decay exponentially. Furthermore, the result obtained above for the moments $\langle |C|^\ell \rangle$ suggests that $P_n(p) > \exp[-\alpha n^{(d-1)/d}]$ for any $p > p_c$. In fact, if $P_n(p)$ behaves exactly as $\exp(-\alpha n^\xi)$ for some ξ, when n goes to infinity, then Equation 6 would ensure that $\xi \leq (d-1)/d$. However, the exact result one can obtain from Equation 6 is slightly weaker because we do not know the $\exp(-\alpha n^{\xi})$ behavior of the $P_n(p)$.

Indeed, for large p, we can[8] prove the exact result,

$$\exp[-\alpha_1 n^{(d-1)/d}] \leq P_n(p) \leq \exp[-\alpha_2 n^{(d-1)/d}], \tag{8}$$

for large p, a result that had been predicted in Reference 9.

The proof is long and will not be given here. Let us just indicate that one first proves the upper bound on the $P_n(p)$. One then proves that $Q_n = \sum_{m \geq n} P_n/n$ satisfies a lower bound of the desired type. From these two properties, one obtains a subsequence of n such that the associated P_n behaves in the desired way. Finally, one uses the upper-multiplicativity, $P_{n+m}/n + m \geq P_n/n \cdot P_m/m$, to prove that not only does a subsequence satisfy Equation 8, but also that such a result is true for any n.

Extension of the Results

The above methods and ideas can be extended to describe much more general situations. We shall briefly mention which extensions have been obtained.

First of all, if one keeps in mind the usual percolation, such an approach can be applied to quantities other than the size of clusters. In particular, one can study[12] the distribution of bonds in the clusters, that is, the number of bonds of the lattice for which both ends of the bond are occupied sites in the cluster containing the origin. Or one can study the cyclomatic number associate to the cluster, that is, the number of independent cycles of the clusters. The distribution function of these quantities can be proved to behave exponentially for small concentrations, and not exponentially for $p > p_c$. These results are a part of an analysis on compactness of clusters which is at work.

Our method can also be extended to the interacting-site percolation problem. Let us consider the distribution of particles associate to an equilibrium state on a lattice in statistical mechanics: for a given box Λ, we have

$$\text{Probability } (X \text{ occupied}, \Lambda \backslash X \text{ empty}) = z^{|X|} \frac{\exp(-\beta U(X))}{Z_\Lambda(\beta, z)},$$

where $\Lambda \backslash X$ denotes the set of points in Λ, except those of X, and

$$U(X) = \sum_{Y \subset X,\, |Y| \geq 2} \Phi(X)$$

is the energy of the configuration X for the given potential Φ, and, finally,

$$Z_\Lambda(\beta, z) = \sum_{X \subset \Lambda} z^{|X|} \exp\{-\beta U(X)\}$$

is the partition function for the given activity z and inverse temperature β. Boundary conditions can also be added. Then one goes to the thermodynamic limit $\Lambda \uparrow \mathbb{Z}^v$.

This construction leads us to a probability distribution on configurations of occupied sites on the lattice, and we can, therefore, study percolative problems on the lattice. We can obtain the following results, which extend those indicated in the previous sections (note that the case of independent percolation is just the case where $\Phi(X) = 0$ for any X and $p = z/(1+z)$): for any integrable potential such that $\sum_{X \ni 0} |\Phi(X)| < \infty$; without any sign condition on the potential, the function P_n

does not decay exponentially in the percolative region, whereas it decays exponentially for small activity. If the potential is negative, which is the case for ferromagnetic systems, again in the percolative region the moments of the cluster-size distribution function satisfy Equation 6.

Again, if P_n behaves as $\exp(-\alpha n^{\xi})$, the latter result means that for ferromagnets in the percolative region, one has $\xi \leq (d-1)/d$, a result proposed on the basis of numerical analysis in Reference 13 for the low-temperature Ising model in a zero magnetic field.

For the latter model, the exact behavior of P_n has just been obtained[14] in various situations. Before giving the results, let us first mention that the Ising model is better expressed as a model of spins taking a value of $\pm\frac{1}{2}$ and interacting via an interaction $J\sigma_i\sigma_j$ between nearest-neighbor spins, and $h\sigma_i$ with the magnetic field h. In such a system, one is interested in the behavior of clusters of plus spins. The exact behavior (8) has been obtained for low enough temperature for $h \geq 0$, including the $h = 0$ plus phase and minus phase, and for large positive magnetic field and arbitrary temperature.

References and Notes

1. Flory, P. J. 1941. J. Am. Chem. Soc. **63:** 3083, 3091, 3096; Coniglio, A., H. E. Stanley & W. Klein. 1979. Phys. Rev. Lett. **42:** 518.
2. Broadbent, S. R. & J. M. Hammersley. 1957. Proc. Cambridge Philos. Soc. **53:** 629.
3. Harris, T. E. 1960. Proc. Cambridge Philos. Soc. **56:** 13.
4. Hammersley, J. M. 1966. J. R. Stat. Soc. Ser. B **28:** 491.
5. Smythe, R. T. & J. C. Wierman. 1978. First-passage percolation on the square lattice. Lecture Notes in Math. No. 671. Springer-Verlag.
6. Coniglio, A., C. R. Nappi, F. Peruggi & L. Russo. 1977. J. Phys. A: Math. Nucl. Gen. **10:** 205.
7. Reich, G. R. & P. L. Leath. 1978. J. Stat. Phys. **19:** 611, Delyon, F. Unpublished.
8. Kunz, H. & B. Souillard. 1978. Phys. Rev. Lett. **40:** 133; 1978. J. Stat. Phys. **19:** 77.
9. Stauffer, D. 1976. Z. Phys. **B25:** 391; Flamang, A. 1977. Z. Phys. **B28:** 47 for the two- and three-dimensional cases, respectively.
10. Fortuin, C. M., J. Ginibre & P. W. Kasteleyn. 1971. Commun. Math. Phys. **22:** 89. The idea and first proof of this inequality actually appeared in Reference 3 precisely for the percolation problem.
11. Before being proved in Reference 8, this result was conjectured, or derived under various approximations, in: Stauffer, D. 1975. J. Phys. C **8:** 172; Reich, G. R. & P. L. Leath. 1978. J. Phys. C **11:** 1155; Stauffer, D. 1978. J. Stat. Phys. **18:** 125.
12. Delyon, F., M. Duneau & B. Souillard. To be published.
13. Binder, K. 1976. Ann. Phys. (N.Y.) **98:** 390; 1976. J. Stat. Phys. **15:** 267.
14. Delyon, F. 1979. J. Stat. Phys. **21:** 727.

ON THE SURFACE TENSION OF LATTICE SYSTEMS*

Jean Bricmont,† Joel L. Lebowitz,†‡ and Charles E. Pfister†§

Departments of †Mathematics and ‡Physics
Rutgers University
New Brunswick, New Jersey 08903

There are various microscopic definitions of the surface tension $\beta^{-1}\tau$ in the literature, but it is far from obvious (or known), in general, that they are all equivalent.[1–4] A proof[5] that for the two-dimensional Ising model (on a square lattice with nearest-neighbor interactions), many different definitions give the same answer as that obtained explicitly by Onsager is therefore encouraging. Here, we use a "grand canonical"[2,4] definition of surface tension that seems natural to us. It is particularly simple when the two pure phases are related to each other by a symmetry of the Hamiltonian, as is the case for the Ising models we shall consider. To make things easy, we deal first with the simplest cases and leave all generalizations to the end.

We consider the d-dimensional ($d = 2, 3$) Ising model with nearest-neighbor ferromagnetic interactions on a simple cubic lattice. At each point, $i \in \mathbb{Z}^d$, there is a spin variable $\sigma_i = \pm 1$, and the Hamiltonian in a finite region $\Lambda \subset \mathbb{Z}^d$ is:

$$\mathscr{H}_{\Lambda,\,\mathrm{b.c.}} = -J\left(\sum_{\substack{\langle ij\rangle \\ i,\,j \in \Lambda}} \sigma_i\sigma_j + \sum_{\substack{\langle ij\rangle \\ i \in \Lambda \\ j \notin \Lambda}} \sigma_i\tilde{\sigma}_j\right), \tag{1}$$

with $\langle ij\rangle$ = nearest-neighbor pair. Here, $J > 0$, and $\tilde{\sigma}_j$ is some fixed value of the spin outside Λ, that is, some boundary condition (b.c.) on Λ. Let $\Lambda_{L,M} \subset \mathbb{Z}^d$ be a parallelepiped of height $2M$ and base $(2L+1)^{d-1}$, $i = (i_1, \ldots, i_d) \in \Lambda_{L,M}$ if $-M \le i_1 \le M - 1$, $-L \le i_2, \ldots, i_d \le L$. We shall generally write Λ for $\Lambda_{L,M}$. We introduce three types of b.c.

(a) the + b.c. (respectively, − b.c.):
$\tilde{\sigma}_j = +1 (\tilde{\sigma}_j = -1)$;

(b) the ± b.c.: $\tilde{\sigma}_j = +1$ if $j_1 \ge 0$,
$\tilde{\sigma}_j = -1$ if $j_1 < 0$;

(c) the free b.c. where the second sum in Equation 1 is set equal to zero.

The Gibbs measure in Λ is, for a given b.c. and inverse temperature β,

$$\mu_{\Lambda,\,\mathrm{b.c.}} = \exp(-\beta\mathscr{H}_{\Lambda,\,\mathrm{b.c.}})/Z_{\Lambda,\,\mathrm{b.c.}} \tag{2}$$

with

$$Z_{\Lambda,\,\mathrm{b.c.}} = \sum_{\sigma_i = \pm 1} \exp(-\beta\mathscr{H}_{\Lambda,\,\mathrm{b.c.}}). \tag{3}$$

* Supported in part by a grant (PHY 78-15920) from the National Science Foundation.
§ Supported by the Swiss National Foundation for Scientific Research.

We write $\langle\ \rangle_{\Lambda,+}$, $\langle\ \rangle_{\Lambda,\pm}$, $\langle\ \rangle_{\Lambda}$ for the expectation value with respect to the b.c. (a), (b), and (c). As L, $M \to \infty$, the Gibbs measures with these b.c. converge to infinite-volume Gibbs states denoted, respectively, ρ_+, (ρ_-), $\rho_\pm$, ρ; the states ρ_+, ρ_-, ρ are translation invariant in all directions,[6,7] and $\rho_\pm$ is translation invariant in the $i_2, \ldots, i_d$ directions.[8]

The surface tension $\beta^{-1}\tau$ is now defined[4] as the suitable thermodynamic limit of the excess free energy per unit cross section of the system with "mixed" $\pm$b.c. over the system with "pure" +b.c.; that is,

$$\tau(K; d) = \lim_{L\to\infty} \frac{1}{(2L+1)^{d-1}} \lim_{M\to\infty} \tau_\Lambda, \tag{4}$$

where $\tau_\Lambda = -\log(Z_{\Lambda,\pm}/Z_{\Lambda,+})$ and $K = \beta J$. This definition of τ is based on the following reasoning:

(a) For $\Lambda \nearrow \mathbb{Z}^d$, the state of the system with +b.c., ρ_+, always corresponds to a pure phase (translation-invariant extremal Gibbs state). For $T \geq T_c$, the critical temperature for the onset of the spontaneous magnetization (which, as is well known, equals $\rho^+(\sigma_0)$), this infinite-volume Gibbs state is the same as that obtained with any b.c. since the system can exist only in one phase (Gibbs state). For $T < T_c$, however, ρ_+ is different from the state obtained with pure −b.c., ρ_-. Moreover, for these systems, it is known that there are only two pure phases at low temperatures[6] (all $T < T_c$ for $d = 2$, almost all for $d \geq 3$[9]).

(b) When $M \to \infty$, the state of the system in the infinite cylinder B_L with base area $(2L+1)^2$ resembles very closely, as $i_1 \to \infty$ (or $-\infty$), the state of a system in B_L with pure + (or pure −) b.c.[10] This can be interpreted to mean that the system with ±b.c. in B_L is spatially segregated into a + and − phase (vapor and liquid phase in the lattice-gas language). τ then measures the excess free energy due to the interface thus created. Although this interface may fluctuate wildly as $L \to \infty$, depending on d and T,[10–12] it is present somewhere (with, presumably, a finite thickness at all $T < T_c$; this will be discussed elsewhere).

As mentioned earlier, Equation 4 can be shown[5] to give, for $d = 2$, the value, found by Onsager,[1]

$$\tau(K; 2) = \begin{cases} 2K + \log[\tanh K], & \text{for } T < T_c \\ 0, & \text{for } T \geq T_c \end{cases}. \tag{5}$$

Although there is unfortunately no explicit formula for τ in other systems, it will be shown that Equation 5 is a lower bound to τ for $d \geq 3$. This is based on the monotonicity property of τ (Theorem 1): when the strength of ferromagnetic interactions is increased, τ increases too. The consequent monotonicity of τ in the temperature is certainly what we would expect of the physical surface tension. We would also expect the surface tension to vanish whenever there is only one phase present ($T \geq T_c$). This is the content of Theorem 2.

We remark here that it has been shown that for a large class of ferromagnetic spin systems:

(a) the limit τ in Equation 4 exists[4] and, for a subset of these spin systems,

(b) $\tau > 0$ for sufficiently low temperatures.[13]

These results apply to spin systems in which the different low-temperature phases (there can be more than two) are related by some symmetry of the Hamiltonian (as in our case where $Z_{\Lambda,+} = Z_{\Lambda,-}$). It is an interesting problem to extend these results to systems without symmetry of the kind considered by Pirogov and Sinai.[14]

RESULTS

Inequalities

Theorem 1. $\tau(K; d)$ is monotone increasing in K and d.

Proof. Consider a general ferromagnetic Ising spin system with (free b.c.) Hamiltonian

$$\mathcal{H}_\Lambda = \sum_{A \subset \Lambda} J_A \sigma_A, \qquad J_A \geq 0.$$

Then, by well-known arguments,[7]

$$\frac{\partial}{\partial J_B} \ln \left[\frac{Z_{\Lambda,+}}{Z_{\Lambda,\pm}}\right] \equiv \langle\sigma_B\rangle_{\Lambda,+} - \langle\sigma_B\rangle_{\Lambda,\pm} \geq 0.$$

The inequality survives in the limit $L \to \infty$ in Equation 4, and so we have monotonicity of $\tau(K; d)$ in K. The monotonicity in d follows from the observation that the d-dimensional system can be obtained from the $(d+1)$-dimensional one by "cutting" the bonds between (hyper-) planes. (In particular, $\tau(K; 3) > 0$ for $T < T_c$ of the $d = 2$ system.)

Remark. The monotonicity of τ is analogous to the well-known monotonicity of the spontaneous magnetization.

We shall now prove that the surface tension vanishes above T_c, the critical temperature for the spontaneous magnetization.

To do this, we introduce a modified Hamiltonian, $\mathcal{H}^s_{\Lambda,\text{b.c.}}$, defined as in Equation 1 but with a coupling sJ instead of J for those $\langle ij\rangle$ with $i_1 = 0$ and $j_1 = -1$. $\tau_\Lambda(K; d, s)$ is defined as in Equation 4 but with $\mathcal{H}^s_{\Lambda,+}$, $\mathcal{H}^s_{\Lambda,\pm}$ (thus, $\tau_\Lambda(K; d, 1) = \tau_\Lambda(K; d)$), and $\langle\ \rangle^s_{\Lambda,\text{b.c.}}$, ρ^s_+, ρ^s_-, $\rho^\pm_s$, denote the corresponding expectation value and Gibbs states. ρ^s_+ and ρ^s_- are now translation invariant only in the $i_2, \ldots, i_d$ directions.

For $s = 0$, there is no coupling between the top and bottom parts of Λ, and the system splits into two uncoupled systems with free b.c. on the spins with $i_1 = 0$ and $i_1 = -1$. In the thermodynamic limit, we obtain two uncoupled "semi-infinite" lattices. We use the subscript s.i. for Gibbs states of this semi-infinite system. The b.c. refer, then, to the b.c. put on the $(2d - 1)$ other sides of Λ.

Theorem 2. For any K and d,

(i) $\tau(K; d) = K \int_0^1 [\rho^s_+(\sigma_0\sigma_{-1}) - \rho^s_\pm(\sigma_0\sigma_{-1})]\, \mathrm{d}s,$

where $-1 = (-1, 0, \ldots, 0)$.

(ii) $\tau(K; d) \leq 2K(\rho_+(\sigma_0))^2$.

(iii) $\tau(K; d) \leq 2K\rho_{+,\text{s.i.}}(\sigma_0)$.

(iv) $\tau(K; d) \geq 2K \int_0^1 \rho^s_\pm(\sigma_0)\rho^s_+(\sigma_0)\, ds.$

Proof. (i) We first remark that $\tau_\Lambda(K; d, 0) = 0$. Therefore,

$$\tau_\Lambda(K; d, 1) = \int_0^1 \frac{\mathrm{d}}{\mathrm{d}s} \tau_\Lambda(K; d, s)\, \mathrm{d}s$$

$$= K \int_0^1 \sum_{\langle ij\rangle \subset \Lambda}' [\langle\sigma_i\sigma_j\rangle^s_{\Lambda,+} - \langle\sigma_i\sigma_j\rangle^s_{\Lambda,\pm}]\, \mathrm{d}s,$$

where the prime indicates $i_1 = 0, j_1 = -1$. We claim that

$$\lim_{L\to\infty} \frac{1}{(2L+1)^{d-1}} \lim_{M\to\infty} \sum_{\langle ij\rangle\subset\Lambda}' \langle\sigma_i\sigma_j\rangle^s_{\Lambda,+} = \rho^s_+(\sigma_0\sigma_{-1})$$

and similarly for $\langle\sigma_i\sigma_j\rangle_{\Lambda,\pm}$. Indeed, we know, from monotonicity in Λ, the existence of the limit and its translation invariance (in the $i_2, \ldots, i_d$ directions) and that, for any $\varepsilon > 0$, there exists Λ_0 such that for every $\Lambda \supset \Lambda_0 + i$, $i_1 = 0$, $|\langle\sigma_i\sigma_j\rangle^s_{\Lambda,+} - \rho^s_+(\sigma_0\sigma_{-1})| < \varepsilon$, $j = (-1, i_2, \ldots, i_d)$. It is clear that

$$\lim_{L,M\to\infty} \frac{1}{(2L+1)^{d-1}} (\#\{i \in \Lambda, i_1 = 0 | \Lambda_0 + i \not\subset \Lambda\}) = 0$$

for any given Λ_0. To conclude the proof of (i), we may simply use the Dominated Convergence Theorem.

(ii) From Lebowitz' inequalities on duplicate variables,[15] one concludes that

$$\rho^s_+(\sigma_0\sigma_{-1}) - [\rho^s_+(\sigma_0)]^2 \leq \rho^s_\pm(\sigma_0\sigma_{-1}) - \rho^s_\pm(\sigma_0)\rho^s_\pm(\sigma_{-1}).$$

On the other hand, by Griffiths' inequalities,[7]

$$\pm\,\rho^s_\pm(\sigma_0) \leq \rho^s_+(\sigma_0) \leq \rho_+(\sigma_0),$$

and this, together with (i), shows (ii).

(iii) We use another modification of the Hamiltonian Equation 1 by putting $s = 1$ but adding an external field $h \geq 0$ on all sites with $i_1 = 0$ or $i_1 = -1$. Then, by Theorem 1, $\frac{d}{dh}\tau_\Lambda(h) \geq 0$, and

$$\tau_\Lambda(K; d) = \tau_\Lambda(K; d, h = 0) \leq \lim_{h\to\infty} \tau_\Lambda(K; d, h) = \log\left(\frac{Z^{++}_{\Lambda'}}{Z^{-+}_{\Lambda'}}\right),$$

where $\Lambda' = \{i \in \Lambda | i_1 > 0\}$ and the superscript $+\ +$ refers to $+$b.c. on Λ' and $-\,+$ to $+$ b.c. on the line $i_1 = 0$ and $-$ everywhere else. Introducing now a factor λ that multiplies the coupling between the spins (equal to $+1$) on $i_1 = 0$ and on $i_1 = 1$, one sees that our last expression equals, after taking the limits of Equation 4,

$$K \int_0^1 [\rho_{+,\lambda}(\sigma_1) - \rho_{-,\lambda}(\sigma_1)]\, d\lambda,$$

$1 = (1, 0, \ldots, 0)$. The subscript λ refers here to the coupling, that is, to the strength of the external field imposed on the spins with $i_1 = 1$. By use of the duplicate variables as in (ii), one sees that the derivative with respect to λ of $\rho_{+,\lambda}(\sigma_1) - \rho_{-,\lambda}(\sigma_1)$ is negative; that is, the integral is bounded from above by its value at $\lambda = 0$. But for $\lambda = 0$, the integrand is just $2\rho_{+,\text{s.i.}}(\sigma_0)$, and this shows (iii).

(iv) We now use (i) and the inequalities of Lebowitz,[9] which show that

$$\rho^s_+(\sigma_0\sigma_{-1}) - \rho^s_\pm(\sigma_0\sigma_{-1}) \geq |\rho^s_+(\sigma_0)\rho^s_\pm(\sigma_{-1}) - \rho^s_+(\sigma_{-1})\rho^s_\pm(\sigma_0)|.$$

By symmetry, $\rho^s_\pm(\sigma_{-1}) = -\rho^s_\pm(\sigma_0)$, which concludes the proof.

Remarks. (1) Although we have used Lebowitz' inequalities to prove (ii) and (iii), we remark that F.K.G. inequalities and (i) give $\tau(K; d) \leq 2K\rho_+(\sigma_0)$ for Ising spins. In fact, we can prove that $\tau(K; d) = 0$ above T_c for systems with ferromagnetic two-body interactions and even a priori measure with compact support on the real line. However, the stronger results (ii) and (iii) hold only for some measures, for example, the uniform measure on $[-1, +1]$ or $\exp(-\lambda\sigma_i^4 + h\sigma_i^2)\, d\sigma_i$ that satisfy

Lebowitz' inequalities. The restriction to nearest-neighbor interactions is not important. Analogous results hold for general ferromagnetic two-body interactions.

(2) (iii) shows that $\tau(K; d)$ vanishes if there is no spontaneous magnetization in the semi-infinite system. However, the critical temperatures for the infinite and the semi-infinite systems are expected to coincide, as they do in two dimensions.[16] On the other hand, if we could show that, whenever $\rho_{+,\text{s.i.}}(\sigma_0) \neq 0$, there exists an $s \neq 0$ such that $\rho^s_{\pm}(\sigma_0) \neq 0$, it would follow from (iii) and (iv) that $\tau(K; d) \neq 0$ if, and only if, $\rho_{+,\text{s.i.}}(\sigma_0) \neq 0$. Indeed, Lebowitz' inequalities in the form used by Messager and Miracle-Sole[8] show that $\rho^s_{\pm}(\sigma_0)$ is monotone decreasing with s and, of course, $\rho^s_{+}(\sigma_0) \geq \rho^s_{\pm}(\sigma_0)$.

Low Temperatures

Let $z = \exp(-K)$.

Theorem 3 (proven by Bricmont *et al.*,[17] Part III). There exists an $r > 0$ such that $\tau(K; 3) - 2K = \mathrm{f}(z)$ is analytic in z for $|z| < r$.

Remark. For $d = 2$, it follows from Formula 5 that $\tau(K; 2) - 2K$ is analytic in z for $|z| < \exp(-K_c)$, $K_c = \beta_c J$, and has an analytic continuation for all K. However, even if we did not have Formula 5, we could prove analyticity of $\tau(K; 2)$ for large K by use of the results of Gallavotti.[11] This proof is slightly more difficult than that of Theorem 3 and proceeds in two steps. We define

$$\bar{\tau} = -\lim_{L\to\infty} \frac{1}{(2L+1)} \lim_{M\to\infty} \log\left(\frac{\sum_{i\in\mathbb{Z}} Z^i_{\Lambda,\pm}}{Z_{\Lambda,+}}\right),$$

where $Z^i_{\Lambda,+}$ is defined with $\pm$ b.c. but where on one side the separation between $+$ and $-$ is put at the height i. One shows that $\bar{\tau}$ is analytic in z and then that for K, real $\tau = \bar{\tau}$. Both results are contained implicitly in Gallavotti.[11]

Let $e(i) = -\frac{1}{2}K\sigma_i \sum'_j \sigma_j$, where the sum $\sum'$ is over the $2d$ nearest-neighbor sites, be the energy of site i (times β) in a given configuration.

Theorem 4 (proven by Bricmont *et al.*,[17] Part II). There exists a $K_0 < \infty$ such that for $K > K_0$,

$$\frac{\mathrm{d}}{\mathrm{d}K}\tau(K; 3) = \sum_{i_1=-\infty}^{+\infty} K^{-1}\{\rho_{\pm}[e(i_1, 0, 0)] - \rho_{+}[e(i_1, 0, 0)]\}(K)$$

$$= -\frac{1}{2}\sum_{i_1=-\infty}^{+\infty}\sum_j{}' [\rho_{\pm}(\sigma_i\sigma_j) - \rho_{+}(\sigma_i\sigma_j)](K),\ i = (i_1, 0, 0). \quad \textbf{(6)}$$

Remarks. (1) Clearly, for fixed Λ, we always have:

$$(2L+1)^{-(d-1)}\frac{\mathrm{d}}{\mathrm{d}K}\log\left(\frac{Z_{\Lambda,\pm}}{Z_{\Lambda,+}}\right) = \frac{[\langle H_{\Lambda,\pm}\rangle_{\Lambda,\pm} - \langle H_{\Lambda,+}\rangle_{\Lambda,+}]}{(2L+1)^{d-1}}$$

$$= \frac{1}{2}\sum_{i\in\Lambda}\sum_j{}' \frac{[\langle\sigma_i\sigma_j\rangle_{\Lambda,\pm} - \langle\sigma_i\sigma_j\rangle_{\Lambda,+}]}{(2L+1)^{d-1}}, \quad \textbf{(6')}$$

where in the sum $\sum'$ the σ_j for $j \notin \Lambda$ are fixed $= \pm 1$ by the b.c.

What has to be proven to get Equation 6 is the validity of the interchange in Equation 6′ of the limits $L \to \infty$, $M \to \infty$, with the summation. Whereas the first interchange, $M \to \infty$, is valid also in two dimensions, the second one, $L \to \infty$, is certainly not; due to the fluctuations of the interface as $L \to \infty$, the Gibbs state ρ_{+} is a superposition of the pure-phase Gibbs states ρ_{+} and ρ_{-};[8,11,18] that is,

$$\rho_{\pm}(\sigma_A) = \tfrac{1}{2}[\rho_{+}(\sigma_A) + \rho_{-}(\sigma_A)]. \quad \textbf{(7)}$$

In particular, $\rho_{\pm}(\sigma_i\sigma_j) = \rho_{+}(\sigma_i\sigma_j)$, so the right-hand side of Equation 6 would be equal to zero for $d = 2$. In fact, this happens also presumably in three dimensions for temperatures above the roughening temperature (see part 2 in the discussion below).

(2) Adapting Proposition 4.2 of Bricmont *et al.*[17] (Part II) to the Ising model, one shows that $\tau - 2K$ is composed of two terms, both exponentially small as $K \to \infty$. The first term is the difference in the free energy (per unit interface, in the limit $L \to \infty$) between a system in B_L with + b.c. and one in B_L with $\pm$b.c. and the additional constraint that all the spins $\sigma_i = +1$ for $i_1 = 0$ and $= -1$ for $i_1 = -1$. This separates B_L into two semi-infinite cylinders with all + or all $-$b.c. The second term results from the fluctuation of the interface separating the + and the $-$ phases in B_L, under $\pm$b.c.

Duality

For a large class of ferromagnetic Ising models, it is possible to construct dual models.[19] Writing the Hamiltonian of the system with "free" b.c. in the form (see Equation 1)

$$-\beta\mathscr{H}_\Lambda = K \sum_{B \subset \Lambda} \sigma_B, \tag{8}$$

where B is a lattice bond and $\sigma_B = \prod_{i \in B} \sigma_i$, the dual model is constructed on a dual lattice with dual bonds B^*, such that

$$-\beta^*\mathscr{H}^*_{\Lambda^*} = K^* \sum_{B^* \subset \Lambda^*} \sigma_{B^*},$$

the coupling K^* is given by

$$K^* = -\tfrac{1}{2} \log \tanh K. \tag{9}$$

In Equation 8 we have free b.c., but the dual of a model with free b.c. has to be taken with +b.c. and vice versa.

Our interest in this duality comes from the observation that by "flipping the spins" in the bottom half of Λ for a system with $\pm$b.c., we obtain, with $\mu_B = exp(-2K\sigma_B)$,

$$Z_{\Lambda,\pm}/Z_{\Lambda,+} = \Big\langle \prod_{\substack{\langle ij\rangle \subset \Lambda \\ i_1 = 0,\, j_1 = -1}} \mu_{\langle ij\rangle} \Big\rangle_{\Lambda,+}$$

$$= \Big\langle \prod_{\langle ij\rangle^* \subset \Lambda^*} \sigma_{\langle ij\rangle^*} \Big\rangle^*_{\Lambda^*}. \tag{10}$$

The first product here is over all bonds crossing the plane ($d = 3$) or line ($d = 2$) $i_1 = -\frac{1}{2}$, whereas the second product is over the dual bonds. Thus, the surface tension of an Ising system at reciprocal temperature β is directly related to the asymptotic behavior of certain spin correlations in the dual model at β^*. This leads to useful relations, as we shall now see.

We note first that for Ising ferromagnets with free boundary conditions,

$$-\lim_{k \to \infty} \frac{1}{k} \rho(\sigma_0\sigma_k) = m. \tag{11}$$

In Equation 11, 0 and k are the lattice sites i with $i_2 = \cdots = i_d = 0$ and $i_1 = 0$ or $i_1 = k$, m is the mass gap, or inverse correlation length. The existence of the limit in

Equation 11 can be shown in the same way as the existence of τ in Equation 4.[4] It is known[19] that, for $d = 2$, our Ising model is self-dual; the dual square lattice $(\mathbb{Z}^2)^*$ has its vertices at the centers of the squares of $\mathbb{Z}^2$, and $\langle ij \rangle^*$ is the bond crossing $\langle ij \rangle$. For $d = 3$, the dual of our model is the Ising gauge model, which is constructed as follows.[20] We take as lattice $\mathscr{L}$, instead of $\mathbb{Z}^3$, the set of centers of the faces of the unit cubes in $\mathbb{Z}^3$. For each $i \in \mathscr{L}$, we have a spin $\sigma_i = \pm 1$. Instead of the *n.n.* pairs, we take as bonds, B^*, the four-point sets given by the centers of all the faces to which a given bond of the *n.n.* $\mathbb{Z}^3$ Ising model (i.e., an edge of an elementary cube) belongs.

A distinctive feature of the gauge model is that $\mathscr{H}$ is invariant under symmetry transformations that involve only the flipping of finitely many spins; for example, changing the sign of σ_i, for all $i \in \mathscr{L}$, which belong to the faces of a given unit cube, does not change the energy. However, this symmetry has no role here (see point 1 in the discussion). In the gauge model, one introduces the "Wilson loop"[20,21]: $\sigma_c = \prod_{i \in c} \sigma_i$, where c is a square of size $L \times L$ in $\mathscr{L}$ for and the coefficient of the (area-law) decay of this loop is

$$-\lim_{L \to \infty} \frac{1}{L^2} \log \rho(\sigma_c) = \alpha. \tag{12}$$

This limit exists and is nonnegative.

We now state the result of this section.

Theorem 5. (1) For the two-dimensional Ising model,

$$\tau(K; 2) = m(K^*). \tag{13}$$

(2) For the three-dimensional Ising model and for the gauge model,

$$\tau(K; 3) = \alpha(K^*), \tag{14}$$

with K^* given by Equation 9.

Proof. We start with Equation 13. By use of Equations 4, 10, and 11 and the self-duality of the model, we see that we have only to show:

$$\lim_{L \to \infty} \frac{1}{(2L+1)} \lim_{M \to \infty} \log\langle \sigma_{-L} \sigma_L \rangle_\Lambda = \lim_{L \to \infty} \frac{1}{(2L+1)} \log \rho(\sigma_{-L} \sigma_L); \tag{15}$$

that is, the decay of the long-long range order $\langle \sigma_{-L} \sigma_L \rangle_\Lambda$ equals the decay of the short-long range order $\rho(\sigma_{-L} \sigma_L)$.[22]

It is clear that the left-hand side is less than or equal to the right-hand side by Griffiths' inequalities:[7] for fixed A, $\langle \sigma_A \rangle_\Lambda$ is monotone increasing in Λ. To show the converse inequality, we take $L = nk$, for n, k integers and k fixed (since we know that the limits exist in Equation 13, we may use subsequences). Then, since $\sigma_i^2 = 1$, $\sigma_{-L} \sigma_L = \prod_{i=-n}^{n-1} \sigma_{ik} \sigma_{(i+1)k}$ and, by Griffiths' inequalities,[7]

$$\log\langle \sigma_{-L/2} \sigma_{L/2} \rangle_\Lambda \geq \sum_{i=-n}^{n-1} \log\langle \sigma_{ik} \sigma_{(i+1)k} \rangle_\Lambda .$$

We claim that

$$\lim_{M, L \to \infty} \frac{1}{2n} \sum_{i=-n}^{n-1} \log\langle \sigma_{ik} \sigma_{(i+1)k} \rangle_\Lambda = \rho(\sigma_0 \sigma_k), \tag{16}$$

by the same argument as the one used in the proof of Theorem 2. Since $(2nk+1)/2n \to k$, we see that the left-hand side of Equation 15 is larger, for any k, than $1/k \log \rho(\sigma_0 \sigma_k)$. Letting $k \to \infty$ finishes the proof.

The proof of Equation 14 is similar. With Equation 9, we obtain

$$\frac{Z_{\Lambda,\pm}}{Z_{\Lambda,+}} = \left\langle \prod_{\substack{i_1=-1/2 \\ i\in\partial\Lambda}} \sigma_i \right\rangle_{\Lambda}^{*},$$

where the expectation is in the dual (gauge) model, and the product runs over all the spins in Λ adjacent to the exterior of Λ, which are in the plane $i_1 = -\frac{1}{2}$ crossing the bonds in Equation 10. Thus, this product is over a square of size $L \times L$, and we write this square as the union of n^2 squares of size $k \times k$, and the rest of the proof goes through.

Discussion

(1) Although we have restricted ourselves to isotropic nearest-neighbor Ising models on $\mathbb{Z}^d$ and their duals, some of our results extend immediately to other systems. Thus, $\tau(K)$ for the triangular lattice with *n.n.* interactions is equal to the mass gap $m(K^*)$ of the honeycomb lattice and vice versa, since these two models are duals.[23] Similarly, our argument, combined with the results of Fontaine and Gruber,[13] shows that some models in three dimensions exhibit the area-law decay of the Wilson loop ($\alpha \neq 0$ in Equation 12). These are all systems obtained through low-temperature–high-temperature duality from a model for which the surface tension is nonzero at low temperature. In fact, the proof in Fontaine and Gruber[13] that $\tau > 0$ at low temperatures for some systems also shows that $\alpha > 0$ at high temperatures for the dual models. In particular, this area law holds also when there is no gauge symmetry. For example, we can take a low-temperature–high-temperature dual of the Ising model where we have "plaquettes," as in the gauge model (4-points bonds) only in planes perpendicular to some axis and *n.n.* two-body bonds between the plaquettes in the different planes. This model has no gauge symmetry but still has a Wilson loop that decays proportionally to the area at high temperatures for the same reason as in the gauge model. This last example was suggested to us by Gruber. Of course, $\alpha = 0$ at low temperatures for all these models because $\tau = 0$ at high temperatures.[13,24]

(2) There is some numerical evidence[25] that, in the three-dimensional Ising model, the nontranslation-invariant Gibbs states that exist at low temperatures do not persist above a roughening temperature T_R strictly less than T_c. It has been questioned[26] whether the surface tension exhibits nonanalytic behavior at T_R.

On the other hand, combining Theorem 5 and Theorem 3, one sees that $\alpha + \log\tanh(\beta J)$ is the restriction to real temperatures of a function of β, analytic around $\beta = 0$. One may ask: What is the interpretation of the corresponding, possibly nonanalytic, behavior of α at T_R^* or, rather, how do the nontranslation-invariant Gibbs states reflect themselves in the gauge model at high temperatures? We remark first that in two dimensions, taking some fixed *n.n.* pair $\langle ij\rangle$, the difference

$$\langle \mu_{\langle ij\rangle}\rangle_{\Lambda,\pm} - \langle \mu_{\langle ij\rangle}\rangle_{\Lambda,+} \tag{17}$$

goes to zero as $\Lambda \to \infty$ (see Equation 7)[8,11,18] for any temperature (this is equivalent in Lebowitz[9] to saying that the state $\langle\ \rangle_{\Lambda,\pm}$ is translation invariant in the thermodynamic limit).

By duality, this means, as pointed out to us by D. Merlini,

$$\frac{\langle \sigma_{-L/2}\sigma_i\sigma_j\sigma_{L/2}\rangle_{\Lambda}^{*}}{\langle \sigma_{-L/2}\sigma_{L/2}\rangle_{\Lambda}^{*}} - \langle \sigma_i\sigma_j\rangle_{\Lambda}^{*} \xrightarrow[\Lambda\to\infty]{} 0 \tag{18}$$

(we denote by $\langle ij \rangle$ also the pair dual to $\langle ij \rangle$). At high temperatures, this is a nontrivial cluster property of the high-temperature state. At low temperatures, Equation 18 still holds but is less striking since then both the numerator and denominator of the first term in Equation 18 tend to a nonzero value.

In three dimensions, duality applied to Equation 17 gives

$$\frac{\langle \sigma_B \sigma_C \rangle^*_\Lambda}{\langle \sigma_C \rangle^*_\Lambda} - \langle \sigma_B \rangle_\Lambda, \tag{19}$$

where B is the bond that crosses $\langle ij \rangle$, and $C = \{i \,|\, i_1 = -\frac{1}{2}, i$ adjacent to the exterior of $\Lambda\}$. Therefore, we know that, as soon as $\langle\ \rangle_{\Lambda,\pm}$ is nontranslation invariant in the thermodynamic limit [and this holds at least up to T_c (two dimensions)[27]], Equation 19 does not go to zero at the dual temperature [i.e., for temperatures higher than T_c (two dimensions) since T_c (two dimensions) $= T_c^*$ (two dimensions) (with $J = 1$)]. So, if there is a $T_R < T_c$ (three dimensions), Equation 19 does not go to zero above the dual temperature T_R^* and goes to zero below. This is another kind of transition in the gauge model.

Summary

We collect here some exact results concerning the surface tension $\beta^{-1}\tau$ of two- and three-dimensional Ising ferromagnets. Some of these results are new: the monotonicity of τ in the coupling constants, the fact that $\tau = 0$ above the critical temperature, the analyticity in $z = \exp(-\beta J)$ of $\tau - 2\beta J$ at low temperatures for the three-dimensional system with nearest-neighbor interaction J, and an expression for τ in terms of correlation functions. Other results are already known, for example, the equality of τ in the two-dimensional square lattice to the mass gap m (inverse correlation length) at the dual temperature β^*; this is proven here simply by means of inequalities. The proof of $\tau(\beta) = m(\beta^*)$ then extends immediately to other lattices, for example, the triangular-honeycomb dual lattices.

References

1. Onsager, L. 1944. Phys. Rev. **65:** 117.
2. Abraham, D. B., G. Gallavotti & A. Martin-Löf. 1973. Physica **65:** 73.
3. Fisher, M. E. & A. E. Ferdinand. 1967. Phys. Rev. Lett. **9:** 163.
4. Gruber, C., A. Hintermann, A. Messager & S. Miracle-Sole. 1977. Commun. Math. Phys. **56:** 147.
5. Abraham, D. B. & A. Martin-Löf. 1973. Commun. Math. Phys. **32:** 245.
6. Gallavotti, G., A. Martin-Löf & S. Miracle-Sole. 1973. Statistical mechanics and mathematical problems. *In* Lecture Notes in Physics. Vol. 20: 162–204. Springer-Verlag. Berlin, Heidelberg, New York.
7. Griffiths, R. B. 1977. Les Houches Lectures, 1970. Gordon and Breach. New York, N.Y.
8. Messager, A. & S. Miracle-Sole. 1977. J. Stat. Phys. **17:** 245.
9. Lebowitz, J. L. 1977. J. Stat. Phys. **16:** 463.
10. Dobrushin, R. L. 1972. Theory Probab. Its Appl. **17:** 582; 1973. **18:** 253.
11. Gallavotti, G. 1972. Commun. Math. Phys. **27:** 103.
12. Abraham, D. B. & P. Reed. 1976. Commun. Math. Phys. **49:** 35.
13. Fontaine, J. R. & C. Gruber. 1978. Surface tension and phase transition for lattice systems. Preprint.
14. Pirogov, S. A. & Y. G. Sinai. 1975. Theor. Mat. Fiz. **25:** 358; 1976. **26:** 61.
15. Lebowitz, J. L. 1974. Commun. Math. Phys. **35:** 87.

16. MacCoy, B. & T. T. Wu. 1973. The Two-Dimensional Ising Model. Harvard University Press. Cambridge, Mass.
17. Bricmont, J., J. L. Lebowitz & C. E. Pfister. 1979. Non translation invariant Gibbs states with coexisting phases: II. Cluster properties and surface tension. Commun. Math. Phys. **66**: 21: III. 1979. Analyticity properties. Commun. Math. Phys. **69:** 267.
18. Russo, L. 1978. The infinite cluster method in the two-dimensional Ising model. Modena University preprint.
19. Gruber, C., A. Hintermann & D. Merlini. 1977. Group analysis of classical systems. *In* Lecture Notes in Physics. Vol. 60. Springer-Verlag. Berlin, Heidelberg, New York.
20. Wegner, F. J. 1971. J. Math. Phys. **12:** 2259.
21. Balian, R., J. M. Drouffe & C. Itzykson. 1974. Phys. Rev. D **10:** 3376; 1975. **11:** 2038; 1975. **11:** 2104.
22. Lieb, E., D. Mattis & T. Schultz. 1964. Rev. Mod. Phys. **36:** 856.
23. Wannier, G. H. 1945. Rev. Mod. Phys. **17:** 50.
24. Gallavotti, G., F. Guerra & S. Miracle-Sole. 1978. Mathematical problems in theoretical physics. *In* Lecture Notes in Physics. Vol. 80: 436–438. Springer-Verlag. Berlin, Heidelberg, New York.
25. Weeks, J. D., G. H. Gilmer & J. H. Leamy. 1973. Phys. Rev. Lett. **31:** 543.
26. van Beijeren, H. 1977. Phys. Rev. Lett. **38:** 993.
27. van Beijeren, H. 1975. Commun. Math. Phys. **40:** 1.